IET TELECOMMUNICATIONS SERIES 94

Intelligent Wireless Communications

Other volumes in this series

Volume 9	**Phase Noise in Signal Sources** W.P. Robins
Volume 12	**Spread Spectrum in Communications** R. Skaug and J.F. Hjelmstad
Volume 13	**Advanced Signal Processing** D.J. Creasey (Editor)
Volume 19	**Telecommunications Traffic, Tariffs and Costs** R.E. Farr
Volume 20	**An introduction to Satellite Communications** D.I. Dalgleish
Volume 26	**Common-Channel Signalling** R.J. Manterfield
Volume 28	**Very Small Aperture Terminals (VSATs)** J.L. Everett (Editor)
Volume 29	**ATM: The broadband telecommunications solution** L.G. Cuthbert and J.C. Sapanel
Volume 31	**Data Communications and Networks, 3rd Edition** R.L. Brewster (Editor)
Volume 32	**Analogue Optical Fibre Communications** B. Wilson, Z. Ghassemlooy and I.Z. Darwazeh (Editors)
Volume 33	**Modern Personal Radio Systems** R.C.V. Macario (Editor)
Volume 34	**Digital Broadcasting** P. Dambacher
Volume 35	**Principles of Performance Engineering for Telecommunication and Information Systems** M. Ghanbari, C.J. Hughes, M.C. Sinclair and J.P. Eade
Volume 36	**Telecommunication Networks, 2nd Edition** J.E. Flood (Editor)
Volume 37	**Optical Communication Receiver Design** S.B. Alexander
Volume 38	**Satellite Communication Systems, 3rd Edition** B.G. Evans (Editor)
Volume 40	**Spread Spectrum in Mobile Communication** O. Berg, T. Berg, J.F. Hjelmstad, S. Haavik and R. Skaug
Volume 41	**World Telecommunications Economics** J.J. Wheatley
Volume 43	**Telecommunications Signalling** R.J. Manterfield
Volume 44	**Digital Signal Filtering, Analysis and Restoration** J. Jan
Volume 45	**Radio Spectrum Management, 2nd Edition** D.J. Withers
Volume 46	**Intelligent Networks: Principles and applications** J.R. Anderson
Volume 47	**Local Access Network Technologies** P. France
Volume 48	**Telecommunications Quality of Service Management** A.P. Oodan (Editor)
Volume 49	**Standard Codecs: Image compression to advanced video coding** M. Ghanbari
Volume 50	**Telecommunications Regulation** J. Buckley
Volume 51	**Security for Mobility** C. Mitchell (Editor)
Volume 52	**Understanding Telecommunications Networks** A. Valdar
Volume 53	**Video Compression Systems: From first principles to concatenated codecs** A. Bock
Volume 54	**Standard Codecs: Image compression to advanced video coding, 3rd Edition** M. Ghanbari
Volume 59	**Dynamic Ad Hoc Networks** H. Rashvand and H. Chao (Editors)
Volume 60	**Understanding Telecommunications Business** A Valdar and I Morfett
Volume 65	**Advances in Body-Centric Wireless Communication: Applications and state-of-the- art** Q.H. Abbasi, M.U. Rehman, K. Qaraqe and A. Alomainy (Editors)
Volume 67	**Managing the Internet of Things: Architectures, theories and applications** J. Huang and K. Hua (Editors)
Volume 68	**Advanced Relay Technologies in Next Generation Wireless Communications** I. Krikidis and G. Zheng
Volume 69	**5G Wireless Technologies** A. Alexiou (Editor)
Volume 70	**Cloud and Fog Computing in 5G Mobile Networks** E. Markakis, G. Mastorakis, C.X. Mavromoustakis and E. Pallis (Editors)
Volume 71	**Understanding Telecommunications Networks, 2nd Edition** A. Valdar
Volume 72	**Introduction to Digital Wireless Communications** Hong-Chuan Yang
Volume 73	**Network as a Service for Next Generation Internet** Q. Duan and S. Wang (Editors)
Volume 74	**Access, Fronthaul and Backhaul Networks for 5G & Beyond** M.A. Imran, S.A.R. Zaidi and M.Z. Shakir (Editors)
Volume 76	**Trusted Communications with Physical Layer Security for 5G and Beyond** T.Q. Duong, X. Zhou and H.V. Poor (Editors)

Volume 77 **Network Design, Modelling and Performance Evaluation** Q. Vien
Volume 78 **Principles and Applications of Free Space Optical Communications** A.K. Majumdar, Z. Ghassemlooy and A.A.B. Raj (Editors)
Volume 79 **Satellite Communications in the 5G Era** S.K. Sharma, S. Chatzinotas and D. Arapoglou
Volume 80 **Transceiver and System Design for Digital Communications, 5th Edition** Scott R. Bullock
Volume 81 **Applications of Machine Learning in Wireless Communications** R. He and Z. Ding (Editors)
Volume 83 **Microstrip and Printed Antenna Design, 3rd Edition** R. Bancroft
Volume 84 **Low Electromagnetic Emission Wireless Network Technologies: 5G and beyond** M.A. Imran, F. Héliot and Y.A. Sambo (Editors)
Volume 86 **Advances in Communications Satellite Systems Proceedings of the 36th International Communications Satellite Systems Conference (ICSSC-2018)** I. Otung, T. Butash and P. Garland (Editors)
Volume 89 **Information and Communication Technologies for Humanitarian Services** M.N. Islam (Editor)
Volume 90 **Communication Technologies for Networked Smart Cities** S.K. Sharma, N. Jayakody, S. Chatzinotas and A. Anpalagan (Editors)
Volume 91 **Green Communications for Energy-Efficient Wireless Systems and Networks** Himal Asanga Suraweera, Jing Yang, Alessio Zappone and John S. Thompson (Editors)
Volume 92 **Flexible and Cognitive Radio Access Technologies for 5G and Beyond** H. Arslan and E. Başar (Editors)
Volume 93 **Antennas and Propagation for 5G and Beyond** Q. Abbasi, S.F. Jilani, A. Alomainy and M.A. Imran (Editors)
Volume 95 **ISDN Applications in Education and Training** R. Mason and P.D. Bacsich

Intelligent Wireless Communications

Edited by
George Mastorakis, Constandinos X. Mavromoustakis, Jordi Mongay Batalla and Evangelos Pallis

The Institution of Engineering and Technology

Published by The Institution of Engineering and Technology, London, United Kingdom

The Institution of Engineering and Technology is registered as a Charity in England & Wales (no. 211014) and Scotland (no. SC038698).

First published 2021

The Institution of Engineering and Technology
Michael Faraday House
Six Hills Way, Stevenage
Herts, SG1 2AY, United Kingdom

www.theiet.org

British Library Cataloguing in Publication Data
A catalogue record for this product is available from the British Library

ISBN 978-1-83953-095-1 (hardback)
ISBN 978-1-83953-096-8 (PDF)

Typeset in India by MPS Limited

Contents

About the editors

George Mastorakis is a director of the e-Business Intelligence Laboratory and an associate professor in the Department of Management Science and Technology at Hellenic Mediterranean University in Greece. His research interests include cognitive radio networks, IoT applications, IoE architectures, radio resource management, artificial Intelligence applications, networking traffic analysis, 5G mobile networks, dynamic bandwidth management, and energy-efficiency networks. He has actively participated in many EC funded research projects (FP6, FP7, and Horizon2020), and acted as a technical manager in several research projects funded by GSRT (General Secretariat for Research & Technology, Ministry of Development, Greece). He is an editor and a co-editor of several edited books. He received his PhD in Telecommunications from the University of the Aegean, Greece.

Constandinos X. Mavromoustakis is a professor in the Department of Computer Science at the University of Nicosia, Cyprus, where he is leading the Mobile Systems Lab. (MOSys). His research covers the design and implementation of hybrid wireless testbed environments, high-performance cloud and mobile cloud computing (MCC) systems, IoT/IoE, modeling and simulation of mobile computing environments, and protocol development and deployment for large-scale heterogeneous networks as well as new "green" mobility-based protocols. He has authored/co-edited several books. He is a management member of the IEEE Communications Society (ComSoc) Radio Communications Committee (RCC), and has served as track Chair and co-Chair of several IEEE International Conferences (including AINA, IWCMC, and IEEE Internet of Things). He received his PhD degree from the Department of Informatics at Aristotle University of Thessaloniki, Greece.

Jordi Mongay Batalla is a professor at Warsaw University of Technology, and deputy director of Research at the National Institute of Telecommunications, Warsaw, Poland. His research is aimed at developing new infrastructures for the future Internet and its applications. He is an editor of four books and author of more than 150 papers published in books, international journals and conference proceedings, and patents (Polish and European Patent Offices). He is/has been a guest editor and member of Editorial Board in more than 10 international journals. Currently, Prof. Mongay Batalla is a technical adviser of the Polish Government for 5G cybersecurity law and a technical adviser of Polish Ministry of Infrastructure for autonomous automotive and others. He is also a technical member of the European Blockchain Services Infrastructure (European Commission) and is a member of Hyperledger platform (a The Linux Foundation project).

Evangelos Pallis is a professor in the Department of Electrical and Computer Engineering at Hellenic Mediterranean University, Greece, and a director of the Research and Development of Telecommunication Systems Laboratory "PASIPHAE". His research interests include intelligent wireless networks, mobile communication systems, and emerging technologies. He has coordinated many European funded R&D projects and participated in standardization bodies for future wireless networking infrastructures. He is general chairman of International Conferences on Telecommunications and an active contributor to standardization bodies for wireless networks. He has co-edited several books in the above research fields.

Chapter 1

An overview of the intelligent big data analytics and their technological presence in the modern digital age

Reinaldo Padilha França[1], *Ana Carolina Borges Monteiro*[1], *Rangel Arthur*[2] *and Yuzo Iano*[1]

Big data (BD) refers to the huge volume of data, both structured and unstructured, generated by our society as a large number of people are connected through information technology and the Internet. This is essential and necessary to have technologies capable of monitoring and interpreting this great flow of information that travels through computerized environments. So, companies optimize their solutions and manage to improve the daily lives of the population through the intelligent and efficient processing of these data. This is where the BD and big data analytics (BDA) come into the picture, which together with artificial intelligence (AI) techniques, analyze this large volume of data in real-time and extract information and knowledge. Big data is a process of analyzing and interpreting a large volume of data stored remotely, integrating any data collected on a subject or a company, such as purchase and sale records and even nondigital interaction channels. Big data analytics is a data analysis process with a specific purpose, forming analysis strategies aiming at a large number of data enabling the study of consumer behaviors and expectations, in addition to observing market trends. Artificial intelligence, as its main focus, is the processing of data in order to make the device or technology more intelligent and capable of reproducing human abilities. Therefore, this chapter aims to provide an overview of intelligent BDA, showing its relationship and technological integrations, approaching its success relationship, with a concise bibliographic background, categorizing and synthesizing the potential of both technologies.

[1]School of Electrical and Computer Engineering (FEEC), University of Campinas (UNICAMP), Campinas, SP, Brazil
[2]Department of Telecommunications Engineering, University of Campinas (UNICAMP), Limeira, SP, Brazil

1.1 Introduction

Reliable data alone provide analytics and insights for business owners to decode, innovate, and earn their customer's trust. For efficient data analysis, digital technologies such as artificial intelligence (AI) and its dominant form play an inevitable role in business. Machine learning (ML) has become today's most sought-after technology to help innovate and transform businesses. Businesses generate a huge amount of data each day, generating as much as 20 petabytes a day such as Google, for example. This volume grows rapidly and there is a need for analysis of this large mass of data. Big data solutions are designed to accelerate decision making, cost reduction, and server processing time to understand consumer behavior and business needs [1].

Increasingly, organizations are realizing the importance of analytics for analyzing data in their business to increase their effectiveness and gain competitive advantage. To achieve redesigned business goals, organizations are considering implementing ML and AI, along with analytics, in their daily tasks. This is increasingly true today, leveraging and correlating various data sets and applying advanced AI techniques. Through deep learning (DL) techniques, data can be automatically recognized to classify images, text, and speech more accurately. This has led to impressive application development in text and speech recognition, image analysis, natural language processing (NLP) and many other advances in different industries [2].

Solving complex problems of increasing volumes of data (structured and unstructured) is a major challenge that requires proper analysis techniques for proper information modeling, where it is necessary to generate appropriate information quickly and intelligently through extremely large databases through ML. Predictive analytics is a set of technologies that focus on anticipating future behaviors or estimating unknown outcomes based on current and historical data. These tools employ a variety of statistical modeling, ML, data mining (DM), and AI techniques to gather all the technical information to make predictions for the future [2–4].

To advance the competition and gain an advantage over others in the industry, organizations are involved in predictive analytics based on ML. Neural networks and DL algorithms successfully discover and use patterns hidden in unstructured datasets, and reveal new information from those datasets. Today, where companies can collect data about their customers in seconds, it is very important to quickly process and extract real-time information from the data. Organizations must devise a strategy to leverage large volumes of BD almost immediately and reposition many of their business processes accordingly [5].

Though descriptive analysis, being the basic form of analysis, aggregates large data and provides useful information from previous records, predictive analytics, now a popular concept, uses historical data, AI, and ML to predict future outcomes while prescriptive analysis uses a combination of business rules, ML, and computer modeling to recommend the best course of action for any prespecified result, typically implemented through workflows [6].

In short, predictive analytics uses a huge amount of data to determine the probabilities of future behaviors. And that is where BD comes in. As is well known, data are always being generated by digital technologies, through the use of apps on smartphones, social media, online shopping, search engine searches, etc. All of this information combines with other data sources and becomes BD. Companies combine BD with technologies such as ML and AI to further improve their performance. In addition, business patterns and behaviors can be determined, enabling the company to offer its customers faster and more personalized experiences [7,8].

Predictive models use these data and technologies, among others, to better understand customers, products, and partners and identify potential future risks and opportunities for business. Companies apply all three types of analytics while working on a dataset, but using predictive analytics creates more value for companies by helping them predict future outcomes. Organizations will need to strike a balance within their data, technology, and employees to completely transform their business into an AI-driven predictive analytics model that is smarter and helps make faster decisions. Implementing AI requires migrating to a data-driven culture coupled with leveraging advanced enterprise-level technologies. Features that control AI and predictive analytics can be applied to virtually any business domain in any industry you can think of, such as protecting IT desktops, detecting security fraud, and cyber data security and theft, among others [9,10].

Combining BD with predictive analytics that leverages cloud computing can do wonders and deliver important information immediately and accurately. Still considering it, the possibilities of AI are endless. This will allow machines to learn, reason, solve problems and understand language instantly. But these capabilities may be years away. Big data analytics and ML, for example, allow marketers to use intelligence to detect, learn, and optimize operations. Big data analytics and AI solutions are extremely powerful tools in an increasingly competitive economic landscape where you almost always have to do more, with fewer resources and in a shorter amount of time. And that is exactly what BD and AI deliver [3,11].

It is the fastest and most efficient decision making, because of the speed of these technologies to process and analyze different data sources and memorize the best paths with AI, companies are able to understand and analytically address information immediately, making decisions better and more assertive in less and less time, since adopting such tools is vital to any organization [12].

Therefore, this chapter aims to provide an updated review and overview of intelligent BDA, showing its relationship and technological integrations, approaching its success relationship, with a concise bibliographic background, categorizing and synthesizing the potential of both technologies.

1.2 Big data

Big data is a process that systematizes the large flow of information that is generated today, by everyone, online or offline, every second. This concept deals with the process of identifying and interpreting this information, in order to favor different strategies, in a more simplistic way. It can be seen as a set of techniques capable of analyzing large amounts of data for the generation of important results that, in smaller volumes, would hardly be possible [13].

Big data works as a huge database. It is the data with a greater variety that arrives in increasing volumes and with increasing speed. Big data is the term in information technology (IT) that deals with large sets of data that need to be processed and stored. The BD concept started with three Vs: speed, volume and variety [14].

Simply put, BD is a larger and more complex data set, especially from new data sources, where these data sets are so massive that traditional data processing software simply cannot manage them; however, these large volumes of data can be used to solve business problems that were not possible to solve before. Since more extensive data sets allow new discoveries to be made, through BD, it is possible to gain an understanding of how to filter web records by understanding the behavior of e-commerce, deriving feelings from social media and customer support interactions, and understanding the statistical methods of correlation and their relevance to product data, manufacturing, customers, and engineering [14,15].

Since many things currently happen quickly, a lot of data are generated all the time, so the use of this database is almost unlimited, and this volume of information only grows exponentially over time and can/should be used for analysis. As long as it is based on that, it is possible to define the concept of BD as an extremely large set of data that, for this reason, needs special tools to support this large volume of data that is found, extracted, organized, and transformed in information that enables a broad and timely analysis. Thus, the use of BD is dependent on the solutions that those who use it are currently looking for [13,16].

For example, there is a difference in distinguishing the sentiment of all customers from that of only your best customers, whether you are capturing customer data, products, equipment, or environmental BD, the goal will be to add more relevant data points in analytical and key summaries, getting better conclusions, which is seen as an integral extension of its existing business intelligence, data warehousing platform, and information architecture resources [16,17].

In order for a data analysis process using technology to be categorized as BD, it is important to pay attention to **Volume**, which in the current hyperconnected and digitized context, companies and various types of organizations collect data from a huge variety of sources, from commercial transactions to social media or even information from sensors and data on the interaction between machines. Since the amount of data matters, processing large volumes of unstructured, low-density data, still considering that it may be data of unknown value, such as clickthrough's on a web page, social data feeds, or an application for mobile devices, or a sensor-enabled device. In reality BD, it is still possible to carry out this cloud storage and processing architectures with new technologies such as Hadoop, being an open-source software structure for data storage and execution of applications in common hardware clusters, providing massive storage for any data type, great processing power, and the ability to handle almost unlimited tasks and jobs occurring at the same time, considered one of the preferred tools of those who work with large volumes of data and need to process them quickly. Thus, the volume is no problem, but rather a solution, even considering that with data-driven solutions, large volumes mean great possibilities [18].

Velocity is the fastest rate at which data are received and perhaps managed. In the BDA ecosystem, data flows happen at an unprecedented speed and the steps for storage, processing, and analysis must take place in the shortest possible time, yet considering data such as Radio-Frequency IDentification (RFID) tags, sensors and the smart measurements they require are driving the need to deal with torrents of data in practically real-time, and in general, the highest data speed is transmitted directly to memory instead of being recorded on the disk, since some smart Internet-enabled products operate in real-time or near real-time requiring real-time evaluation and action. Thus, for some analyses, the closer to real-time, the greater the business increase can be [18,19].

Variety is related to the various types of data available, considering that the traditional data types were structured and are perfectly suited to a relational database, which with BD, the data come in new types of unstructured and semi-structured data, such as text, audio, and video, requiring additional preprocessing to get meaning and support metadata. Big data could also be considered as any data since it has the ability to capture and analyze structured and unstructured data, text, sensors, web navigation, audio, video, log files, turnstiles, central air conditioning, among others. Another feature of BDA processes is that data can come in all types of formats, from structured and numeric data in traditional databases to unstructured text documents, e-mail, video, audio, stock data, and financial transactions, which after processing the possible insights are extracted [18,20].

Thus, over the years, two more Vs have emerged in recent years. Corresponding to **Value** related to the data, it has intrinsic value; however, this is useless until this value is discovered. In addition to the value related to the investment necessary to generate a return for the companies, it can improve the quality of services and increase their revenue. In the same sense as related to other Vs with regard to Volume, in fact, the BD is based on the huge volume of data generated each day as well as Speed because the way in which this information is managed must be dynamic, otherwise it loses its value. Thus, finding value in BD is not just a matter of analyzing it, it is a complex discovery process that requires insightful analysis, considering that business users and executives ask the right questions, thereby making it possible to recognize the patterns, making assumptions informed and predicting behavior [18,21].

Equally, **Veracity** is related to how reliable these data are, and how much it is possible to confirm in them, referring to the fact that the data originated from a multiplicity of different channels, among them, e-mails, social media, sensors, and many others; and finally, veracity because if the data are not real, it is useless [18,22].

And recently, two more Vs were considered, forming the seven Vs of BD, related to **Variability**. In the context of BD, it concerns the inconsistencies in the data or the change of dimensions between them, since they come from various types and origins. Also it can be referred to the inconsistent speed at which large data are loaded into your bank. **Visualization** is explained how the data are presented to facilitate the making of important decisions and can be presented in many ways, such as Excel spreadsheets, Word documents, graphics, among other forms.

Regardless of format, the information must be easily readable, understandable, and accessible, and that is why data visualization is an essential component for BD [18,23,24].

Big data analytical processes and models can be based on humans and machines, since BD analytical resources include statistics, spatial analysis, semantics, interactive discovery, and visualization, and using analytical models, it is possible to correlate different types and sources of data to make meaningful associations and discoveries [25,26].

With respect to **structured data** are those that have some standard or format that can be used in reading and extracting the data, which can be data from databases, text files such as CSV, text or XML, even among the structured data are the GPS information or even the clicks that a given user made on a given website, for example, considering that this information is stored in a fixed format or in a database called structured data, where it is formatted so that it is easily accessed and thus can be used for analysis [27].

The **unstructured data** do not have a standardized format for reading, it can be Word files, Internet/intranet pages, audios, videos, among others, making reference to the information which are not stored in a database or in any other type of structure, considered as "loose" information, generated in a medium, such as MP3 audio files, e-mail messages, video files, and images, among others. This way, these data have a much higher volume when compared to the other two categories [24,28].

With regard to **semistructured data**, it is that which contains both structured and unstructured data, relating to data that are not organized in specific databases, but that still have associated information that makes it easily accessible, as if it were a point of reference. This type of categorization concerns the most technical part of the concept; however, BD has specific tools for that [29].

In addition to this categorization, there are still three types of data, which are not related to their structural character, being **Personal data** or data of things, which are related to the type of information generated through objects, such as a TV and/or a connected mobile device on the same Internet network and communicating with each other, in addition to real-time traffic information based on application data. **Social data** with respect to how a person searches on Google shares or comments on a social network publication, serving as a basis for analyzing user behavior to ensure a better future experience. In the same way, the **Enterprise data** with respect to this type of data are provided by the companies themselves, such as financial data, on the productivity of the team, among other aspects [30].

Big data can be used in various routines of a business, being necessary not only to have knowledge about the technology but also to identify which are the points of the company that will be impacted with its implementation, provided that with that knowledge, it is possible to more efficiently target resources by increasing the return on investment in these data analysis solutions. Taking into account another factor related to the implementation of a data-driven culture, making the company must act so that all its professionals understand the importance of data in the definition of strategies and working based on that [31].

With BD, you get more complete answers, because you have more information. Its relevance also appears in relation to what is done with this amount of data and how they are used to make analyses, since it is essential that they revert in benefits for the business, getting more complete answers that mean more confidence in the data, that is, a completely different approach to dealing with problems. The usefulness and importance of BD appear in pain points common to any enterprise, such as cost, time, product development and decision making, considering that the information collected and interpreted serves as a basis to make the business more efficient, productive, and differentiate itself from the competition, taking into account the modern reality considering that the markets are highly competitive [32].

When associated with other tools, such as analytics, BD helps in identifying the main causes of business failure, contributing to the analysis of sales trends, based on the evaluation of the customer's purchase history, for example, considering the increasing capacity to identify behaviors. But storing data is not enough, it must be used to be useful, and it depends on curation, as long as data are clean or relevant to the client and organization allowing meaningful analysis. Finally, BD technology is changing as the world sees information at an accelerated pace, considering that it is possible through the data collected through the Internet to obtain a greater knowledge of a user than he has of himself [33].

1.3 Artificial intelligence

Artificial intelligence is a branch of computer science that studies the development and creation of machines and algorithms capable of learning on their own, concept consists of the viability of machines thinking such as human beings, developing the ability to learn, reason, perceive, deliberate and logically decide on the facts. The concept is popularized more recently due to the volume of available data that grows exponentially, being necessary more advanced algorithms and improvements in the computational power and storage for processing these data [34].

Another important aspect of AI is that, due to the ability to learn, it needs to be constantly fed so that it can continue to evolve, similar to what happens to a person, allowing systems to simulate human-like intelligence, going beyond order programming making decisions autonomously, based on standards from huge databases. In essence, it allows these systems to make decisions independently, accurately, and supported by digital data, which means multiplying the rational capacity of the human being to solve practical problems, simulating situations, thinking with greater processing power in responses or, more broadly, enhancing the ability to be intelligent [35].

A comparison for better visualization of concepts is that traditional computing works with accurate information, while AI works with inaccurate information. Since a traditional computer system in the context of a photo with a person without glasses and another with glasses will be able to affirm that the two photos are relative to the same person, in contrast, the AI algorithms have the capacity to affirm with a great degree of certainty that the two photos are relative to the same person. In this sense, the AI has the capacity of machines thinking as human beings learning, perceiving, and rationally in the face of certain situations deciding which paths to follow [36–38].

Simple computing for the current one shifted from three main pillars to evolve to AI are as follows: access to large amounts of unprocessed data; good data models for classification, processing, and analysis; and powerful, affordable computing for fast and efficient processing. Considering that currently, an AI solution involves a grouping of various technologies, such as algorithms, learning systems, artificial neural networks, among others, which has the ability to simulate human capacities linked to intelligence, such as reasoning, environmental perception, and the ability analysis for decision making. Thus, through this evolution, they reached three segments based on AI, enabling the formula: BD + cloud computing + good data models [39].

It is possible to make an analogy that AI learns like a child, based on the premise that teaching computers to "think" is not so simple. So depending on the purpose for which it was created little by little, a system absorbs, analyzes, and organizes the data in order to understand and identify what are objects, people, patterns, and reactions of all kinds. Just as current systems are trained to do very specific tasks, that is, a system that plays poker cannot play chess, a tool that detects tax fraud cannot do the same in warranty claims, and the feature used in an autonomous car does not serves to give legal advice. However, AI still has an important limitation with respect to the only way to incorporate knowledge into it is by entering data, meaning that additional layers of forecasting or analysis must be added separately, just as any inaccuracy is reflected in the results. Thus, an AI is the combination of large amounts of data, intelligent algorithms, and fast processing, and from this equation, the system learns, delimits patterns, and processes information [35,40].

Artificial intelligence is also a field of science, whose purpose is to study, develop, and use machines to carry out human activities in an autonomous way, conceptualizing the capacity of technological solutions to carry out activities in a way considered intelligent, being able to “learn by themselves” due to learning systems that analyze large volumes of data, consisting of the simulation of reasoning, the perception of the environment, and the most varied forms of problem-solving, enabling them to expand their knowledge. It is also linked to robotics, ML, voice and vision recognition, facial recognition, speech and the location of a user, the possibility of predicting weather or traffic conditions, and the creation of more real games, among other technologies. Since from this fusion, there are already agricultural machines to cancer diagnosis equipment that use it successfully, for example, as the classification of a large number of resonance images, which increases the possibility of cancer recognition by means of standards delimited and observed, still considering features that are even closer to the common reality, such as personal assistants, found on smartphones [41].

Another important point is that AI algorithms have the ability to learn and evolve, called ML, that is, they are not static algorithms (traditional computing) but algorithms that learn through data, considering that the more data and tests are done, the more these systems evolve, and so when this "artificial brain" is properly "tuned," it can be copied and replicated. In environments with human-supervised learning, AI is usually programmed with the right answers during a training period, and then it will follow these rules. But in unsupervised environments, the AI itself is responsible for analyzing, grouping, and making its own decisions and conclusions, and many times the models generated are not understood by humans, although they are often more efficient [35,42].

In summary, the objective is to create software that reason better with respect to data inputs and explain, as human beings, their outputs, through actions or reports. Unlike automation by robots, AI allows the performance of frequent and bulky reliably and without fatigue. Basically, in relation to an AI, a base code is installed on the machines together with some type of training, and from that, it can learn and evolve alone, according to its interaction with the environment in which it is inserted, that is, the AI will be able to predict some results and make its own decisions, without the interference of human beings, considering that it will need a large amount of information to guarantee comparative power, analyzing the pattern in the commands themselves to learn what is being executed and anticipate actions [43].

Predictive systems aim to analyze data and, based on this data, in general with respect to past history, make predictions for the future. In pattern recognition systems, they have the objective of identifying patterns in data, which can be images, sounds, information, and data. The AI algorithms are able to analyze MRI images of people with and without Alzheimer's identifying the disease patterns and then relating to people without the disease, still considering the possibility of identifying people with future potential to have the disease [44].

Cognitive systems aim to understand and interact with humans through language, needing to do voice recognition, interpreting what is being written or spoken in relation to semantics, abbreviations, ambiguities, grammar, ironies, among other aspects as well as learning to speak even considering details such as accent, in certain cases. Classification systems have the objective of classifying data by similarity, such as the example given of systems with a biological algorithm allowing the identification of a person wearing glasses, even if they have their hair cut, or even changed their eye color [45].

In general, AI has the potential to transform all industries, since it can improve from operational efficiency to productivity, directly reflecting on the company's efficiency and the quality of service, mainly due to the agility of execution and the reduction of the time of implementation. Considering that the collection and storage of data can be carried out in a traditional way, however, its processing needs to be intelligent, where the results are analyzed continuously, making it possible to obtain insights about processes and customers. And from this information about users, it is possible to direct them directly to their demand, eliminating the waiting time and the need to inform personal data whenever the service is performed by a different person, as well as analyzing the data strategically collected, where the information is cross-referenced in a more organized way, streamlining and simplifying the service, creating a consumption profile, through the pattern of behavior and historical data, suggesting options that best fit that moment, as can be seen and felt in the suggestions for complementary or similar items in e-commerces [34].

In short, AI allows companies to gain new insights about their customers and processes, improving strategic aspects, mainly related to efficiency, allowing the creation of more suitable products and smarter services to remain competitive. The trajectory of modern AI is considered in three distinct cycles, such as AI; ML and cognitive computing (CC). Since the period of the initial advent of AI was between 1950 and 1980, beginning the studies and the possibility of carrying out intellectual activities performed by humans, this period is marked as weak AI or application of AI techniques to limit problems. From the 1980s, there was the advent of ML with the aim of offering systems to computers that make them able to learn and develop models for the execution of activities and from the end of the 2000s, there was the advent of CC with the emergence of the first example in order to meet and interact with human beings in the most natural way possible [38,41].

In this sense, what can be seen through AI is that it has the potential to free people from tasks that do not challenge creativity and to easily automate monotonous functions, such as call center operation, document classification, content moderation, operations and support of production processes, opening bank accounts, among others. In this context, companies start investing in AI solutions, such as chatbots, which are conversation platforms that allow interaction between machines and people. Considering that instead of the presence of a collaborator behind the service of their customers, through a chat, it is possible to leave this function to a robot, solving simple questions or performing the first call, in order to understand the customer's need. In addition, the number of services focused on creating content and experience for AI is also growing. Still considering that AI can optimize the service offered to the public in different aspects, such as delivering more efficient communication, increasing transaction agility, and extended personalization [43].

Artificial intelligence improves decision making by helping to simplify analysis processes, especially for a company that values data-based decision making, as long as it is able to organize and check with greater clarity, especially if it is linked to a solution of BD, dealing with a gigantic volume of unstructured data, cloudy or confusing data, which make it difficult to establish strategies. A virtual AI solution employs algorithms that perform more precise segmentation, suggesting goods tuned to the profiles of consumers analyzed, increasing the chances of developing good commercial strategies, together with a high level of process replicability. These systems are capable of performing the same analysis several times, ensuring that any workflow becomes scalable, in the same sense that obtaining relevant information from reports is faster since text mining algorithms are able to analyze a document by finding information [46].

Artificial intelligence contributes to the automation of logical, analytical and cognitive activities, generating greater speed in the treatment of information, replacing operational tasks, such as tightening a screw with precision, optimizing processes and improving business performance, serving as a complement to the automation of physical tasks, especially in production, generally provided by robotic machines, as well as automating bulky computational processes, avoiding the need for people to perform tasks or identifying patterns [47].

Artificial intelligence is able to obtain predictability of markets, behaviors, and processes due to the large analysis of data identifying patterns and establishing predictions from past events, through predictive analysis using ML. It is possible to consider infinite data and scenarios identifying the most likely events, helping in a more effective and strategic decision making. What through BD is made possible is a systematic analysis of the data. Along with AI, this interpretive capacity is deepened, generating more intelligence from the analysis of information, allowing to extract more complex and valuable information from the data [43,46].

Through the use of AI to monitor machines and computer systems of the company, considering the possibility of using Internet of things (IoT) to capture information from the equipment that will feed the AI, which will be grouped with the data generated by the solutions that manage them, until even through the use of data from sensors, cameras, records in monitoring software, among other sources, it is possible to reduce errors, risks, and operational costs in relation to systems which have information coming from databases, reports, history, among others. From the AI's ability to discover bottlenecks, failures and other weaknesses in the company's processes, to reducing errors and increasing operational efficiency, enabling a constant evolution in the use of data, since it uses neural networks with several layers that are used to build more complex and effective interpretive structures, adopting DL which needs BD so that the model can learn from this information. The more data are inserted in the model, the more effective it becomes [48].

For an effective implementation of AI, a lot of data are necessary, since they are important for the process of integration of the AI platform with the workflows and software of the business, making possible the increase of the precision of the information generated by it. These data can be obtained in the internal business management solutions, and a BDA solution is important to generate insights and information from unstructured data, which will be used by the AI solution. This is because processes have a predictable pattern of repetitive activities that can be replicated by ML algorithms, taking into account even complex activities, such as those that require the processing of large data sets in real-time, using AI to observe, decide, or act from well-defined optimization functions [27].

Artificial intelligence is already used in almost all market segments, due to the flexibility and customization of the technology that enables it to contribute both to the development of operational tasks and to its strategic use. In this way, AI is a broad technology with applications so diverse that, in modern days, it is present in the daily lives of all people connected to the Internet, whether in an e-mail, accessing a social network, news site, among others. There is a beneficial potential for organizations that have adopted this technology, regardless of the area in which they find themselves [24].

Though it is possible to obtain insights and intelligence so that there is greater diffusion of resources, with simple computing processes, such as data structures to process, categorize, and analyze data in an intelligent way, BD makes these large amounts of data available for processing and operational processing power to process this information quickly and efficiently. Thus, AI goes beyond mechanical automation, encompassing cognitive processes, which generate a learning capacity, managing to carry out activities that are not only repetitive, numerous and manual but also those that require analysis and decision making [46,48].

1.3.1 Machine learning

With ML, it is possible to automate the construction of analytical models, where through mathematical models, statistics, and neural networks, patterns are found hidden in the data, since it is not programmed in a limited way, just to give a certain conclusion. It is the ability that machines have to learn from a large amount of data, since from that, they can either make decisions autonomously or help humans to do so [49].

A **neural network** is composed of interconnected units, similar to human neurons, that process information through passages in their networks until they find connections or extract meanings that do not exist yet. Instead of programming rules for a machine and waiting for the result, you can let the machine learn these rules on its own from the data, reaching the result autonomously. An example is the personalized recommendations on the streaming platforms indicating the titles according to what the user watches, that is, the system includes data, as the user watches and learns he likes. Therefore, human inference is practically zero, making the machine self-managing [50].

All of this occurs through algorithms and BD. Algorithms are the basis of everything, they are sequences of rules and operations that, when applied to a set of data, bring about a certain result, so that it is possible for machines to learn, the algorithms they are subjected to certain methods, identifying data patterns and creating connections with each other, for the purpose of execution, based on statistical analysis, where any tasks without human interference predicting responses in a more precise way. They are minimum parameters that need to be met and that aim at the better performance of activities, serving to make decisions autonomously or assist in decision making, which is made by a human being [49].

Machine learning involves a method of data evaluation that automates the development of analytical standards, based on the conception that technological systems can learn using data, in order to discover patterns, make decisions, and improve with little human interference carrying out an activity over time [47].

Supervised learning is the method in which the trained algorithms use data input in which the desired output is already foreseen; in it, the learning of the algorithm occurs because it receives data that contain the correct answer, for the detection of a transaction banking system suspected of fraud, it is necessary to have had some kind of previous experience so that the pattern can be identified and this becomes the source of ML [51].

Unsupervised learning, on the other hand, is the method in which the data that the algorithm receives are not labeled, so the effects of the variables are unpredictable. So it is more complex and advanced, because in it the machine itself finds the desired patterns and improves its filters as the use, unlike the previous one, occurs when there are no previous events to be able to base a decision, being necessary that the algorithm needs to have sufficient intelligence to discover what is being requested, occurring through the exploration of a lot of data. **Semisupervised learning** is a category that differs in one aspect from supervised learning, related to the use of labeled data for training, seen in the identification of a person's faces through a webcam, serving to create classifications, categories, and also predictions [52].

Reinforcement learning is a method based on trial and error in search of the actions that produce the best results. In general, it is used in games, navigation, and robotics, with a simple objective of finding the best strategy to be employed presenting three components: who are the agents, who learn or make the decisions; the environment, the respective one where this agent interacts; and the actions, related to what the agent does [53].

Machine learning is present in applications that analyze the route to reach the ideal one, with several patterns, variables, and points; in the same way, as a user's interest in a type of series, film, documentary, music, and whatever is observed by digital streaming platforms; in e-commerce, machine learning is doing it all the time. The most well-known statistical methods used by machine learning to process the data are **Regression** with the objective of determining its characteristics to be able to predict the output results, using the supervised learning algorithms, allowing a mapping of the input variables. **Classification** is also a method that uses supervised learning, working with responses with two or more variables, allowing the grouping of results in several categories. **Clustering** is a method that uses unsupervised learning algorithms, making it possible to find patterns in a database in which the effects of variables cannot be perceived [47,49].

Just like other learning approaches used in ML, according to the objectives proposed by the systems, **Decision Trees** is a structured, systematic and logical way of solving problems, with respect to the graph that presents decisions and their possible consequences. **Bayesian classification** uses a simple probability formula based on Bayes' theorem. **Logistic regression** is related to a statistical form capable of modeling a binomial result with explanatory variables, considering one or more [49,50].

Machine learning platforms are capable of providing computational capacity, as well as data, algorithms, application programming interfaces (APIs). Among other solutions to design, train, and apply models to the area in machines, applications, and processes related to the current world, considering more and more computerization nowadays, in a single day, more information is produced than in the whole past centuries. In this sense, through BD, such processing is possible, since in the face of so much data it is humanly impossible to take advantage of all this data, and for this reason, it has resorted to machines. With the creation of algorithms made to take advantage of this immense amount of data, for what purpose ML is related to it, machines can learn on their own to process this data and give it usefulness. Thus, man-made software will be able to analyze increasingly complex and numerous data automatically and quickly [42].

Artificial intelligence is a somewhat broad concept but it encompasses ML as one of its resources, since AI is understood as computational mechanisms for problem-solving, based on human behavior; therefore, the learning of a machine is nothing more than a subset of AI, so ML has AI, but not all AI has automatic learning [41].

1.3.2 Natural language processing

Natural language processing is the computational capacity to analyze, understand, and develop an almost human language, even considering the stage of communication with normal language between humans and computers. It is an aspect of AI that helps computers understand, interpret, and manipulate human language, resulting from several disciplines, including computer science and computational linguistics, which seek to bridge the gap between human communication and understanding of computers. This processing uses ML techniques to find patterns in large sets of pure data and recognize natural language, since one of the examples of application is the analysis of feelings, in social networks, where the algorithms can search for patterns in posts to understand how customers feel about specific brands and products [54].

It incorporates diverse techniques to interpret human language, from algorithmic and rule-based approaches to statistical and ML methods, taking into account the need for a good variety of approaches, because the text or voice-based data diverge a lot, as well as their practical applications [55].

Basic tasks of natural language processing include stemming, labeling of speech components, tokenization and parsing, language detection, and identification of semantic relationships. In general terms, natural language processing tasks segment the language into smaller, essential parts, trying to understand the relationships between them and exploring how these pieces work together to create meaning [54].

The technology helps computers communicate with human beings in their own language and scales other language-related tasks, related to the possibility that computers hear and interpret speech, read texts, identify feelings, and determine which passages are important. Reflecting that today's machines analyze more data based on language than humans, without fatigue, in a consistent and impartial way, since this gigantic amount of unstructured data are generated every day, from social media to medical records, among others, the automation will be essential for a complete and efficient speech and text analysis [56].

Through it, machines can better understand the texts, which involves context recognition, information extraction, development of summaries, among other characteristics, making it possible to compose texts based on data obtained by computers, being applicable in areas such as customer service and in production of corporate reports [54].

Human language is surprisingly complex and diverse, and people express themselves in many ways, both verbally and in writing, where not only are there hundreds of languages and dialects, but there are also expressions and slang, that is, a unique set of grammatical rules and syntax within each of them, and when writing, people also tend to shorten words and even make mistakes, or omit punctuation; when we speak, regional accents are perceived, as is the mixture of terms from other languages. Although natural language processing is not a new science, this technology is advancing rapidly due to the growing interest in human–machine communication, in addition to the availability of BD. Computing being more powerful and with improved algorithms, the technology aims to study and attempt to reproduce development processes linked to the functioning of human language, using software, programming, and other solutions [57].

Although supervised and unsupervised learning and, specifically, in the same way as DL, are currently widely used for modeling human language, there is also a need for syntactic and semantic understanding, which are not necessarily present in these approaches to ML. So natural language processing is important because it helps to resolve ambiguity in the language and adds a useful numerical structure to the data for many applications such as text analysis or speech recognition [52].

1.3.3 Deep learning

Deep learning is the sum of BD and ML, which through a large volume of information and neural networks with several layers of processing, have the potential to learn complex patterns and transform them into CC, so that machine simulates, in a natural way, the interpretation of lines and images. Enabling to imitate the activity in layers of neurons in the neocortex, linked in 80% of the brain in the region where thought occurs, it is an advanced technique that learns through artificial neural networks, making the computer "think, learn, and act "as a human being, making machines able to recognize objects and translate the voice in real-time [56].

Deep learning refers to a part of ML that uses complex algorithms to "mimic the neural network of the human brain" and learn an area of knowledge with little or no supervision. The advantage of DL algorithms is their ability to learn in large amounts of data in a way that do not need to be supervised, thus being a valuable tool for BD, where most of the data are of this nature, also called unstructured data, since machines can learn complex data abstractions through of a hierarchical learning process very similar to what happens with the brain, especially in the visual cortex. It is a special type of ML that involves articular neural networks with several layers of abstraction, being applied for pattern recognition and classicization applications supported by data sets [51].

With DL, computers perform tasks as if they were human beings, being able to perform speech recognition, image identification, and make behavior predictions; however, the technique does not organize data for execution using previously defined equations. The system itself sets basic parameters on this information and conducts training on the machine to learn on its own using standard recognition in different layers of processing [49].

Deep learning algorithms present an innovative approach since they dispense with a large part of preprocessing by automatically generating invariant properties in their hierarchical representation layers, where such methods have produced excellent results in different applications including natural language processing, being the base technology behind numerous tools such as online language translators and personal assistants for mobile devices [47].

Deep learning is one of the techniques used by ML, consisting of training computers to perform activities similar to human beings, such as speech recognition, predictions, and image identification, pattern recognition, in the same way, that it improves the opportunities that the algorithms have to learn through the use of neural networks. Thus, ML is one of the developments of AI that allows the system to be able to analyze data, make predictions, and even learn without human interference during the process; however, DL is a deepening of ML, having a differentiated learning process that allows the analysis of voice and images, expanding the capacity of AI.

For medicine, the DL algorithms allow the implementation of detection tools guaranteed in AI, very useful for the areas of tomography, magnetic resonance, hematology, and x-rays, which need to evaluate results in a short time. Since a machine trained based on the principles of DL can identify from traits that indicate diseases such as cancer and tumors in imaging exams to faces for facial recognition on smartphones and security systems [40,44,58].

In a way, machines are already taught to understand past documents and information and can answer questions about content; however, knowledge is limited to the extent of these documents, and just as the increasing amount of online data allow models computational, DL can make use of a significantly greater number of natural language, attributing to the computer in-depth knowledge on various topics, using methods based on this technology in different applications, such as computer vision, speech recognition, and language processing [48].

All of this is possible due to the processing based on neural networks in "layers" on which the algorithms build their representation in natural language. Since the learning process occurs between its layers of mathematical neurons, where the information is transmitted through each layer, whereas this scheme, the output of the previous layer is the input of the posterior layer. Thus, DL “trains” machines to perform activities as if they were human, and processes this data in a certain way, so that it can, for example, decipher the language used, and as it is fed and exposed to data, the computer relates terms, words, and images to infer their meaning. In a nutshell, DL, like ML, is an offshoot of AI; however, it is a new concept that uses artificial neural networks in a deep way, with more layers of training to understand user behavior and, thus, teaches the machine [43].

1.4 Big data analytics

Big data analytics can be understood as a concept of clusters of structured and unstructured data that are generated all the time. These procedures help companies to explore their data, analyzing and obtaining answers almost in real-time, and using them to know the new market trends in order to create new opportunities. It is the science of examining raw data in order to find patterns and draw conclusions about that information, applying an algorithmic or mechanical process to obtain information. This data analysis may be divided into three categories such as social data, enterprise data, and personal data, which explores the data in order to create principles to optimize the understanding of scenarios and patterns. It is the enhanced analysis of large amounts of raw data for extracting information and insights for a given business, used in the most varied segments allowing organizations to make correct decisions, in addition to testing processes that already exist, enabling anticipation of trends and considering methods to answer questions through the application of an algorithmic process [3,59–64].

Big data analytics is a process that involves examining data by drawing useful business conclusions, using specialized software and technologies such as Qlik, consisting of an end-to-end platform that includes data integration, user-oriented business intelligence, and conversational analytics, which are used extensively in industries to help companies make decisions. Since all these standards are developed with the objective of filtering and taking all useful information to the company, containing a whole cycle that begins with data extraction, organization, treatment, and understanding of the data [4,65–67].

The term data analytics refers to a number of applications, such as business intelligence tools; however, the difference between the two is that while the latter is focused on the use of data within the business, BDA allows companies to use information already available to identify opportunities and gain insights, with a broader focus, which can also be used in academic research [6,62–68].

The main benefits that the implementation of BDA brings are related to the identification of precise patterns occurring through personalized segmentation, studying the behavior, whether by age, behavior and/or region, of the company's target audience; reduction of costs in relation to cloud data storage is an opportunity that companies have to conduct more efficient business, reflected with the speed and intelligence of data analysis, enabling all necessary information in real-time; development of products and services based on the needs and desires of customers, made with the collection of concrete data from consumers, and subsequently, the realization of DM, recognizing standards and rules; and competition analysis, being able to anticipate the customer's needs and desires, making it possible to identify different and better paths than those followed by the competition [13,63–69].

Considering that alone, BD is not an efficient strategy to visualize the answers to the questions, but BDA can organize and put all this data in perspective, since it deals with visualization tools, such as intuitive dashboards with responses in real-time, which can be implemented regardless of the size of the company, as the implementation is not complex, but arduous. Unlike Business Analytics, which has a low learning curve, BDA makes it possible for all employees, even those less informed about data, to gain insights from them, essential for a company to be able to incorporate processes guided by your operations [15,61–66].

Big data and BDA are two essential tools by which errors within an organization can be known instantly, making the real-time perception of errors is what helps companies to react quickly to mitigate the effects of operational problems; yet considering that it is possible to monitor products that are not used by its customers and to respond promptly to failures. Another great benefit is being able to observe the competition's strategies in real-time, allowing you to stay one step ahead of the competition or receive notifications when a direct competitor decides to change its strategy by reducing its prices [17,64–69].

Fraud can also be detected the moment it happens and the necessary measures can be taken quickly to limit the damage, as well as assessing the implementation of BDA tools reduce the burden on a company's overall IT landscape, freeing up resources and saving money with respect to that there will be no time wasted with poorly functioning databases. Through real-time analysis indicating exactly how sales are being made can lead to increased revenue and this is another major advantage of BDA [13,61–68].

Tracking customer trends in real-time is the ultimate benefit of BDA, and gaining insights into competitive offers, promotions, and customer movements are valuable insights into acquisition and retention trends. In this way, BDA is the collection, organization, and analysis of large data sets to discover patterns and other useful information, helping companies of all sizes to better understand the information contained in the data they store and identifying the most important data for business decisions [4,61–67].

Big data analytics is much more than the concern of managing analytics since there are several aspects of BD such as the varied new data sources that keep growing and proliferating; the volume of data created each day increases exponentially; and the speed of data creation influencing decisive elements. With BDA, it is possible to help in all stages in which business managers eliminate what was previously done under a combination of intuition, experience, and some level of analysis to make business decisions in modern times through technology by the development of science instead of the intuition based on facts, patterns, and relationships that exist in the behavior of the client/user [22,67–72].

1.4.1 Descriptive analysis

Descriptive analysis is perhaps the most popular modality of data analysis within the concept of BD. It is the real-time understanding of events that creates panels, notifications, and alerts to tell what happened in the past, and this information is usually displayed in graphs bars, in tables, among other forms of visualization, allowing the manager a global view of the monitored processes; but it is not enough to point out the reasons why something happened or what can change from there. This is the best-known model as it helps to predict future scenarios based on the analysis of database patterns. So it is possible to make more precise decisions with this type of analysis of data referring to a specific process, and it becomes possible to make immediate decisions with a huge safety margin since it is made and based on hard data [17].

Descriptive or exploratory data analysis is the most basic process for any type of data analysis, although simple on several occasions it is a process that, when combined with a consistent methodology of analytics answers important management questions almost immediately, can be related to univariate investigating the behavior of a single variable (or column); bivariate related to the investigation of how two variables are related; and it can be multivariate regarding the investigation of the behavior of several variables, multicolumnar analysis [8].

Descriptive analysis is seen as a way of visualizing data, understanding how a database is organized and what it means at the present time with regard to the need to relate it to past patterns or future projections. on the contrary, it is more useful and uses past data to model what might happen in the future, in the sense of how a customer can respond to a marketing campaign, for example [18].

The purpose of this model is to allow the understanding of events in real-time, widely used in situations such as credit analysis, where the bank evaluates the information of the individual, checking the risk involved in the process, defining interest rates, for example, through techniques such as regression and prediction analysis, pattern matching, predictive modeling, and multivariate statistics. In addition to its industrial use, it is also widely used by companies in the stock market and financial investments, providing specific data on variants at a given time and the market [19].

As it does not issue a judgment of value, it is indicated to visualize the data and understand the impact on the present, but without having any relation with the past or the future, it helps in making immediate decisions with tranquility and security. The central element of the predictive analysis is the predictor considering a variable that can be measured by an individual entity predicts future behaviors [21].

1.4.2 Predictive analysis

The methods used by predictive analysis are statistical and historical data, in addition to DM and AI, to find meaning in large amounts of data to project future behaviors of the public and the market, in addition to assessing fluctuations in the economy and trends in consumption [3].

This means that, in predictive analysis, after data collection, a statistical model is formulated, predictions are made, and then the model is validated or revised. It is a way to use simple or advanced metrics and statistical techniques to understand and explain how the data are applicable in many areas of research, including meteorology, security, genetics, economics, and marketing [7].

In other words, the predictive analysis combines statistical techniques, such as DM, ML, and AI, ranging from classic regression models to complex models and algorithms that involve DL, dimensionality reduction methods, ensemble learning, and swarm intelligence to find meaning in large amounts of data, making predictions, applicable in many areas of research, including meteorology, genetics, economics, security, and marketing [12].

1.4.3 Prescriptive analysis

The prescribed analyses use techniques such as optimization and A/B testing, being a method used in digital marketing to provide insights into the performance of what is being tested, which may be applicable to an email marketing or landing page, for example, aimed at to guide employees and managers on the best way to fulfill their roles in an organization, and can help a salesperson decide what kind of discount is most interesting to offer the customer to close a sale, for example [19].

The idea is to verify the consequences of the actions taken, making it possible to know what should happen when choosing certain attitudes; it is relevant because it defines the path to be taken so that the action occurs as expected. In other words, a goal is outlined and, based on that, the paths that must be followed to reach it are indicated, is considered a more complex analysis, since you must know data science techniques and the ecosystem in which it is inserted [20].

A little deeper than the predictive analysis, the prescriptive analysis translates the forecasts into viable plans for the business, identifying which are the best measures to be implemented, despite their importance; this can be done by listing standards and applying filters by specifics, allowing to have a real context of the situation and the effects of the actions, as in its application in the health context, being able to outline disease patterns for the patients, verifying how each attitude will impact on this group, and allowing verification of the best alternative [24].

This type of assessment is used to raise hypotheses about possible consequences of the actions taken by the company, identifying system behaviors under different configurations, ensuring that performance indicators such as waiting times, queue length, among others, are satisfied. It also has the optimization that supports the tactical, operational, and strategic planning of an ongoing business, taking advantage of linear programming to identify the best result, given the restrictions and objectives [15].

The prescriptive analysis is uses as statistical tools, both descriptive and predictive, in line with business management, generating recommendations for actions to be taken automatically or semiautomatically, with the aim of providing the strategies adopted by the companies in the shortest possible time to achieve better results. While predictive analysis is limited to saying how the future is likely to be, the prescriptive analysis provides support for making decisions that will change the future, that is, what must be done for the desired future to be realized [10].

1.4.4 Diagnostic analysis

While the previous types of analysis are related to decision making, the diagnostic analysis is done after the work done, discovering the causes and reasons, using techniques such as drill-down, data discovery, DM, and correlations. It is a form of analysis that examines data or content by answering questions related to their events and reasons. Monthly sales reports are very practical examples of diagnostic analysis, fundamental to know what is working, what is not, why and how it affects business, serving as a comparison with other periods. In the same way, more in-depth reports allow the user to decide the best path to follow, identifying possible errors and leading to correct decisions [6].

Diagnostic analysis can provide a report that reveals the details of each split of actions that led to a particular problem in the process, taking into account that it is possible to change nonfunctional strategies and reinforce those that are being effective, even considering that despite being numerous, the data say nothing when an analysis is not carried out according to structured standards, capable of extracting usual indicators from them [73,74].

1.5 Discussion

It is the reliable data that provide analysis and insights for companies to decode, innovate, and win the trust of the customer. Since the technology combined with data analysis, which is generated, in millions and uninterruptedly, every second of the day, is in the online or offline environment, they are revolutionizing the way of doing business in all sectors, considering that companies generate a very large volume of data every day, which grows quickly and there is a need to analyze this large mass of data.

Initially, data preparation is the process of collecting, cleaning, normalizing, combining, structuring, and organizing data for analysis, and this is an initial and fundamental step for successful handling with BD since it increases the quality of data. Consecutively results of DM thus disregard the "poor" data, of poor quality, which generate incorrect and unreliable results at the end of the process using the technologies.

Data mining consists of using a set of technologies and techniques that allow automating the search of large volumes of data by patterns and trends that are not detectable by simpler analyses. Using sophisticated algorithms to segment the data and assess the probability of occurrence, certain events in the future, since this type of analysis, generate a high-value basis for making strategic decisions, allowing the early detection of market trends and making it possible to anticipate actions to respond to new scenarios.

Big data is the technology that allows the processing and analysis of data that uses large volumes, fast speeds, enormous variety, and greater veracity, since this technique allows to transform data into relevant information, which in turn are converted into results for decision making in real-time. Big data solutions were created to accelerate the decision-making process and reduce costs and server processing time to understand consumer behavior and organizations' needs, by investigating the data collected and generating useful insights, generating useful information at the appropriate time which is crucial to the success of any business.

Solving complex problems of increasing volumes of data, both structured and unstructured, is a major challenge that requires adequate analysis techniques for modeling information, generating adequate information quickly and intelligently from extreme databases through ML algorithms.

The possibilities are endless for BD with AI that allow machines to learn, reason, solve problems, and understand the language instantly, since, with BD, companies start to update the way they process information about customers and the market. Considering that this practice is a cultural change in the way of handling information, where it is no longer necessary to place all records in standardized forms or structured bases. Since it currently deals with a mass of data, which are more extensive and tend to be statistically more relevant, because it is more flexible, including unstructured information, and it allows to provide much more complete and comprehensive analyses, in this sense.

Big data analytics and ML allow marketers to use intelligence for the purpose of detecting, learning, and optimizing operations.

For efficient data analysis, digital technologies play an unavoidable role in business, and in AI, through ML, it has become the most sought-after technology today that helps innovate and transform business, gaining an advantage in a particular sector that a company is, since organizations are increasingly involved in predictive analytics based on ML. Predictive analysis is an effective decision-making tool that helps in redefining all business activities, redistributing resources, and initiating an organizational transformation.

In order to fully use the accumulated potential of their data and obtain a competitive advantage, companies quickly integrate new analytical methods into their overall strategy, making it possible to maintain competitiveness and optimize the use of resources, taking advantage of available data, considering and including data quality. Since BD requires cataloging what is available and its associated quality metrics before it can be used for any type of analysis. With BD, it is possible to obtain more information about the business and use it to analyze markets, forecast sales, and identify ways to reduce costs, where this knowledge can be considered a precursor to innovation and, therefore, to differentiate businesses.

Still considering that ML also takes place in scalable cloud-based infrastructure, making it important for organizations to be able to send data to cloud storage, where computing capacity is unrestricted. United with neural networks and DL algorithms, ML is successfully discovering and using hidden patterns in unstructured data sets, revealing new information from these data sets. Since currently with the technological potential that companies have, they can collect data about their customers in seconds, quickly processing and extracting real-time information from the data. Through these technologies comes the ability to identify the real needs and interests of its customers to be able to offer what they really need or want, at the most appropriate time, in the best way, at the lowest possible cost. Organizations are creating strategies to take advantage of large volumes of BD almost immediately and repositioning several of their business processes.

Companies apply all three types of analysis while working on a data set. Descriptive analysis is the basic form of analysis that aggregates large data and provides useful information from previous records. Still considering prescriptive analysis that uses a combination of rules business, ML, and computer modeling to recommend the best course of action for any prespecified outcome, usually implemented through workflows, but using predictive analysis creates more value for companies as it helps them predict future results. Predictive analysis is a popular concept today that uses historical data, AI and ML to predict future results, and it is a way to take advantage of all this information, obtain new tangible insights, and enable a company to be ahead of the competition.

Of the types of BD analysis, predictive analysis is perhaps the best known, since it is possible to define it as an analysis of future possibilities that happens from the identification of past patterns in your database, consisting of a set of technologies focused on anticipating future behaviors or estimating unknown results based on current and historical data, employing various techniques of statistical models, ML, DM, and AI to gather all the technological information to make predictions. This type of analysis allows managers to map possible futures in their fields.

The idea is to stop making decisions based solely on intuition, managing to establish a more solid prognosis for each action, using DM, statistical data, and historical data to know the trends, processing and analyzing different data sources and memorizing the best paths with AI, making companies able to understand and analyze analytically information immediately, making better and more assertive decisions in less and less time.

So it is not enough to just collect millions of data, it is necessary to be able to understand them, considering that one of the predictive analysis modalities is text analysis. It is possible to quickly discover the correlations between some keywords in customer complaints, making it possible to allocate adequate resources to deal with the dissatisfaction of a certain group of customers or users, and can also segment them by location, preferences, age, type of product purchased/contracted service, among other characteristics.

Data are always being generated by digital technologies, whether through the use of applications on mobile devices, social media, online shopping, research on search engines, among others. In this sense, all this information is combined with other data sources and become BD. However, the predictive analysis does not tell you what will happen, but what can happen with an acceptable level of reliability and also including hypothetical scenarios and risk assessment; in short, it uses a huge amount of data to determine probabilities of future behavior.

Companies combine BD with technologies such as ML and AI to further improve their performance, making it possible to determine patterns and behaviors, which allows the company to offer its customers faster and more personalized experiences, where predictive models use this data and technologies to better understand customers, products, and partners to identify potential future risks and opportunities for a company. The combination of BD, with predictive analytics taking advantage of cloud computing, which brings fast processing, can provide important information immediately and accurately.

In recent years, technology giants, as well as most companies in the digital market, have understood the strategic potential of capturing data that flow in their business environments, especially with the increase in investments in digitization leading them to make AI becomes mainstream (taste of dominant popular character) with more focus on harnessing the power of DL techniques as well as accessing synthetic data to simulate meaningful perceptions, and along with the advent of technologies that allow previously performed analyses to happen automatically, in unimaginable volumes, with high speed, and with the highest possible precision indexes, making these organizations able to establish a balance within their data, technology, and collaborators, completely transforming their business into a predictive analysis model guided by AI, provided that what is smarter helps in taking faster decision making.

Allied with the text mining technology, as already described, it is possible to analyze text data from the web, comment fields, books, and other text-based sources, discovering previously unseen insights, using natural language processing technology or language learning technology, machine panning large amounts of information in documents, considering e-mails, twitter feeds, surveys, blogs, competitive intelligence, among others, discovering new topics and term relationships.

With the implementation of AI, migration to a data-driven culture was performed, coupled with the leverage of advanced technologies at the corporate level, where the resources that control AI and predictive analysis can be applied to virtually any business domain in any industry, as well as protecting information technology (IT) work environments by detecting security fraud, and cyber data security and theft, leading to smarter business movements, more efficient operations, greater profits, and happier customers.

Big data is present in organizations of all segments, such as banking. Since in this type of organization, there are large amounts of information flowing in numerous sources, taking into account the application of methods that involve innovative forms of management and analysis of BD, including customers and increasing their satisfaction, while also minimizing risk and fraud and maintaining regulatory compliance. Governments are able to take advantage of and apply analysis to the "mountains" of data to which they have access, increasing public transparency to improve congestion and crime prevention; however, it is necessary to ensure that these institutions safeguard essential rights such as privacy while working with data from entire populations.

In educational institutions, teachers and teaching managers can process student data intelligently, since their school systems and curricula analyze large volumes of data, making it possible to identify students at risk, assessing whether they are making progress enabling the implementation of a more adequate system of evaluation and support for teachers and principals. In healthcare, with respect to patient records, prescription information, treatment plans, through BDA, hospital managers and healthcare providers can discover insights capable of improving infinite indicators, including patient care, as long as everything is needed be done quickly, accurately, and, in some cases, with sufficient transparency to meet strict regulations.

In retail, building relationship with the customer is fundamental and the best way to manage this relationship with the use of BD. Through a data-driven way, technology is able to trace the social profile of its customers, understanding their consumption habits, predicting market trends, and evaluating the most effective way of dealing with transactions. In the same sense, it involves the manufacturing of products on a large scale being benefited by BDA strategies with regard to increasing the quality and volume of production while minimizing waste processes related to a culture based on data analysis in its sales sectors, supply, and logistics.

Thus, the importance of BD does not revolve around the amount of data that are obtained, but what is done with it, through the ability to gather and analyze large volumes of information, quickly and with quality, transforming the way companies understand the market and deal with customers. Since the world becomes more connected, there are enormous growth in man-made data, as well as an exponentially increasing amount of data created by machines.

Finally, through BDA, it is possible to take data from any source and analyze it to find answers that allow the development of new products and optimized offers; cost reduction and time reductions; intelligent decision making, combining large volumes of data with high-power analysis, determining root causes of failures, problems, and defects in almost real-time; recalculating entire risk portfolios in minutes; as well as identifying fraudulent behavior before an organization is affected.

1.6 Trends

Although not very widespread regarding the control of home appliances, smartphones will become a kind of universal control, through IoT, with increasingly intelligent devices, several companies are already developing products that can connect to the network and be operated through the Internet, which relates that as organizations start to provide better IoT applications, new ways of collecting, managing, and analyzing information will be created [75–79].

Intelligence security is another element that many organizations will be incorporating, with BD analysis being noticeable in their security strategy, security log data will provide a treasure trove of information ranging from a history of cyber attacks to information about how to best obtain protection from future attacks.

A major trend is the analysis of data in the cloud, since many other business processes migrate to cloud structures which offer great opportunities to process gigantic data sets. Pushed by AI, chatbots will be increasingly implemented to handle queries and customer service, in the same sense are virtual assistants, dealing with a conversation technology that provides personalized interactions without the need for human involvement, processing large volumes information. So BD will allow these assistants to be programmed to collect and analyze the data obtained during conversations.

The use of predictive analytics related to the fact that technologies become more powerful, larger data sets can be analyzed, and this increases the predictability. Big data analytics solutions will have to be extremely capable not only to analyze but also to process the data and understand the reasons why certain events occur. With predictive analysis using BD, with regard to understanding consumer behavior, it will be possible to predict what may happen in the future.

In-memory computing is the storage of information in random access memory, RAM, and not in complicated relational databases that operate on comparatively slow disk drives, which will help companies quickly detect patterns by analyzing data volumes in real-time and performing its operations quickly, related to the drop in-memory prices in the current market is also an important factor that will contribute to the growing popularity of this technology.

As BDA capabilities advance, some companies will have to invest more and more in ML, focusing on allowing computers to learn new things without being explicitly programmed, meaning they are able to analyze large databases, existing data, and BD, reaching conclusions that change how the application behaves.

The emergence of the semantic graph will become the backbone that supports BD and BDA in a constantly changing data landscape, as will the growing adoption of the next generation of embedded analytics, which references the concise analysis provided in the context of specific applications and interfaces accelerating decision making, being a style of incorporating and curating concise and contextual analyses including no-code and low-code development methods.

Finally, there is DL, being a set of techniques that uses neural networks to detect interesting trends in massive amounts of unstructured data, inferring relationships without the need for models or explicit programming. In this context, each company will use the practice in its operations, in which strategies they will be taking to stand out in comparison to the competition.

Thus, the combination of these analysis tools with BD is an important part of maintaining the competitiveness of companies, and such a step will help in creating the right conditions for data specialists to test theories based on the information they have.

1.7 Conclusions

Big data is the concept that covers not only the data that are collected during this process but also the ways that they are used for a given objective, fulfilling a prominent role for managers who use data analysis to identify bottlenecks and areas of concern for low productivity.

Big data analytics is increasingly present in everyday life being able to change the landscape of companies, especially with the amazing increase in information; however, these benefits extend with the advancement of technologies and Industry 4.0, considering that the world is on the verge of a new technological revolution that will most likely be marked by the emergence of robots, virtual reality, autonomous vehicles, AI, IoT devices, and many other great technologies that are yet to come, but data are at the center of this revolution and how data can be analyzed and processed efficiently is utmost important.

In the past, periodically the vast majority of systems needed to back up and combine data in a database, where reports could be run and all management members could have a view of what was happening; however, this technology could not handle flows multiple and continuous data, nor modify them in real-time, without taking into account the volume of data.

Thus, BD emerged to analyze a large data flow with high performance, low computational cost and high scalability, made according to the business profile, through the use of resources that are aligned with the company's needs, leading to investment capable of generating a high return. And soon, BDA solutions appeared offering highly indexed and optimized data structures, cloud hosting, designed reporting interfaces, and automatic archiving and extraction capabilities to provide accurate analysis that supports decision making.

Thus, BD is the concept to make this happens with technological inclusion enough to store these increasingly larger data volumes, obtaining the necessary information that will drive business intelligently, whether in creating messages with greater relevance for users through campaigns created to increase service returns and expand the spontaneous disclosure of goods even in the best engagement of consumers.

In this way, technologies are available; however, it is necessary to invest time, money, and resources to fully implement a BD solution, since it is the compass of any successful entrepreneur who wants to use privileged knowledge of the market to suppress competition, regardless of the economic moment.

References

[1] Brynjolfsson, E. and Mcafee, A. "The business of artificial intelligence." *Harvard Business Review*. 2017; 1: 1–20.

[2] Ransbotham, S., Gerbert, P., Reeves, M., Kiron, D., and Spira, M. "Artificial intelligence in business gets real." *MIT Sloan Management Review*. 2018; 1: 60280.

[3] Gunasekaran, A., Papadopoulos, T., Dubey, R., et al. "Big data and predictive analytics for supply chain and organizational performance." *Journal of Business Research*. 2017; 70: 308–317.
[4] Pandya, D., Srivastava, A., Doherty, A., et al. "Increasing production efficiency via compressor failure predictive analytics using machine learning." *Offshore Technology Conference*. 2018.
[5] Laha, A. K. (ed.). *Advances in analytics and applications*. Springer; 2019.
[6] Balali, F., Nouri, J., Nasiri, A., and Zhao, T. "Data analytics." In A. Nasiri, J. Nouri, T. Zhao, and F. Balali (eds.), *Data intensive industrial asset management*. Cham: Springer; 2020. pp. 105–113.
[7] Chipidza, W., George, J., and Koch, H. "Chartering predictive analytics: A case study." *Twenty-Second Americas Conference on Information Systems, San Diego, 2016*. 2016.
[8] Sun, Z., Sun, L., and Strang, K. "Big data analytics services for enhancing business intelligence." *Journal of Computer Information Systems*. 2018; 58 (2): 162–169.
[9] Breuker, D., Matzner, M., Delfmann, P., and Becker, J. "Comprehensible predictive models for business processes." *MIS Quarterly*. 2016; 40(4): 1009–1034.
[10] Kibria, M. G., Nguyen, K., Villardi, G. P., Zhao, O., Ishizu, K., and Kojima, F. "Big data analytics, machine learning, and artificial intelligence in next-generation wireless networks". *IEEE Access*. 2018; 6: 32328–32338.
[11] Shah, N. D., Steyerberg, E. W., and Kent, D. M. "Big data and predictive analytics: Recalibrating expectations." *JAMA*. 2018; 320(1): 27–28.
[12] Dubey, R., Gunasekaran, A., Childe, S. J., Blome, C., and Papadopoulos, T. "Big data and predictive analytics and manufacturing performance: integrating institutional theory, resource-based view and big data culture." *British Journal of Management*. 2019; 30(2): 341–361.
[13] Erl, T., Khattak, W., and Buhler, P. *Big data fundamentals: concepts, drivers and techniques*. Prentice Hall Press; 2016.
[14] Zomaya, A. Y. and Sakr, S. (eds.). *Handbook of big data technologies*. Berlin: Springer; 2017.
[15] Furht, B. and Villanustre, F. *Big data technologies and applications*. Berlin: Springer; 2016.
[16] Stucke, M. E. and Grunes, A. P. "Introduction: Big data and competition policy." In *Big data and competition policy*. Oxford University Press; 2016.
[17] Grable, J. E. and Lyons, A. C. "An introduction to big data." *Journal of Financial Service Professionals*. 2018; 72(5): 17–20.
[18] Younas, M. "Research challenges of big data." *Service Oriented Computing and Applications (SOCA)*. 2019; 13: 105–107.
[19] Johnson, J. S., Friend, S. B., and Lee, H. S. "Big data facilitation, utilization, and monetization: Exploring the 3Vs in a new product development process." *Journal of Product Innovation Management*. 2017; 34(5): 640–658.
[20] Al-Mekhlal, M. and Khwaja, A. A. "A synthesis of big data definition and characteristics." *2019 IEEE International Conference on Computational*

Science and Engineering (CSE) and IEEE International Conference on Embedded and Ubiquitous Computing (EUC). IEEE; 2019. pp. 314–322.

[21] He, W., Wang, F.-K., and Akula, V. "Managing extracted knowledge from big social media data for business decision making." *Journal of Knowledge Management*. 2017; 21(2): 275–294.

[22] Skourletopoulos, G., Mavromoustakis, C. X., Mastorakis, G., *et al*. "Big data and cloud computing: a survey of the state-of-the-art and research challenges." In C. X. Mavromoustakis, G. Mastorakis, and C. Dobre (eds.), *Advances in mobile cloud computing and big data in the 5G era*. Cham: Springer; 2017. pp. 23–41.

[23] Erraissi, A., Belangour, A., and Tragha, A. "A big data Hadoop building blocks comparative study." *International Journal of Computer Trends and Technology*. 2017; 18.

[24] Lee, I. "Big data: Dimensions, evolution, impacts, and challenges." *Business Horizons*. 2017; 60(3): 293–303.

[25] Miller, J. D. *Big data visualization*. Packt Publishing Ltd.; 2017.

[26] Sivarajah, U., Kamal, M. M., Irani, Z., and Weerakkody, V. "Critical analysis of big data challenges and analytical methods." *Journal of Business Research*. 2017; 70: 263–286.

[27] Uma, K. and Kannan, A. "The challenges and issues on big data analytics." *Journal of Advanced Research in Dynamical and Control Systems*. 2017; 2017 (Special Issue 2): 1242–1258.

[28] Erevelles, S., Fukawa, N., and Swayne, L. "Big data consumer analytics and the transformation of marketing." *Journal of Business Research*. 2016; 69 (2): 897–904.

[29] Tsirogiannis, D., Binkert, N. A., Harizopoulos, S., et al. Patent No. 9,613,068. Washington, DC: U.S. Patent and Trademark Office; 2017.

[30] Cheng, L., Liu, F., and Yao, D. "Enterprise data breach: Causes, challenges, prevention, and future directions." *Wiley Interdisciplinary Reviews: Data Mining and Knowledge Discovery*. 2017; 7(5): e1211.

[31] Hilbert, M. "Big data for development: A review of promises and challenges." *Development Policy Review*. 2016; 34(1), 135–174.

[32] Leclerc, B. and Cale, J. *Big data*. Routledge; 2020.

[33] Vijayakumar, V., Neelanarayanan, V., Rao, P., and Light, J. *Proceedings of 6th International Conference on Big Data and Cloud Computing Challenges*. Singapore: Springer; 2020.

[34] Russell, S. J. and Norvig, P. *Artificial intelligence: A modern approach*. Malaysia: Pearson Education Limited; 2016.

[35] Krishnamoorthy, C. S. and Rajeev, S. *Artificial intelligence and expert systems for engineers*. CRC Press; 2018.

[36] França, R. P., Peluso, M., Monteiro, A. C. B., Iano, Y., Arthur, R., and Estrela, V. V. "Development of a kernel: A deeper look at the architecture of an operating system." In *Brazilian Technology Symposium*. Cham: Springer; 2018. pp. 103–114.

[37] França, R. P., Iano, Y., Monteiro, A. C. B., and Arthur, R. "Potential proposal to improve data transmission in healthcare systems." In *Deep learning techniques for biomedical and health informatics*. Academic Press; 2020. pp. 267–283.
[38] Chatterjee, S. 2062: The world that AI made [Book Review]. *IQ: The RIM Quarterly*. 2019; 35(1): 14.
[39] Seyedmahmoudian, M., Horan, B., Soon, T. K., et al. "State of the art artificial intelligence-based MPPT techniques for mitigating partial shading effects on PV systems: a review." *Renewable and Sustainable Energy Reviews*. 2016; 64: 435–455.
[40] Du-Harpur, X., Watt, F. M., Luscombe, N. M., and Lynch, M. D. "What is AI? Applications of artificial intelligence to dermatology." *British Journal of Dermatology*. 2020; 183(3): 423–430.
[41] Li, B. H., Hou, B. C., Yu, W. T., Lu, X. B., and Yang, C. W. "Applications of artificial intelligence in intelligent manufacturing: a review." *Frontiers of Information Technology and Electronic Engineering*. 2017; 18(1): 86–96.
[42] Wilson, H. J., Daugherty, P., and Bianzino, N. "The jobs that artificial intelligence will create." *MIT Sloan Management Review*. 2017; 58(4): 14.
[43] Flasinski, M. *Introduction to artificial intelligence*. Springer; 2016.
[44] Gunter, J. L., Spychalla, A. J., Ward, C. P., et al. "Automating cerebral microbleed detection in support of Alzheimer's disease trials using a convolutional neural network AI." *Alzheimer's and Dementia*. 2018; 14(7): P1530–P1531.
[45] Reid, D. J., Pilling, M. J., and Ivancevic, V. G. *Mathematics of autonomy: Mathematical methods for cyber-physical-cognitive systems*. World Scientific Publishing Co Pte Ltd.: 2017.
[46] Goodfellow, I., Bengio, Y., and Courville, A. *Deep learning*. MIT Press; 2016.
[47] Buduma, N. and Locascio, N. *Fundamentals of deep learning: Designing next-generation machine intelligence algorithms*. O'Reilly Media, Inc. 2016.
[48] Al-Garadi, M. A., Mohamed, A., Al-Ali, A., Du, X., and Guizani, M. "A survey of machine and deep learning methods for Internet of things (IoT) security." 2018. arXiv preprint arXiv:1807.11023.
[49] Mohri, M., Rostamizadeh, A., and Talwalkar, A. *Foundations of machine learning*. MIT Press; 2018.
[50] Raschka, S. and Mirjalili, V. *Python machine learning: machine learning and deep learning with Python, scikit-learn, and TensorFlow 2*. Packt Publishing Ltd.; 2019.
[51] Zhou, Z. H. "A brief introduction to weakly supervised learning." *National Science Review*. 2018; 5(1): 44–53.
[52] Yi, X., Walia, E., and Babyn, P. "Unsupervised and semi-supervised learning with categorical generative adversarial networks assisted by Wasserstein distance for dermoscopy image classification." 2018. arXiv preprint arXiv:1804.03700.
[53] Sutton, R. S. and Barto, A. G. *Reinforcement learning: An introduction*. MIT Press; 2018.

[54] Goldberg, Y. "Neural network methods for natural language processing." *Synthesis Lectures on Human Language Technologies*. 2017; 10(1): 1–309.
[55] Aletras, N., Ash, E., Barrett, L., et al. *Proceedings of the Natural Legal Language Processing Workshop 2019*. 2019.
[56] Deng, L. and Liu, Y. (eds.). *Deep learning in natural language processing*. Springer; 2018.
[57] Thanaki, J. *Python natural language processing*. Packt Publishing Ltd.; 2017.
[58] Monteiro, A. C. B., Iano, Y., França, R. P., and Arthur, R. "Methodology of high accuracy, sensitivity and specificity in the counts of erythrocytes and leukocytes in blood smear images." *Brazilian Technology Symposium*. Cham: Springer; 2018. pp. 79–90.
[59] Skourletopoulos, G., Mavromoustakis, C. X., Chatzimisios, P., Mastorakis, G., Pallis, E., and Batalla, J. M. "Towards the evaluation of a big data-as-a-service model: A decision theoretic approach." *2016 IEEE Conference on Computer Communications Workshops (INFOCOM WKSHPS)*. IEEE; 2016. pp. 877–883.
[60] Papadokostaki, K., Mastorakis, G., Panagiotakis, S., Mavromoustakis, C. X., Dobre, C., and Batalla, J. M. "Handling big data in the era of internet of things (IoT)." *Advances in mobile cloud computing and big data in the 5G era*. Cham: Springer; 2017. pp. 3–22.
[61] Skourletopoulos, G., Mavromoustakis, C. X., Mastorakis, G., et al. "Towards mobile cloud computing in 5G mobile networks: applications, big data services and future opportunities." *Advances in mobile cloud computing and big data in the 5G era*. Cham: Springer; 2017. pp. 43–62.
[62] Vakintis, I., Panagiotakis, S., Mastorakis, G., and Mavromoustakis, C. X. "Evaluation of a web crowd-sensing IoT ecosystem providing big data analysis." *Resource management for big data platforms*. Cham: Springer; 2016. pp. 461–488.
[63] Mavromoustakis, C. X., Mastorakis, G., and Dobre, C. (eds.). *Advances in mobile cloud computing and big data in the 5G era*. Springer International Publishing; 2017.
[64] Vavouranakis, P., Panagiotakis, S., Mastorakis, G., and Mavromoustakis, C. X. "Smartphone-based telematics for usage based insurance." In *Advances in mobile cloud computing and big data in the 5G era*. Cham: Springer; 2017. pp. 309–339.
[65] Kopanakis, I., Vassakis, K., and Mastorakis, G. "Big data in data-driven innovation: The impact in enterprises' performance." *Proceedings of 11th Annual MIBES International Conference*. 2016. pp. 257–263.
[66] Skourletopoulos, G., Mavromoustakis, C. X., Mastorakis, G., *et al.* "Game theoretic approaches in mobile cloud computing systems for big data applications: a systematic literature review." In G. Skourletopoulos, G. Mastorakis, C. X. Mavromoustakis, C. Dobre, and E. Pallis (eds.), *Mobile big data*. Cham; Springer; 2018. pp. 41–62.
[67] Skourletopoulos, G., Mastorakis, G., Mavromoustakis, C. X., Dobre, C., and Pallis, E. (eds.). *Mobile big data: A roadmap from models to technologies*.

Vol. 18. Cham, Switzerland: Springer International Publishing AG; 2018. pp. 1–347.

[68] Skourletopoulos, G., Mavromoustakis, C. X., Mastorakis, G., Chatzimisios, P., and Batalla, J. M. "A novel methodology for capitalizing on cloud storage through a big data-as-a-service framework." *2016 IEEE Globecom Workshops (GC Wkshps)*. IEEE; 2016. pp. 1–6.

[69] Mihaita, A. E., Dobre, C., Pop, F., Mavromoustakis, C. X., and Mastorakis, G. "Secure opportunistic vehicle-to-vehicle communication." In G. Mastorakis, C. X. Mavromoustakis, and C. Dobre (eds.), *Advances in mobile cloud computing and big data in the 5G era*. Cham: Springer; 2017. pp. 229–268.

[70] Granat, J., Batalla, J. M., Mavromoustakis, C. X., and Mastorakis, G. "Big data analytics for event detection in the IoT-multicriteria approach." *IEEE Internet of Things Journal*. 2019.

[71] Krawiec, P., Batalla, J. M., Bruneau-Queyreix, J., Negru, D., Mastorakis, G., and Mavromoustakis, C. X. "Collaborative media delivery in 5G mobile cloud networks." *Advances in mobile cloud computing and big data in the 5G era*. Cham: Springer; 2017. pp. 341–360.

[72] Mavromoustakis, C. X., Mastorakis, G., and Batalla, J. M. (eds.). *Internet of things (IoT) in 5G mobile technologies*, Vol. 8. Springer; 2016.

[73] Monteiro, A. C. B., Iano, Y., França, R. P., Arthur, R., and Estrela, V. V. "A comparative study between methodologies based on the Hough transform and watershed transform on the blood cell count." *Brazilian Technology Symposium*. Cham: Springer; 2018. pp. 65–78.

[74] Gudivada, V. N. "Data analytics: Fundamentals." In M. Chowdhury, A. Apon, and K. Dey (eds.), *Data analytics for intelligent transportation systems*. Elsevier; 2017. pp. 31–67.

[75] Goşman, C., Cornea, T., Dobre, C., Mavromoustakis, C. X., and Mastorakis, G. "Secure model to share data in intelligent transportation systems." *2016 18th Mediterranean Electrotechnical Conference (MELECON)*. IEEE; 2016. pp. 1–8.

[76] Bourdena, A., Mavromoustakis, C. X., Kormentzas, G., Pallis, E., Mastorakis, G., and Yassein, M. B. "A resource intensive traffic-aware scheme using energy-aware routing in cognitive radio networks." *Future Generation Computer Systems*. 2014; 39: 16–28.

[77] Batalla, J. M., Mastorakis, G., Mavromoustakis, C. X., and Zurek, J. "On cohabitating networking technologies with common wireless access for home automation system purposes." *IEEE Wireless Communications*. 2016; 23(5): 76–83.

[78] Kryftis, Y., Mastorakis, G., Mavromoustakis, C. X., Batalla, J. M., Pallis, E., and Kormentzas, G. "Efficient entertainment services provision over a novel network architecture." *IEEE Wireless Communications*. 2016; 23(1): 14–21.

[79] Dimitriou, C. D., Mavromoustakis, C. X., Mastorakis, G., and Pallis, E. "On the performance response of delay-bounded energy-aware bandwidth allocation scheme in wireless networks." *2013 IEEE International Conference on Communications Workshops (ICC)*. IEEE; 2013. pp. 631–636.

Chapter 2

Artificial intelligence in IoT and its applications

Mona Bakri Hassan[1], *Elmustafa Sayed Ali*[1,2], *Nahla Nurelmadina*[3] *and Rashid A. Saeed*[1,4]

Artificial intelligence (AI) and Internet of things (IoT) are both become most amazing technologies nowadays. Artificial intelligence deals with intelligent reasoning and speedy data analysis which would cover the smartest future applications. Internet of things promises the deployment of these smart applications everywhere in the world. Both AI and the IoT interact in a unique way to increase increased human–machine interaction efficiency, increased operational efficiency, and true digital transformation. In IoT, the AI can ensure to collect only adequate data in order to reduce the processing time of big data amount sent by sensors. Artificial intelligence processes such as data integration, data selection, data cleaning, data transformation, data mining, and pattern evaluation all will provide the best solution to manage huge data flows and storage in the IoT network. The use of AI in IoT will extract the unique capability in different intelligent aspects such as smart decisions, smart metering, and forecasting and these will generate a new intelligent application in industrial security, health care, and smart homes. In this chapter, we will provide a brief concept about the AI and its relationship to the IoT, in addition to the benefits of AI to solve many challenges in IoT operations such as sensing, computing, energy management, and security. The chapter will also provide different types of AI methodologies related to the IoT applications.

2.1 Introduction

Artificial intelligence (AI) is one of the most amazing technologies nowadays dealing with intelligent thinking and rapid data analysis that will cover the smartest future applications of the Internet of things (IoT). Artificial intelligence and the IoT interact in a unique way to increase the efficiency of human–machine interaction, increase operational efficiency and true digital transformation in IoT networks

[1]Department of Electronics Engineering, Sudan University of Science and Technology, Khartoum, Sudan
[2]Department of Electrical and Electronics Engineering, Red Sea University, Portsudan, Sudan
[3]Department of Computer Science and Engineering, Taibah University, Medina, Saudi Arabia
[4]Department of Computer Engineering, Taif University, Al-Taif, Saudi Arabia

enabling sufficient data to be collected only in order to reduce the processing time of the large amount of data sent from the sensor. In IoT applications, AI technologies can provide intelligent processes that represent, integrate, select, and clean data integrity, as well as convert or extract data to create patterns for a better solution for managing large data flows and storing them in the IoT. The use of AI in the IoT will extract the unique power in different smart aspects such as smart decisions and intelligent measurement and prediction, which will lead to the creation of a new intelligent application in industrial and health applications as well as various welfare applications.

2.1.1 Internet of things

The IoT is becoming the most attractive intelligent technology promising to develop smart human beings' life by enabling people to be connected with different kinds of smart things. Moreover, IoT will provide an intelligent method allowing communications between various objects and machines together which is considered one of the most important technologies for many applications such as in industry and medical areas. In IoT networks, sensors boarded over things can connect to Internet due to many wireless network structure and technologies such as Wi-Fi, Bluetooth, or ZigBee; moreover, it can be implemented over wide-area connectivity using many technologies such as GSM, GPRS, 3G, and LTE and LPWAN technologies. The IoT will allow to share different big data of surrounding environment with people or data related to software systems and machines. The communication of data exchange in IoT networks will be handled based on interoperable protocols, operating in heterogeneous environments and platforms [1]. Internet of things in this context is a generic term and all objects can play an active role to their connection through the Internet by creating smart environments that are able to provide many intelligent applications.

The concept of IoT is defined by IEEE as a network that connects uniquely identifiable "things" to the Internet. These things may be sensors or actuators which have conceivable programmability capabilities. By exploiting the unique identity and sensing in the IoT, you can collect data about the "thing" and modify it from anywhere, anytime. The IoT is considered as an envisions a self-configuring adaptive and complex network that interconnects "things" to the Internet through the use of fashionable conversation protocols [2]. Internet of things provide services through the use of intelligent interfaces those interacted between people to things, things to machine, or people to humans, as indicated in Figure 2.1, which are easy to be used at anywhere, anytime while achieving safety in use. These interfaces are developed according to the IoT orientations interdisciplinary such as middleware Internet oriented, sensors things oriented, and semantic oriented [3].

Because of the IoI characteristics, its acting rapidly to penetrate almost all areas of our lives. The most important indispensable traits in IoT that will govern the degree of IoT application intelligence are related to network connectivity, heterogeneity, scale, dynamic changes, autonomous agency, distributed control, expression, and safety. The connectivity will make the IoT network possible by being accessible and compatible and giving the ability to facilitate connecting anything to

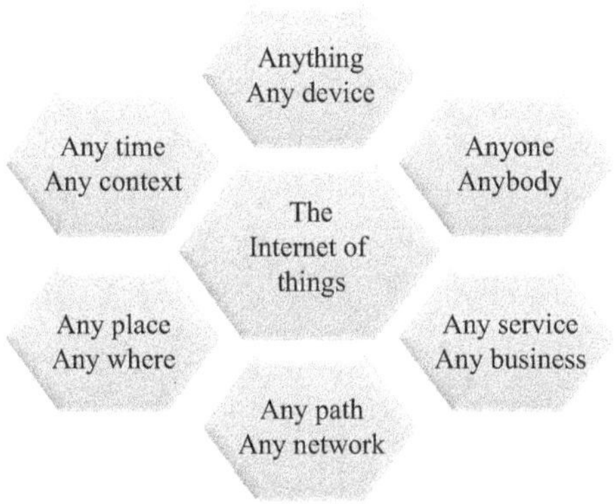

Figure 2.1 Intelligent IoT services. Adapted from [3]

the IoT in general with compatibility to create and devour data. In addition, it has the ability to deal with heterogeneous IoT modules as they are based on exclusive platforms and networks and provides an intelligent mechanism to interact with different modules and server structures with the help of a range of networks [4].

Moreover, IoT offers a range of smart methods to connect many IoT objects and modules that communicate, interact with and manage each other in a way that ensures the handling of big data and the dynamic changes that occur on IoT networks. For the enhanced human environment, IoT can provide immediate and spontaneous interaction capabilities associated with human applications. It also ensures the wide spread of the provision of a number of services and computational functions in everyday objects and makes them speak correctly and perform favorite tasks in a way that reduces human desire to interact with computer systems as computers, and opens new horizons for society, the economy, and the individual. By taking account of expression and safety, IoT design will provide intelligent data security to ensure private and human application security [5]. They are based on securing devices, networks, and information that are transmitted through them, which implies that they create a sophisticated security model. Even for distributed control, IoT will be able to manage a large number of IoT nodes that are distributed in networks and enable emerging aspects and behaviors that require acceptable distributed control allowing expression to interact with humans and the physical world [6]. In all cases, expression will help to create products and things that interact intelligently with the real world and the environment.

2.1.2 Intelligent wireless communications

In wireless communications, system design and optimization may face challenges due to the extreme key performance indictors related to user experience and interacting efficiency in addition to the network environment. The use of AI in wireless communication systems will enable to learn from environment and massive dataset to solve complex problems related to those systems such as decision making, resource optimization, and network management to achieve intelligence to wireless

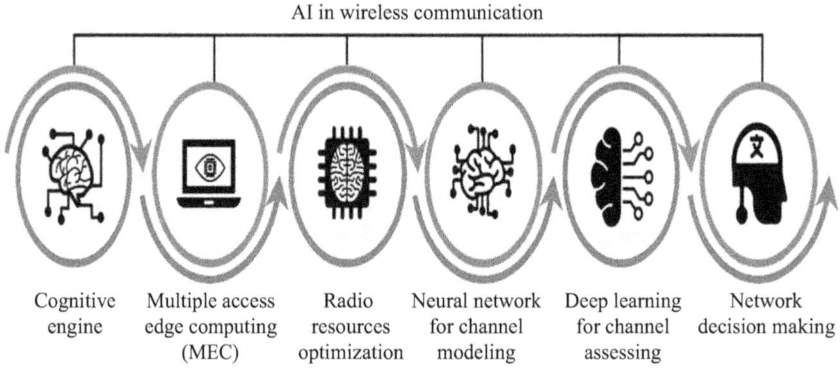

Figure 2.2 Artificial intelligent applications in wireless communications

communication networks. Artificial intelligence can be used to provide wireless network segment with AI capabilities and employ a data-driven model in which the network node is able to determine the best policy for use based on experience gained from previous data processing so that it helps reduce dependence on complex mathematical models with respect to network design and operation [7]. Many AI technologies are developed to act as mechanism in solving the most difficult applications and problems related to wireless communications, and such methodologies are game theory, theorem proving, general problem solving, AI-based expert control, and machine/de0ep learning [8]. Figure 2.2 illustrates a map of AI applications in wireless communications.

In recent years, AI technology has provided unique solutions in building robot phases by building reasonable algorithms to plan tasks, navigate, and control intelligent knowledge. Moreover, it also plays an important role in IoT applications by introducing intelligent management in uniquely identifying machines and linking them to IoT networks and developing complex processing strategies on the machine to be able to learn and improve overall performance through learning, especially in industrial IoT applications [9]. Artificial intelligence with IoT will provide solutions for search and optimization by monitoring and correcting errors just as humans do.

2.1.3 Intelligent IoT

The IoT offers a new type of communication environment that supports ubiquitous connectivity, which makes every object and device in our daily lives accessible to the Internet through and with help of many sensors and engines. Internet of things is shown in a way that makes the connected devices have a sense of purpose and intelligence to respond to their surroundings, making the Internet of industrial things (IoIT) concept in real addition to intelligent communication. The challenges of IoT connectivity and energy management are the most important things to improve as billions of things need to deal with the IoT, which will become a huge burden, requiring intelligent technologies adding the robot behavior to IoT devices to make them acting autonomously to take an accurate decisions in order to get rid of such challenges and moreover to achieve and exhibit complex behaviors, integrating mechanisms for social interactions, autonomous navigation, and object analysis.

Artificial intelligence and the IoT are now very mature and their synergy promises many benefits and is seen as the engine of the new generation for a variety of technological developments and modifications that protect a wide range of fields. Internet of things is the future for the development of AI. Most IoT applications in the near future will become more widely used for AI technologies, algorithms, logic tools, and programming tools. The cooperation between the IoT and AI is that the IoT collects statistics from vast amounts of data [10]. Artificial intelligence acts as an appropriate tool for understanding and analyzing vast amounts of data, making decisions based on these data, as well as understanding patterns, helping to make more informed decisions. With the rapid increase of IoT devices, the data collected by these devices will present a new challenge of how to analyze this huge amount of data. Collecting this data will not be beneficial to anyone unless there is a way to analyze and understand it. With the analytic capabilities of AI, IoT data can be analyzed and different organizations can classify and understand patterns and make more informed decisions. As shown in Figure 2.3, IoT generates and collects a huge amount of data for analysis and meaningful information is extracted using AI approaches.

The use of machine learning (ML) in processing IoT big data has opened up new opportunities for IoT applications by providing technologies capable of gathering information, sorting, analyzing and making decisions. Obviously, in order to have greater benefit in the IoT, AI needs to develop more accurate and faster algorithms and equipment. The IoT-backed AI can provide an efficient way for businesses to get more store operations and ensure their sustainability in a long running time. Using IoT/AI stores can, among other things, reduce theft and maximize purchases through direct sales [11].

2.1.4 Challenges and solutions in building AI with IoT

The implementation challenge related to the building of the AI with the IoT deals with the privacy, complexity, and compatibility. Recently, the IoT data are becoming very faster in speed with different forms which exceed the ability of

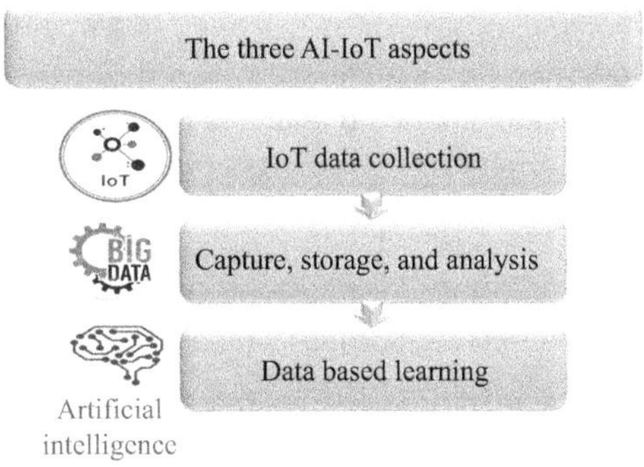

Figure 2.3 AI in analyzing big data created by IoT devices. Adapted from [10]

current information system to handle store, analyze, and process. In addition, the capability of saving big data of some petabytes from waste is considered one of the main challenges of IoT applications. Moreover, the ability of extracting information patterns and structure is much important as processing and store. Artificial intelligence can fulfill the need of storage and processing operation by providing new algorithms such as pattern recognition, ML, and data mining. Developing these techniques in IoT are required to address concerns such as limitations related to complex adaptation, big data scalability, and context awareness, in addition to security and privacy. The use of AI systems in some IoT applications may lead to malignant self-sustaining development such as in medical system applications that can mimic the growth of disease activity in the human body. Further research is needed to combat such systems using sophisticated AI systems. In addition, adapting AI systems to the IoT to be under control is even more important because of the existence of electronic warfare technologies. Hacking gaps in AI systems must be verified so that they remain protected from probable attacks and being fortified by intelligence protection strategies [12].

Due to massive devices and systems in IoT, there are enormous amounts of data have been generated every day even those big data come from IoT social media applications. Machine learning can likewise help machines, humans, devices, etc. in IoT get together to comprehend what important information individuals may need from the data. But with massive big data, it can give IoT a cerebrum to think, which is called self-intelligence. This self-intelligence can be a real threat in some sensitive applications that require to not exceed a certain degree of control, because the machines associated with the IoT, which work in accordance with the structures of AI and developed ML models may lead to accurate plagiarism of the human mind on a much larger scale and the development of complex systems with sufficient wisdom to start in understanding the things that boggle the superiority of human thinking [13].

Big data scalability is related to five analytics variables denoted by 5Vs and they are data volume, velocity, variance, veracity, and value. The use of cognitive IoT (CIoT) will offer new opportunities for IoT big data management, aggregation, and analytics, but it may not concern with challenges. For big data management and storage, current AI technologies may not able to fulfill the considerations of efficient work with heterogeneous data because of slow development in storage capacity compared with massive data volume which requires to evaluate some big data platforms and technically advanced analyses [14,15]. The aggregation of big data is also one of the most important challenges related to data synchronization in distributed big data forms in addition to analytics which are both needed to be heterogeneously and scalable with real-time response. The use of CIoT can provide a means of intelligent data processing able to be scaled with the large number of IoT services shared between multiple applications with different processing requirements, but for a much larger scale. Cognitive IoT will face challenges in processing massive sensing data that may be of mixed characteristics that include higher dimensionality and heterogeneity which require mechanism to develop the IoT heterogeneous devices to be automatically integrated into analytical infrastructures. The data collected from

IoT layer and transmitted to the cognitive computing must be maintained to preserve the security and privacy. Continually and automatically learning in cognitive computing layer must be improved in order to understand the environment of IoT devices and users having accurate provided services [16]. This learning must ensure the degree of awareness to discover and distinguish between the secured and unsecured IoT devices. Artificial intelligence can help solve privacy and security issues in IoT architecture. Artificial intelligence-based models are not only trained in data use, they work best as data volumes grow, meaning that the accuracy of the model improves as they learn from larger data volumes. But the AI needs a steady stream of data to optimize the model and provide better security solutions. The solutions in AI not only require to develop and improve their models but constantly require new information or data to ensure security against unknown attacks.

2.2 AI revolution in IoT

The revolution in AI nowadays leads to perform different intelligent tasks related to voice and image recognition, language translation, robotic perceptions, and intelligent decision-making. Moving from these aspects toward intelligent communication will extract an attractive and most highly developed means of communication and with the Internet can lead to a huge revolution to reach the full integration of information and communication technology [17]. In our daily life, IoT devices provide an advanced level of comfort which able to give unimaginable amounts of data related to user behavior, preferences, and personal information. The massive amount of data flooded by the IoT devices can be gathered by the help of AI for data analysis and rationalization. This tremendous shift in the IoT caused by AI is fundamentally reshaping the technology landscape adding a new concept of interaction between physical world, digital world, and virtual cyber world (see Figure 2.4). It is expected that there will be an increasing need for these technologies, especially in the various fields of industry that are adopted and implemented continuously in various scenarios [18].

The studies and continuous developments conducted in AI have led to the development of intelligence concept in the IoT architectures, resulting in the now widely used term as in intelligent machines. Utilizing the power and AI, smart devices will simplify tasks by performing them in minutes that may take weeks or even months to complete in addition to providing an advanced level of persuasion and self-intelligence [18]. The combined impact of AI and IoT will radically restructure our personal and professional lives in a way that cannot be fully imagined. Not only it will replace the cumbersome and monotonous human jobs with machines, it will also significantly reshape the competitive scenario between AI developing and IoT system companies.

2.3 Intelligent sensing

Sensing in IoT is related to people who carry the sensing devices and the devices itself in networks. For intelligent sensing, people sensing is a new area of application that is

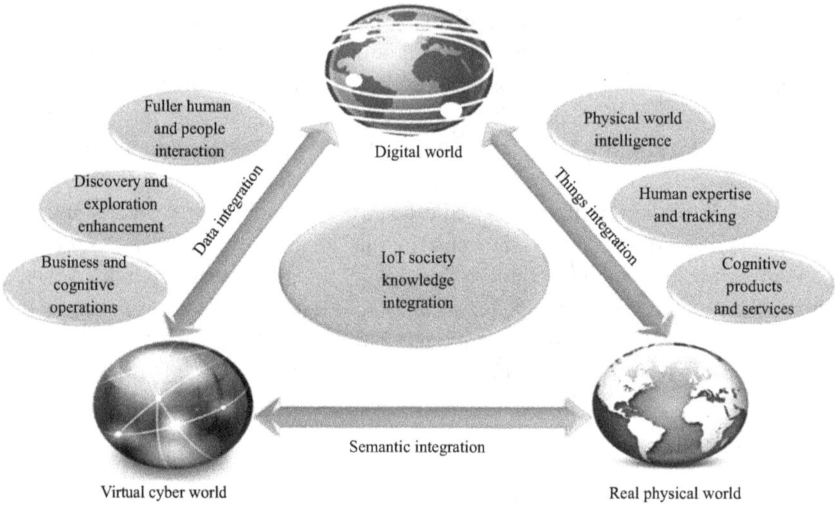

Figure 2.4 Essential AI revolution to realizing intelligence of IoT

expanding the sensor network to monitor the environment and infrastructure. Mobile people sensors can be used to support remote sensing of IoT networks to provide high coverage which can be considered as a new infrastructure with a mobile sensor having potential to logically belonging to more than one IoT network, and it possible to run computations on enormous amounts of data from a very large number of user and indeed it can reach areas where fixed sensors cannot reach, proving to be particularly useful for IoT applications. Distributed sensing systems can be used in IoT as a platform to allow interaction between groups of people and a group of devices. It can depend on the physical location of the distributed devices according to geographical information system (GIS) by sing GPS [19].

In wireless IoT networks, intelligent sensing will reduce the energy consumption due to send all raw sensor data to the cloud and big data analytics. Improving energy efficiency in data transfer is just as critical to making the sensor smarter so that only useful and documented data are delivered to the cloud. The use of generic algorithm will contribute to provide minimum energy consumption when it is used in the IoT wireless network, ensuring to achieves greater energy efficiency in data sensing and collection in addition to balance the residual energy in the IoT devices which will increase the IoT network lifetime [20]. The sensing power consumption in general depends on the IoT sensors placement, network coverage, clustering, and data aggregation. The distribution will impact the IoT wireless network sensing according to the IoT sensors' distributions and the distances between them which is dependent on the type of IoT network and its application concern. For example, in military IoT applications, sensors are usually scattered by airplanes over military zones, while in case of underwater IoT networks, regular distribution is adopted and grid layouts are usually used for urban networks [21].

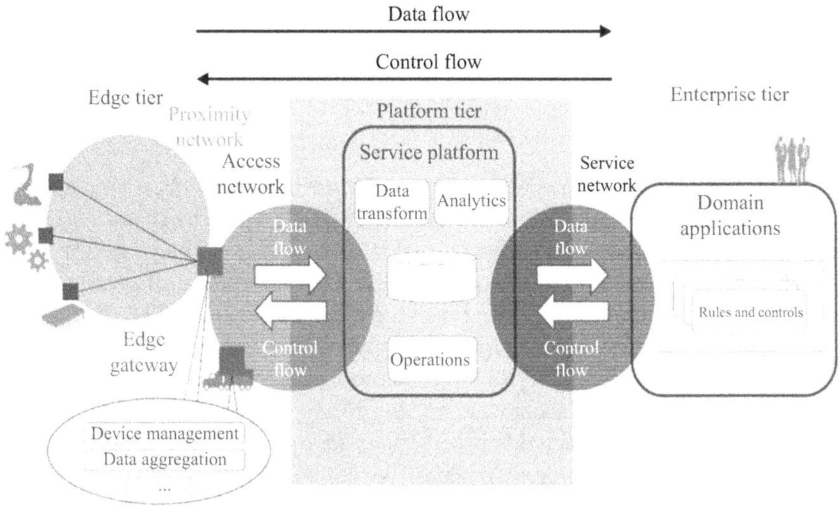

Figure 2.5 Key elements of intelligent sensing domains. Adapted from [22]

Internet of things intelligent sensing can be provided through the virtual sensing which integrates information of multiple sensors under control to configure virtual sensors. Therefore, it will provide a capability to reconfigure sensors functionality and enable to increase the ability of sensing devices at the edge domain of IoT system (see Figure 2.5). Moreover, the calibration and control of IoT devices in network can be enabled according to the functions provided by the edge domain and intelligent IoT platforms for effective runtime and configuration proficiency.

The intelligent platforms in IoT will support and manage the dynamic composition of these devices and they will be capable to adapt them according to environment and context. Intelligent platform capabilities must meet the requirements of network authentication, access control, authorization, and safety of the devices in addition to predictive maintenance. Intelligent and secure IoT platforms will offer possibilities of self-control in IoT devices through hierarchical and collaborative control strategies [22]. The ability of self-control operations in IoT devices and the actuation control in cyber-physical system can be performed by control policies for managing a self-controlled swarm of IoT devices such as collaborative control strategies or context-aware control of devices and adaptation of their behavior according to applications' environment.

2.4 Feature of AI with IoT

A smarter IoT system will have AI and may serve the actual goal of automation and adaptation. Artificial intelligence can play a vital role when combining it with IoT. Artificial intelligence with IoT is a perfect combiner for growth. They will lead to a wide range of characteristics for consumers and companies such as personalized experience, a proactive intervention, and intelligent automation. Here are some of the most popular features of combining IoT with AI technologies that will change the way that businesses work.

2.4.1 Computing

The use of AI in IoT cloud computing will enhance the handling request for cloud data center by deploying the deep learning (DL) services. For low latency characteristic, DL applications can be used on the edge cloud for edge computing purpose. The collaboration of the end level and edge cloud IT levels will perform real-time computation analytics by deep learning for different intelligent IoT applications. The edge level is related to the analysis request for typical IoT applications and maybe responsible for different object actions such as signal or image processing, data compression, or object recognition which are all collaboratively training an adaptive model with a deep model for better computation accuracy. Many distributed edge nodes can be at the edge level cooperate with each other to provide better services and with turning the DL on the edge, it will improve overall performance by significantly reducing the consumption of edge layer resources while ensuring the performance of the analysis [23].

2.4.2 Intelligent energy management

Recently, many researches have investigated AI to produce practical solutions for the smart grid. The expectation of new generation of energy networks will be able to make efficient use of renewable energy sources, support real-time and efficient demand response, in addition to large-scale deployment of electric vehicles (EVs). Artificial intelligence techniques and methodologies could be instrumental in addressing sustainability problems, for example, to increase the efficiency and effectiveness of the way we manage and allocate our natural and societal resources. Combination of the IoT with AI can produce an energy management system. Internet of things solutions can be implemented as narrowly as at the circuit level, and by leveraging and analyzing those data with AI, decision-makers can pull actionable information to significantly reduce waste and further optimize business operations. Also, AI allows real-time alerts and notifications and the automation of key functions such as climate control and lighting [24].

2.4.3 Security

Basically, AI and IoT can enhance the workplace security. For instance, AI can automatically scan the security footage and IoT can control gates when an intruder is on the premises. As beneficial as these emerging technologies are by themselves, combing IoT with AI can offer an extra layer of security. By pairing ML with machine to machine (M2M) communication, companies can predict potential security risks as well as automate rapid response. In the banking sector as an example, ATMs are equipped with the capabilities to detect potential fraudulent behaviors and alert law enforcement about the criminal activities. Several applications that combine IoT with AI are helping companies and organization understand and predict various kinds of risks and automate rapid response to these risks, better manage worker safety and cyber threats [25].

2.5 Intelligent machine and deep learning in IoT

Machine learning is closely related to AI. Machine learning performs duties such as classification, compilation, prediction, sample recognition, and archiving of the identification process by training systems to use different algorithms and statistical processes to analyze sample data. It also uses information obtained at some point in the education department to visualize patterns or make decisions based on new data. Machine learning is ideal for problems such as determining regression rules, grading, and grouping. Deep learning is one of the powerful technologies to recognize the ability of the system to a large amount of unstructured data and to gain a deep knowledge of strategies suitable for dealing with big data and provide intelligent calculation service. Both machine learning and deep learning mechanisms play an important role in the deployment of IoT applications.

2.5.1 IoT machine learning

Machine learning allows learning from large amounts of data and continuous monitoring of IoT data flows generated by intelligent algorithms capable of monitoring system states and behavior patterns, and learning to predict possible system states in the future. These insights may be used to elicit proposals or pioneering actions to the desired future state. The use of ML is possible to automate the conclusion creation process, which is necessary to achieve complex and intensive IoT data use cases in the future. Machine learning can ultimately lead to independent work, a logical extension of existing automated processes and services to increase efficiency and productivity [26]. Machine learning will help to train the IoT network information to identify patterns or make decisions. According to the learning mechanisms, the use of ML algorithms in IoT networks depends on the IoT applications. The supervised learning deals with retraction estimation such as in weather forecasting and monitoring of IoT networks which is related to linear regression algorithms or as in classifying the speech recognition, diagnostics, and identity of fraud detection by using algorithms such as support vector machines. The training of big data visualization, feature elicitation, or the discovery of hidden structures are based on unsupervised learning algorithms which are able to identify patterns on testing data and cluster the data or predict future values [27]. The combination of supervised and unsupervised learning is known as semisupervised learning which works mostly like the unsupervised learning with the improvements that a portion of labeled data can bring.

The integration of ML with IoT can be divided into four levels (see Figure 2.6): the integration with physical endpoint system which is related to ML inference server with IoT endpoints such as a microcontroller-based system or sensor systems. Edge gateways deal with the integration of ML servers to IoT gateway and this will enable ML to inference services for one or more endpoints at a single edge location. The interface with cloud-based IoT platforms such as Amazon Web Services (AWS) IoT and Azure IoT can enable the ML to be interfaced with many endpoints across many IoT edge locations. Machine learning servers also can be integrated with enterprise data centers to facilitate the operations of IoT monitoring and control [27,28].

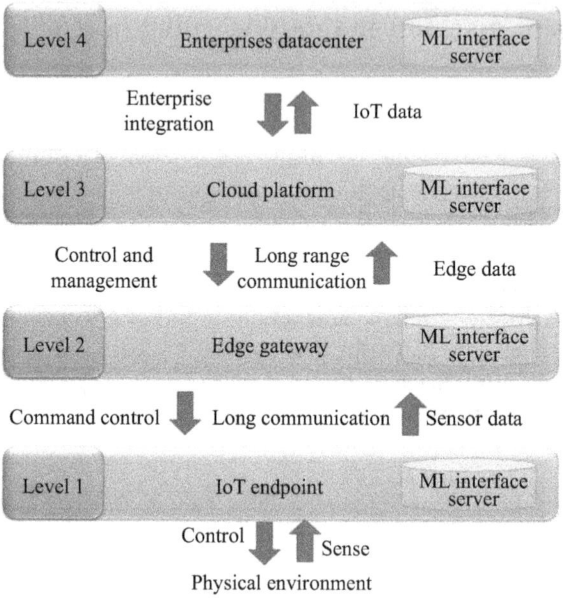

Figure 2.6 ML integration levels with IoT. Adapted from [28]

The use of ML algorithms in IoT wireless networks can lead to significant improvements in the application or infrastructure itself, helping to optimize the network, allocate resources, and avoid congestion. Even in the case of increasing number of devices and increasing amount of data collected, these algorithms deal with "big data" and properly processed data and also it has special techniques for analysis of the huge volume of structured and disorganized data. To support future IoT platforms to be smarter, virtualization techniques can be used by providing virtual machines with a logic of operating systems as well as business logic. The devices can be supported with a virtualization layer and the cloud-connected management layer can download the business logic that allows the device to perform its specified function. This mechanism provides a unique model for making IoT devices connect to the network and obtain identity through a standard interface so that the network can communicate this identity to the cloud, which sets up and anticipates the connection report [28]. The cloud then provides the device with the required software content, which in turn downloads that content and creates an instance of virtual machines that are ready for the public. The management layer can allow business content to communicate securely with the cloud.

2.5.2 IoT deep learning

Artificial neural networks (ANN) and deep learning are both algorithms that can be used in IoT to predict the status of the data in the network due to gathered parameters from different IoT ends to make real-time decisions. In IoT wireless networks, IoT devices need to not only collect data and communicate with other devices but also to be autonomous and able to take context-based decisions and

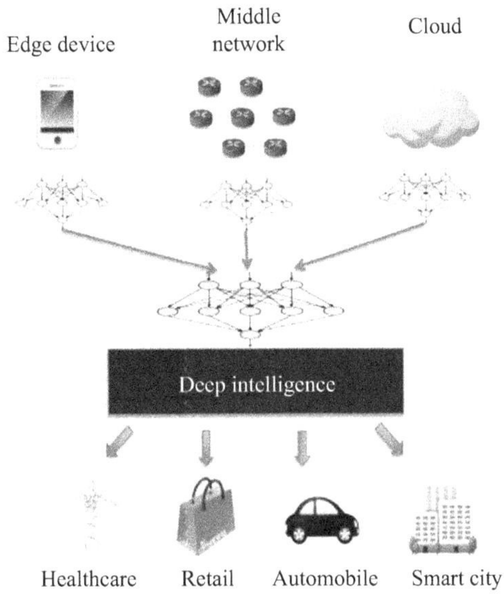

Figure 2.7 IoT network-based deep intelligence learning. Adapted from [30]

learn from their collected data which will turn to develop intelligent IoT devices, able to create automated smart applications with automated resource allocation, communication, and network operation [29]. The use of deep intelligent learning will deploy an artificial IoT system as shown in Figure 2.7 that consists of a device which will be able to connect to base stations or gateways to the cloud servers to interact with different IoT applications ends.

The deep intelligent methods in IoT can able to generate control decisions of IoT devices that enable processing of the data at the edge fog servers or clouds. The control decisions that occurred due to reinforcement learning (RL) method to closed-loop problem of processing the sensory data will let the IoT devices interact with the environment to learn optimal policies that map status or states of actions [31]. This learning type requires that IoT devices must gain the representation state of the environment efficiently and highly sensory data and use this information to learn optimal policies. Therefore, the use of deep reinforcement learning (DRL) can overcome this challenge by combining the RL with deep learning (DL) together.

2.6 IoT-based AI data access and distributed processing

5G technologies with edge computing can enable distributed computing platforms with data resources close to end-users with low-latency connections. The use of AI in the edge ecosystem can concern on performance, cost, and privacy [32]. By making cloud services near or in network edge closed to IoT devices and data sources, it will give an opportunity of achieving different means of communications such as D2D, cellular-based gateway, Wi-Fi-based router, or microdatacenter available for use by

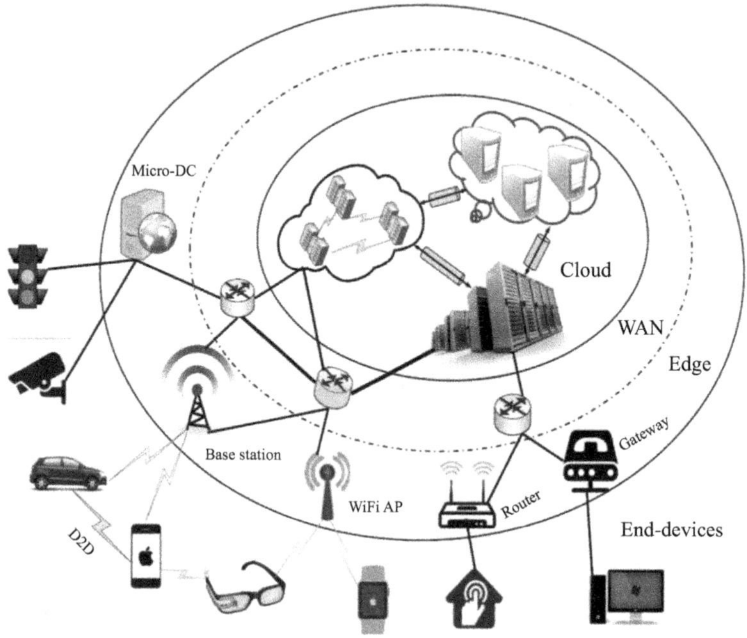

Figure 2.8 IoT edge computing scenario. Adapted from [33]

nearby devices (see Figure 2.8). The convergence of computing and information generation sources can provide several benefits such as energy efficiency, privacy protection, reduced bandwidth consumption, and low latency [33].

For IoT data accessing and processing, edge platforms can be used and become a part of distributed computing IoT systems to train ML models in real-time. Cloud platforms can provide ML services to extract the meanings of IoT data and with the cloud presence, it will be able to handle big data due to its ability to be scalable. To increase the speed of processing response when sensors and devices are far from the cloud, edge platform can overcome such limitation [33,34]. Edge platform is considered as an optimized cloud computing system which can provide data processing at the edge of IoT network enabling computing to the end device in order to reduce the latency of communications. The AI can interact and collaborate with cloud and edge for training and prediction. Some database and cloud companies such as Google are investigating on how to make the edge train itself. The use of deep learning processing algorithms network edge can contribute to achieve collaboration between all heterogeneous IoT platforms and devices over time [35,36]. Figure 2.9 shows a collaborative edge-cloud processing in IoT for heterogeneous raw data coming from different sensors.

In heterogeneous IoT data, the value predication due to ML model from the cloud must be optimized in the edge enabling analysis at edge with assurance of high processing power and storage, such as analysis method like transfer learning. This analysis method can be used to improve the performance of the ML model.

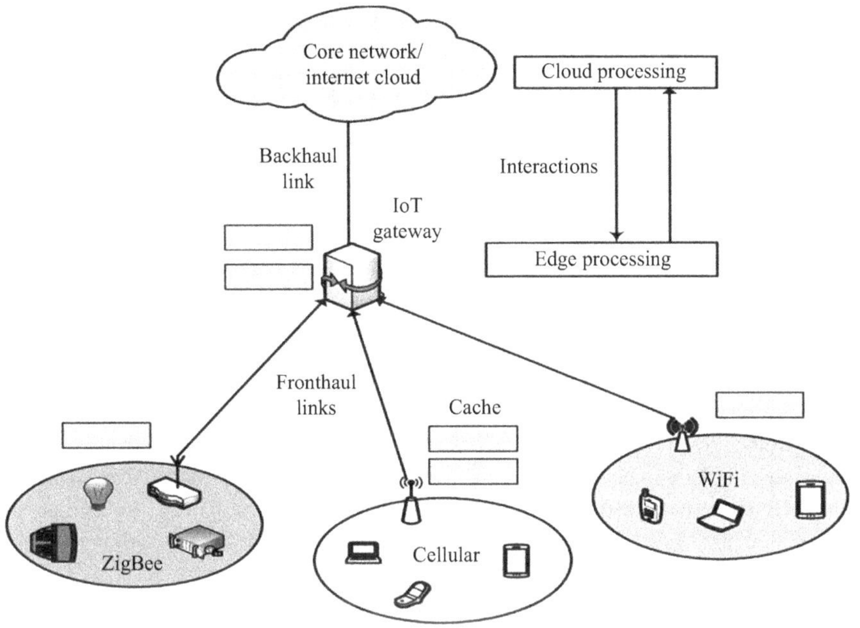

Figure 2.9 IoT-based collaborative edge-cloud processing. Adapted from [35]

The efficient collaboration between the edge and the cloud can also ensure privacy of network traffic by making cloud collecting and preprocessing data before training to let data to be integrated in appropriate format and preprocess it for ML [35]. Such learning process can enable the cloud to deliver the most appropriate model to the edge based on the information transferred from the edge itself.

2.7 AI contribution aspects to Internet of things

AI plays an important role in IoT applications and deployments. Both investments and gain in startups combining AI and IoT have climbed over the past years. Major companies of IoT platforms such as Oracle, Microsoft, Amazon, and Salesforces offer integrated AI capabilities such as ML-based analytics. In this context, several examples will be discussed of existing IoT services with the working of AI behind them.

2.7.1 Data mining and processing

Data mining is a new application and growing rapidly. It shows a great tool for exploring new avenues to automatically examine, visualize, and uncover patterns in data that make possible the decision-making process. Data mining identifies trends within data that go beyond simple analysis. Modern data mining techniques such as association rules, Bayesian networks, regression algorithms, neural networks, decision trees, support vector machines, Gaussian mixture models, etc. are utilized in many region areas to solve diagnosis, association, segmentation, classification, and prediction problems. Data mining techniques are used to find patterns, structure or regularities, and singularities in large and growing data sets. Artificial neural networks (ANN) are gross simplification of real networks of neurons. An ANN is introduced with a systematic step-by-step method which optimizes a criterion commonly known as the learning rule [37].

2.7.2 Voice assistants

These are cloud-based voice services which act as table-top personal assistants for end-users. They provide various tasks through third-party applications and other smart devices in their vicinity. These cloud-based voice services are capable of calling cabs, making restaurant reservations, switching smart lights on/off, playing music, answering queries, and various services based on the user's voice commands. Some of the well-known voice assistants will be maintained such as Alexa which is the voice assistant developed by Amazon used in products such as Amazon Tap, Amazon Echo, etc. There is a specific set of dexterity such as Alexa Skills Kit (ASK) that can be modified and updated to improve or personalize certain skills. Siri is developed by Apple Inc. and is applied in Apple Homepod which serves a similar purpose. Google Assistant also a kind of an AI voice assistance used in Google Home. These voice assistants are able to perform multiple tasks mostly due to the application of various subfields of AI. Wake word detection, automatic far-field voice recognition, natural language processing, speech to text conversion and understanding, contextual reasoning, dialogue management, question answering, conversational AI, etc. are executed continuously to make the voice assistants perform functions in real-time [38].

2.7.3 Smart decisions

The application of AI to smart decision is not new realized. Recently, improvement in this application made an AI technique accessible to a wider audience as seen by increasing in number of applications in some areas such as intelligent decision support systems. Artificial intelligence is used in decision making for tasks such as aiding the decision-maker to select actions in real-time and stressful decision problems, reducing overload information, and enabling up-to-date information providing a dynamic response with intelligent agents, allowing communication needed for collaborative decisions, and dealing with uncertainty in decision problems [39].

2.7.4 Forecasting

For many intelligent IoT applications, AI forecast systems show superiority to traditional models. Yet operational utilization of the AI-based forecast systems remains limited in comparison with older regression-based statistical forecast systems. Besides, development of AI forecast systems by operational forecasters still is uncommon. Given the possibility, adoption of AI-based operational forecasting has not been more widespread. A key factor that restricts the adoption of AI-based meteorology operation is the lack of a pool of forecasters trained in the areas needs to use the new systems wisely. The supply of forecasters and researchers trained in the theory and tools required to develop new AI-based forecast systems is even more limited [40].

2.7.5 Smart meters

Smart meters allow consumption to be able to monitoring data gathered informing decisions and behavior around power efficiency accurately. Artificial intelligence can be applied to smart meters' applications. A smart meter is installed outside of a household or building and sends readings to the company in real-time. Also, if there is an event of tampering or electricity theft, it immediately notifies any power outages and can automatic alert and register the utility companies. They are introduced in every electricity consumption unit and share their information with local control centers. The most cutting-edge smart meters not only have two-way communication but are equipped with real-time sensors that can gather the data on relevant factors consist of frequencies used by different equipment and appliances [41]. The smart metering utilities are now starting to roll out a second generation of these smart meters that will be able to provide even more insights and help with energy efficiency by using AI applications. Recently, an important technique in the grid that makes all the difference and provides invaluable data in real-time to the individual user is the smart meter.

2.8 AI for IoT applications

AI will promise to develop many IoT applications especially those are engaged with variant and myriad data from massive IoT sensors and external sources. For IoT applications, the use of ML to IoT data will provide an ability to automatically improve its algorithms with more data being received and aggregated, even more accurate predictions which can make the decisions artificially answers to various aspects, starting from the machine safety in industries or power control to the increased personalized services and in healthcare applications.

2.8.1 Personal and home devices

Smart sensors play a big role in discovering various physical activities to maintain good health. Many medical companies are investing in medical sensors that can help patients track their activities to improve their health. They can also help monitor levels of vital signs in an emergency. Construction companies also use these sensors to record the load-carrying capacity as well as the position of their workers to avoid any type of injury and increase productivity. For smart home applications, AIoT paved the way for developing smart homes that are enabled with many smart features. Features such as open system mobile applications, enhanced security and control of household equipment such as cooling, telephone, TV, computer systems, fans and other home systems [42]. These are some of the great features that smart homes facilitate. The AI in IoT contributed to the emergence of the concept of smart home where all devices are connected to each other through a shared network. By combining this with AI, all these devices will be able to interpret their owner's instructions and make smart decisions accordingly.

The informative model of data trained for personal or home devices can be processed by the use of ML to provide an intelligent service to the IoT end users. The performance of training is depending on the computation of multiple CPUs interfaced with IoT devices to enabling self-taught knowledge. Self-taught knowledge is considered as an informative model generated from ML techniques which can be performed locally by an IoT device. To share the self-taught knowledge, cloud-based service model can perform such good technical approach that can be supported by knowledge of things (KoT) concept to enable sharing between devices [43,44]. The management services for the KoT between IoT devices can be structured which can be integrated into IoT platforms. The connectivity between IoT home devices and information management can be performed by the IoT platform interacting with the KoT framework to enable IoT home or personal devices to share their own self-taught knowledge with each other at the edge level. Data granulated from different end home devices will then consumed by the smart home or smart wearable applications and managed through the IoT platform layer. In such framework operation, an IoT device with sufficient computing resources can play both roles of a contributor and a catcher [43].

2.8.2 Healthcare

The application of AI in the IoT in healthcare has made it easy to analyze symptoms and treat diseases. Data that is transferred from the medical equipment reinforced system adopt preventive measures. Health is concerned with mobile applications designed by mobile application development companies capable of receiving and processing this data. One of the most important IoT healthcare-based applications issues is trust and security of transmission channel and network resources [45]. The use of preprocessing intelligence prioritizes of incoming data from sensor frameworks which can be handled by using ML and data mining concepts to extract signatures from incoming sensor data and accordingly provide medical interpretation based on the captured data (see Figure 2.10).

Through this concept of AI method, it can provide an assessment of the patient's health condition, in addition enabling remote diagnosis and provide better rural medication when it can have integrated with the IoT and edge computing process. The distributed computing architecture in form of edge computing where some part of the computations can be done at the IoT device or "edge" devices rather than having everything computed in the cloud can provide real-time low latency responses [45]. For such healthcare applications, IoT devices with edge computing can have different decision-making and backend cloud server methods and protocols for IoT endpoints or clients such as MQTT protocol, ML tensor flow or Hadoop databases framework for big data of healthcare [46]. The decision making provided by the AI algorithms to analyze data for diagnosis to ongoing treatment options. The AI algorithms may be one of those used for patient's recognition kinds like facial or motion recognition to recognize a patient's behavior and reaction to take healthcare decision like to reminding patients to take medication or reminding for the critical situation [47].

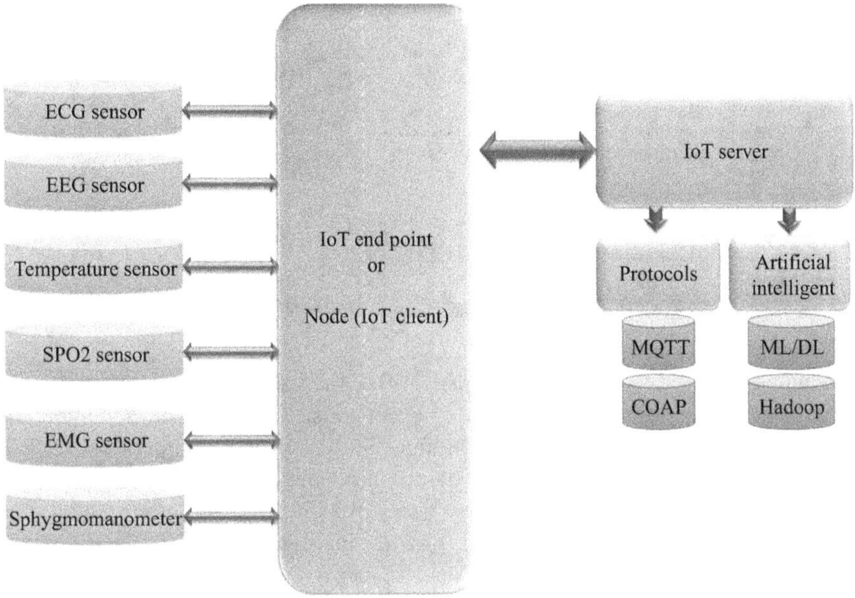

Figure 2.10 Intelligent learning and data mining in IoT healthcare

2.8.3 Agricultural

The use of AI in the IoT helps the agricultural sector to identify and predict crop rotation, risk management, soil characteristics, climate change, crop forecasting, and assessment, etc. It also can contribute to provide improvement in waste reduction, cost-saving, business efficiency, quality, and volume of products. Artificial intelligence and IoT will adopt the smart sustainable agriculture by developing a future intelligent agriculture management system in order to control and maintain various tracking devices and sensors related to agriculture. The devices and sensors able to collect various information and analyze it for complex decision-making tasks to enable an effective agriculture data analysis practices for farmers or farming companies [48]. Such data analysis may be related to soil quality which is collected according to the soil information needed such as its chemical and biological properties. Moreover, it can be used also for monitoring livestock to provide real-time assessment of productivity, health, and welfare. The technologies such as AI and cloud ML will contribute in advance to analytics the smart farming ecosystem. Digital farming-based AIIoT framework (see Figure 2.11) can provide farmers achieve higher average yield and better price control in addition to taking decision support related to how extracts the less resource usage, better agriculture quality, highest yield and precision farming.

Moreover, it can enable to give real-time moisture adequacy data from daily rainfall and soil moisture to build predictability and provide inputs to farmers on ideal planting time. With the advent of technologies such as AI, ML, satellite imaging and advanced analytics, an ecosystem for intelligent, efficient, and

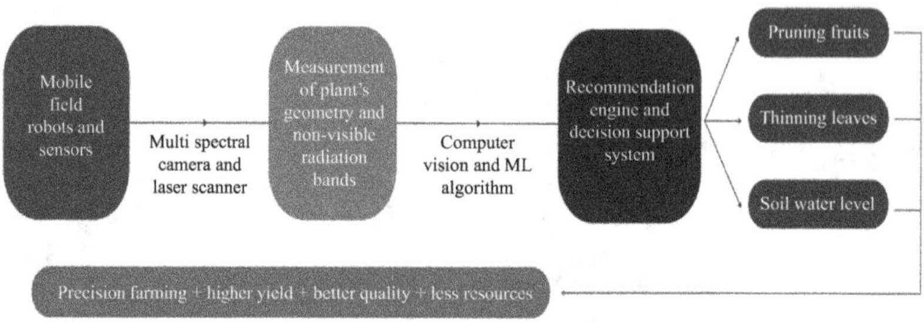

Figure 2.11 Digital farming-based AIoT. Adapted from [49]

sustainable agriculture can be developed [49]. Cloud services with an AI framework designed to monitor the status of greenhouses can provide a number of information, which can be obtained through the human device interface via IoT platforms by pushing data from any device that supports the Internet and contributes to agricultural data management processes [50,51]. In addition, the use of biosensing sensors for some pests can provide intelligent solutions in managing epidemic control and monitoring crop developments.

2.8.4 Intelligent cities

Due to the increase in population density, which is expected to reach 9.7 billion people by the end of 2050 and nearly 70% of their occupations in cities, smart solutions must be devised to meet the challenges of providing resources and energy for the entire population while avoiding environmental degradation and regulation to prevent sanitation issues, alleviate traffic congestion, and thwart crime. Using the IoT with AI can offer technology that facilitates the new residents' experience to make their everyday lives more comfortable and safer. This has led to the concept of smart cities that use information and technologies to improve the quality and performance of urban services such as energy and transport, reducing resource consumption and preventing waste and overall costs [52,53]. The implementation of AI in smart cities not only possess the information technology and communications but also use technology in a way that positively affects the behavior of the population.

The smart sustainable city concept provides a model for a city that uses information and communications technologies (ICTs) and other intelligent means to improve urban services and meet the needs of economic, social, and environmental conditions. Smart Cities are designed by building a flexible, reliable, and scalable ICT infrastructure that is flexible and secure. This concept also includes services based on physical infrastructure such as transport and utilities associated with energy and water services, as well as strengthening the function of prevention and management of natural and man-made disasters, including the ability to address the effects of climate change. The combination of high-speed, flexible, low-latency communication and technologies such as the IoT, ML, and AI will enable the transition to smart, sustainable cities. For such combination, data generated from different city system at high rates can make the intelligent cities concept more difficult in design and development due to the requirement of big

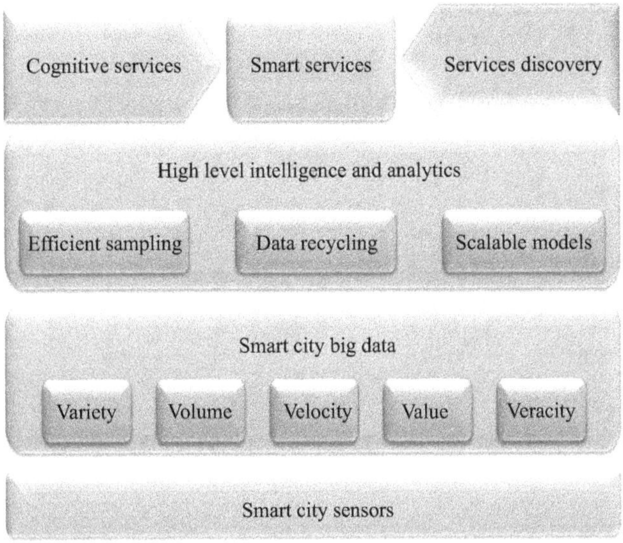

Figure 2.12 Machine learning prospective challenges in intelligent cities. Adapted from [54]

data management, and even more difficult in analytics by using ML perspectives which needs ML algorithms that are able to exploit the availability of unlabeled and labeled data in the context of intelligent cities to overcome the challenges related to services, analytics, and big data (see Figure 2.12). With taking these considerations, smart city infrastructure ML can be used to control the components of the smart city system by an intelligent software agent [52,54]. This ML agent can be deployed in the fog or cloud according to the required analytical properties of the data collected from the different levels of intelligent city systems [55]. The analytical agent returns the appropriate action to the infrastructure devices based on forecasts related to the meaningful application of intelligent city infrastructure.

In hardware infrastructure level, IoT sensors are enabled to be connected with environmental resources. The devices are also connected at the IoT level to fog levels using deep learning to make them more intelligent and capable of analyzing different resource information. In smart cities, AI plays an important role in the design and development of software systems capable of managing some urban flows and services such as traffic liquidity, self-driving cars, logistics, and energy use [56]. Big data-driven ML applications are often used in intelligent cities to adapt markets and citizen behavior, reduce energy consumption, adopt vehicles, allow social and cultural mixing, promote social consistency, and improve safety and health.

2.8.5 AIoT industry

The use of IoT devices with industrial equipment provides a wide range of data. Using AI algorithms applied to the data collected, potential problems can be detected and repaired in advance by gradually training the system to identify external and internal factors that have an impact on machine operation. By improving resources and

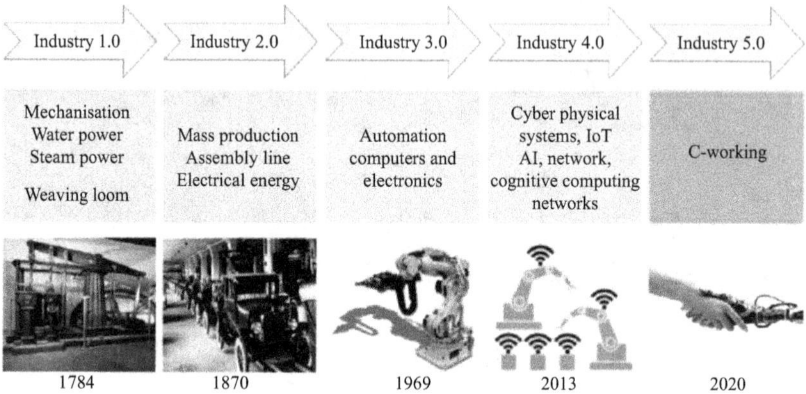

Figure 2.13 Industrial revolution moving to intelligence. Adapted from [57]

increasing industrial safety, the whole production process is simplified. Artificial intelligence in the IoIT performs predictive or perspective maintenance that can be obtained from industrial systems supported by ML algorithms that predict the need for maintenance on factories [57]. Furthermore, AI can help create self-healing and calibrated IoT devices such as sensors, inductors, or transmitters. The greatest benefits AI adds to IoT due to the abilities mentioned are reduced maintenance costs and downtime. The process of automating the complete industrial production system is one of the most important elements of the revolution in the technology of Industry 4 (fourth industrial revolution) and then the evolution has moved to the use of techniques to support the human functions of machines as in the case of the use of industrial robots in addition to connecting to the Internet via IoT devices, which offers a new industrial model known as Industry 5 (fifth industrial revolution) as shown in Figure 2.13.

Industrial companies have tended to design a systematic structure for the implementation of AI in industrial environments, which build intelligent and flexible industrial equipment enabling them to deal with errors more consciously and accurately. Moreover, AI provides intelligent industries on demand with self-regulation [57,58]. The concept of AI can be introduced by developing and deploying various ML algorithms for sustainable performance of industrial applications to provide on-demand manufacturing services to end-users through optimal coordination of distributed manufacturing resources enhanced by AI methodologies [58].

2.9 Summary

According to market statistics, there will be more than 64 billion of IoT devices by 2025 and all of these IoT devices produce a lot of data that must be collected and extracted to get actionable results. To do this, AI comes at the forefront of the intelligent methods used with IoT to collect and process the huge amount of data required from AI algorithms that convert data into actionable results that can be implemented by IoT devices. Internet of things methods in the IoT are expected to lead a large number of intelligent tasks to a series of interconnected devices that transmit data over a network such as voice recognition, language translation, decision-making, etc. without human intervention. Artificial intelligence helps to analyze IoT data and makes it logical. Thus, AI is expected to be the main driver of the unprecedented growth of the IoT revolution. This chapter provides information on making IoT applications intelligent by AI and points to the contribution of AI to the development of the IoT by explaining the methods of learning and its characteristics in data analysis and decision making in IoT computing. General features of AI and its contribution to the development of the IoT are reviewed in this chapter, in addition to the review of some IoT applications that use AI technologies such as machine learning and deep learning and show their impact on the IoT.

References

[1] Mohammed, Z. K. A. and Ahmed, E. S. A. "Internet of things applications, challenges and related future technologies." *WSN*. 2017; 67(2): 126–148.

[2] Mastorakis, G., Mavromoustakis, C. X., and Pallis, E. "Beyond the Internet of things." *Internet of Things Series*. Springer; 2017.

[3] Mavromoustakis, C. X., Mastorakis, G., and Batalla, J. M. "Internet of things (IoT) in 5G mobile technologies." *Modeling and optimization in science and technologies*. Springer, 2016.

[4] Dhivya, V. and Singh, A. K. "Internet of things: A survey of the advancements." *International Journal of Engineering and Technology*. 2018; 7(3.12). doi: 10.14419/ijet.v7i3.12.16036.

[5] Ahmed, E. S. A. and Saeed, R. A. "A survey of big data cloud computing security." *International Journal of Computer Science and Software Engineering (IJCSSE)*. 2014; 3(1): 78–85.

[6] Pathan, A.-S. K., Saeed, R. A., Feki, M. A., and Tran, N. H. "Integration of IoT with future internet." *Journal of Internet Technology (JIT)*, 2014; 15(2): 145–147.

[7] Zappone, A., Renzo, M. D., and Debbah, M. "Wireless networks design in the era of deep learning: Model-based, AI-based, or both?." *arXiv, 13* Jun 2019, pages 1–45.

[8] Wang, F. and Preininger, A. "AI in health: State of the art, challenges, and future directions." *IMIA Yearbook of Medical Informatics*. 2019; 28(1):16–26.

[9] Saeed, M. M., Saeed, R. A., and Saeid, E. "Preserving privacy of paging procedure in 5th G using identity-division multiplexing." *First International Conference of Intelligent Computing and Engineering ICOICE*. Yemen; 2019.

[10] Atlam, H. F., Walters, R. J., and Wills, G. B. "Intelligence of things: Opportunities & challenges." *3rd Cloudification of the Internet of Things (CIoT)*, Paris, France; 2018, pp. 1–6. doi: 10.1109/CIOT.2018.8627114.

[11] Pramanik, P. K. D., Pal, S., and Choudhury, P. "Beyond automation: The cognitive IoT. artificial intelligence brings sense to the internet of things." *Cognitive Computing for Big Data Systems Over IoT*. 2018.

[12] Ghosh, A., Chakraborty, D., and Law, A. "Artificial intelligence in Internet of things." *Institution of Engineering and Technology*. 2018; 3(4): 208–218.

[13] Park, J.-H., Salim, M. M., Jo, J. H., *et al.* "CIoT-Net: a scalable cognitive IoT based smart city network architecture." *Human Centric Computing Information Science*. 2019; 9: 29. Available from https://doi.org/10.1186/s13673-019-0190-9.

[14] Skourletopoulos, G., Mavromoustakis, C. X., Mastorakis, G., and Batalla, J. M. "Big data and cloud computing: A survey of the state-of-the-art and research challenges." *Advances in mobile cloud computing and big data in the 5G era*. Springer. 2016, 23–41.

[15] Skourletopoulos, G., Mavromoustakis, C. X., Mastorakis, G., *et al.* "An evaluation of cloud-based mobile services with limited capacity: A linear approach." *Soft Computing*. 2017; 21: 4523–4530.

[16] Mavromoustakis, C., Mastorakis, G., Bourdena, A., and Pallis, E. "Energy efficient resource sharing using a traffic-oriented routing scheme for cognitive radio networks." *IET Networks*. 2014; 3(1): 54–63.

[17] Davenport, T., Guha, A., Grewal, D., *et al.* "How artificial intelligence will change the future of marketing." *Journal of the Academy Marketing Science*. 2020; 48: 24–42.

[18] Susar, D. and Aquaro, V. "Artificial intelligence: Opportunities and challenges for the public sector." *Proceedings of the 12th International Conference on Theory and Practice of Electronic Governance (ICEGOV 2019)*. Association for Computing Machinery, New York, NY, USA; 418–426. Available from https://doi.org/10.1145/3326365.3326420.

[19] Manogaran, G., Mohamed Shakeel, P., Fouad, H., *et al.* "Wearable IoT smart-log patch: An edge computing-based Bayesian deep learning network system for multi access physical monitoring system." *Sensors*. 2019; 19(13): 3030.

[20] Almeida, A., Mulero, R., Rametta, P., Urošević, V., Andrić, M., and Patrono, L. "A critical analysis of an IoT—aware AAL system for elderly monitoring." *Future Generation Computer Systems*. 2019; 97: 598–619.

[21] Fotios, Z., Koulouras, G., Karabetsos, S., and Kandris, D. "A review of machine learning and IoT in smart transportation." *Future Internet*. 2019; 11(4): 94.

[22] Cheng, C. "How AI at the edge is defining next-generation hardware platforms." *DesignCon - Santa Clara Electronics & Test Artificial Intelligence, IoT, Electronics & Test*. November 2019. Available from https://www.designnews.com/electronics-test/how-ai-edge-defining-next-generation-hardware-platforms/183048850061875. [Accessed in 27 November 2019].

[23] Wang, X., Yiwen, H., Victor C. M. Leung, Niyato, D., Yan, X., and Chen, X. "Convergence of edge computing and deep learning: A comprehensive survey." *arXiv,* to be Appeared in IEEE Communications Surveys & Tutorials; 2019, pages 1–36.

[24] Bassiliades, N. and Chalkiadakis, G. "Artificial intelligence techniques for the smart grid." *Advances in Building Energy Research*. 2018; 12(1): 1–2.

[25] Fang, H., Qi, A., and Wang, X. "Fast authentication and progressive authorization in large-scale IoT: How to leverage AI for security enhancement?" *IEEE Network*. 2020; 34(3): 24–29.

[26] Mohammed, M., Khan, M. B., and Bashier, E. B. M. *Machine learning: Algorithms and applications*. Boca Raton, FL: CRC Press; 2016.

[27] Kubat, M. *An introduction to machine learning*. Cham, Switzerland: Springer; 2017.

[28] DeBeasi, P. "Machine learning integration options." Gartner. 2019. Available from https://blogs.gartner.com/paul-debeasi/2019/01/30/machine-learning-integration options/ [Accessed 02 December 2019].

[29] Gopalakrishnan, K. *Deep learning in data-driven pavement image analysis and automated distress detection: A review*. Ames, IA: Iowa State University; 2018.

[30] Ota, N. "Create deep intelligence in the Internet of things." 2014. Available from http://on-demand.gputechconf.com/gtc/2015/presentation/S5813-Nobuyuki-Ota.pdf. [Accessed 11 November 2019].

[31] Xu, G., Shi, Y., Sun, X., and Shen, W. "Internet of things in marine environment monitoring: A review." *Sensors*. 2019; 19(7): 1711.

[32] Batalla, J. M., Mavromoustakis, C. X., Mastorakis, G., and Sienkiewicz, K. "On the track of 5G radio access network for IoT wireless spectrum sharing in device positioning applications." *Internet of things (IoT) in 5G mobile technologies*, 2016.

[33] Zhou, Z., Chen, X., Li, E., Zeng, L., Luo., K, and Zhang, J. "Edge intelligence: Paving the last mile of artificial intelligence with edge computing." *Proceedings of the IEEE*. 2019; 107(8): 1738–1762.

[34] Karolewicz, K., Beben, A., Batalla, J. M., Mastorakis, G., and Mavromoustakis, C. X. "On efficient data storage service for IoT." *International Journal of Network Management*. 2017; 27: e1932.

[35] Moon, J., Kum, S., and Lee, S. "A heterogeneous IoT data analysis framework with collaboration of edge-cloud computing: Focusing on indoor PM10 and PM2.5 status prediction." *Sensors* 2019; 19: 3038.

[36] Batalla, J. M., Sienkiewicz, K., Latoszek, W., Krawiec, P., and Mavromoustakis, C. X. "Validation of virtualization platforms for I-IoT purposes." *The Journal of Supercomputing*. 2018.

[37] Chamatkar, A. J. "An artificial intelligence for data mining." *Journal of Computer Science (IOSR-JCE)*. 2014; 53–57.

[38] Ghosh, A., Chakraborty, D., and Law, A. "Artificial intelligence in Internet of things." *CAAI Transitions of Intelligence Technology*. 2018; 3(4): 208–218.

[39] Phillips-Wren, G. and Jain, L. "Artificial intelligence for decision making." In B. Gabrys, R. J. Howlett, and L. C. Jain (eds.), *Knowledge-Based Intelligent Information and Engineering Systems. KES 2006*. Lecture Notes in Computer Science; vol. 4252. Springer, Berlin, Heidelberg. Available from https://doi.org/10.1007/11893004_69.

[40] Saima, H., Jaafar, J., Belhaouari, S., and Jillani, T. A. "Intelligent methods for weather forecasting: A review." *2011 National Postgraduate Conference*; Kuala Lumpur, 2011, pp. 1–6. doi: 10.1109/NatPC.2011.6136289.

[41] Kee, K. K., Shahab, S. M. F., and Loh, C. J. "Design and development of an innovative smart metering system with GUI-based NTL detection platform." *4th IET Clean Energy and Technology Conference (CEAT 2016)*; Kuala Lumpur, 2016, pp. 1–8. doi: 10.1049/cp.2016.1293.

[42] Kopytko, V., Shevchuk, A., Yankovska, L., Semchuk, Z., and Strilchuk, R. "Smart home and artificial intelligence as environment for the implementation of new technologies." *International Electronic Scientific Journal*. 2018; 4(9): 2007–2012.

[43] Kryftis, Y., Mastorakis, G., Mavromoustakis, C. X., Batalla, J. M., and Pallis, E. "Efficient entertainment services provision over a novel network architecture." *IEEE Wireless Communications*. 2016.

[44] Batalla, J. M., Mastorakis, G., Mavromoustakis, C. X., and Zurek, J. "On cohabitating networking technologies with common wireless access for home automation system purposes." *IEEE Wireless Communications*. 2016.

[45] Mohan Kumar, S. and Majumder, D. "Healthcare solution based on machine learning applications in IOT and edge computing." *International Journal of Pure and Applied Mathematics*. 2018; 119(16): 1473–1484.

[46] Papadokostaki, K., Mastorakis, G., Panagiotakis, S., and Mavromoustakis, C. X. "Handling big data in the era of Internet of things (IoT)." *Advances in mobile cloud computing and big data in the 5G era*. 2016.

[47] Machorro-Cano, I., Alor-Hernández, G., Paredes-Valverde, M.A., Ramos-Deonati, U., Sánchez-Cervantes, J.L., and Rodríguez-Mazahua, L. "PISIoT: A machine learning and IoT-based smart health platform for overweight and obesity control." *Applied Sciences.* 2019; 9: 3037. doi:10.3390/app9153037.

[48] Alreshidi, E. "Smart sustainable agriculture (SSA) solution underpinned by Internet of things (IoT) and artificial intelligence (AI)." *(IJACSA) International Journal of Advanced Computer Science and Applications.* 2019; 10(5): 93–102.

[49] Dharmaraj, V. and Vijayanand, C. "Artificial intelligence (AI) in agriculture." *International Journal of Current Microbiology and Applied Sciences*. 2018; 7(12): 2319–7706.

[50] Ferrández-Pastor, F. J., García-Chamizo, J. M., Nieto-Hidalgo, M., and Mora-Martínez, J. "Precision agriculture design method using a distributed computing architecture on Internet of things context." *Sensors*. 2018; 18: 1731. doi:10.3390/s18061731.

[51] Ruan, J., Hu, X., Huo, X., *et al.* "An IoT-based E-business model of intelligent vegetable greenhouses and its key operations management issues." *Neural Computing & Applications*. 2020; 32: 15341–15356. Available from https://doi.org/10.1007/s00521-019-04123-x.

[52] Xu, H. and Geng, X. "People-centric service intelligence for smart cities". *Smart Cities*. 2019; 2(2): 135–152. Available from https://doi.org/10.3390/smartcities2020010.

[53] Batalla, J. M., Gajewski, M., Latoszek, W., Krawiec, P., and Mavromoustakis, C. X. "ID-based service-oriented communications for unified access to IoT." *Computers and Electrical Engineering*. 2016.

[54] Mohammadi, M. and Al-Fuqaha, A. "Enabling cognitive smart cities using big data and machine learning: Approaches and challenges." *IEEE Communications Magazine*. 2018; 56(2): 94–101.

[55] Nikoloudakis, Y., Panagiotakis, S., Markakis, E., and Mastorakis, G. "Towards a FOG-enabled navigation system with advanced cross-layer management features and IoT equipment." *Cloud and fog computing in 5G mobile networks: Emerging advances and application*. 2017.

[56] Inclezan, D., and Pradanos, L. I. "Viewpoint: A critical view on smart cities and AI." *Journal of Artificial Intelligence Research*. 2017; 60: 681–686.

[57] Miraz, M. H., Ali, M., Excell, P. S., and Picking, R. "Internet of nano-things, things and everything: Future growth trends." *Future Internet.* 2018; 10: 68. doi:10.3390/fi10080068.

[58] Lee, J., Singh, J., and Azamfar, M. "Industrial artificial intelligence." *ArXiv* abs/1908.02150; 2019, pages 1:10.

Chapter 3
Green energy harvesting protocols for intelligent wireless communication systems

Ganesh Prasad[1] *and Deepak Mishra*[2]

With the exponential growth of economical and sophisticated user devices in the digitized world, the competence of communicating high volume of data such as high-definition videos or high-quality multimedia has become users' requirements nowadays. However, spectral-efficient networking for high-speed data and self-sustainability in energy are two major trade-offs in design of energy-aware next-generation wireless communications. Efficient utilization of energy resources reduces the cost of power consumption that can improve the economic characteristics of the network. In this chapter, we investigate the energy and quality-of-service (QoS) aware smart services using artificial intelligence (AI) assisted automated wireless communications, describes the novel energy state prediction model, online optimization algorithms for energy-harvesting (EH), and enhancement of autonomy by maximizing the throughput based on self-sustainability. The intelligent power management technological developments while satisfying the desired QoS paving the way for AI-powered mechanisms in future wireless networking applications.

3.1 Background and motivation

With the exponential growth of data traffic due to multimedia applications in next-generation communications, the power dissipation in underlying network infrastructures is also increasing at the same rate [1]. To investigate this coexistence problem, today green communication has not only gained interest in academic researches but also important for a self-sustainable network in a realistic scenario. Specifically, the fundamental green tradeoff in the problem exists between spectral efficiency and energy efficiency. The spectral efficiency of a wireless network is defined as the throughput achieved per unit bandwidth, whereas energy efficiency is throughput achieved per unit power dissipation.

[1]Department of Electronics and Communication Engineering, NIT Silchar, Assam, India
[2]School of Electrical Engineering and Telecommunications, UNSW Sydney, Sydney, Australia

The tradeoff between them can be mathematically expressed in an additive white Gaussian noise (AWGN) channel as [2]:

$$\eta_{ee} = \frac{\eta_{se}}{(2^{\eta_{se}} - 1)\Omega_0} \tag{3.1}$$

where Ω_0 is the power spectral density of AWGN, η_{ee} is the energy efficiency, and η_{se} is the spectral efficiency. For a given bandwidth B_W and transmit power P_t, η_{se} is given as:

$$\eta_{se} = \log_2\left(1 + \frac{P_t}{\Omega_0 B_W}\right) \tag{3.2}$$

Further, the relationship between energy efficiency and spectral efficiency can be described using the plot as depicted in Figure 3.1. It can be observed that η_{ee} decreases at a rate with η_{se} and vice versa and η_{ee} converges to $\frac{1}{\Omega_0 \ln 2}$ for $\eta_{se} \to 0$ and $\eta_{ee} \to 0$ for $\eta_{se} \to \infty$. However, the above tradeoff is obtained in an ideal condition where the channel consists only of AWGN with transmission power P_t. In a real system, where modulation, source coding, and resource management are also considered, the tradeoff is not simple and which is portrayed as bell-shaped in Figure 3.2 [3] that may further vary depending on the types of the networks. In this chapter, to optimize this tradeoff, we describe the green spectral efficient access protocols in self-sustainable machine type communications.

Through optimization of the tradeoff using the underlying system parameters that reduce the power dissipation while satisfying the desired throughput, the resultant saved energy due to it does not fulfill the requirement to make the operating network

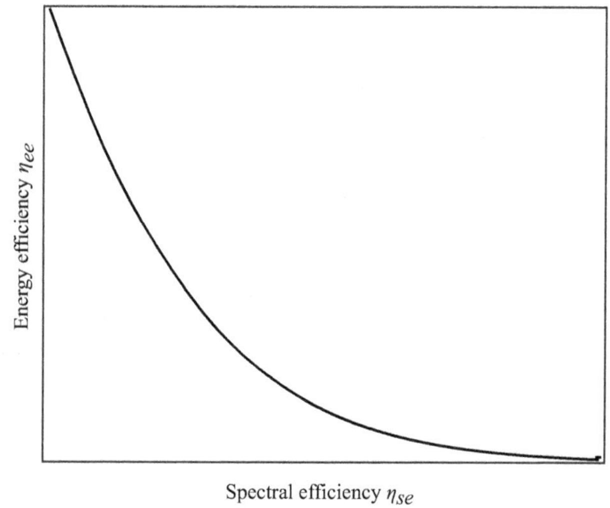

Figure 3.1 Tradeoff between energy efficiency and spectral efficiency obtained. Adapted from [31]

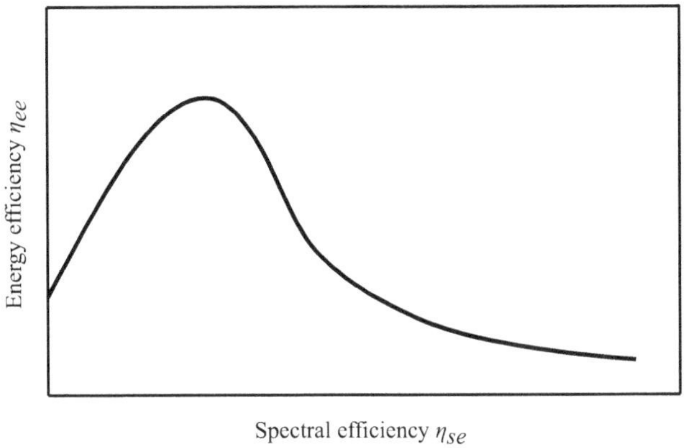

Figure 3.2 Tradeoff between energy efficiency and spectral efficiency in a practical system

self-sustainable [4]. Also, today the devices are restricted in their sizes and have limited energy storage capability [5] that cannot provide sufficient power to the network for its indispensable lifetime [6]. Therefore, laterally, a regenerative mechanism is required to replenish the dissipated energy which can be conceived using energy harvesting (EH). Nevertheless, EH rely on statistics of ambient energy sources such as solar, wind, vibration, and radio frequency (RF). Therefore, essentially, we require a predication model for the statistical knowledge of available energy sources as well as for energy-efficient power-management solutions [7]. In fact, these ambient sources are site-specific and quite unpredictable using the conventional predication models. Inevitably, we need learning techniques based on artificial intelligent (AI) tools to get the availability of ambient sources with high accuracy [8,9].

Moreover, the protocol used in EH wireless network is entirely different than the communication in conventional network run by external supply or battery storage. Consequently, we need to develop an optimal algorithm to enhance the throughput of the EH wireless network. From the existing works, offline and online algorithms are adopted to maximize the system performances [10] which have been explored in this chapter in detail.

3.2 Architecture of energy harvesting wireless networks

An architecture of EH wireless networks is shown in Figure 3.3 that comprises access points, energy harvesters, and user devices as main components. The harvesters use the sources such as solar and RF to harvest the energy. The access points get the energy from solar harvesters and smart grid, whereas user devices rely only on the EH from sunlight and RF. The received energy by the network nodes is stored in a battery of finite capacity associated with each of them. The information can be transmitted through device-to-device communication or channel between an access point (AP) and a device. In contrast, RF energy harvesting (RFEH) takes place only

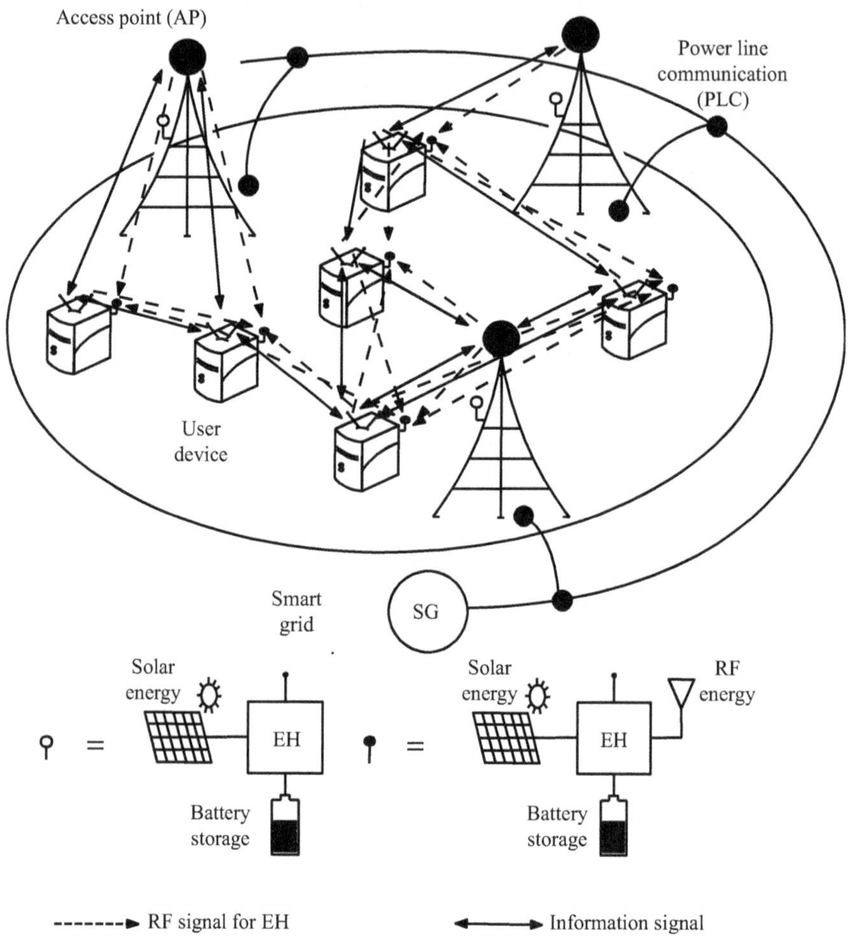

Figure 3.3 A typical architecture of EH wireless network

at devices either using the received signals from an AP or from other devices as portrayed in Figure 3.3. Here the solid lines show the information transmissions, while the dashed lines indicate the RF energy transmissions. Note that typically, the received signal power for the EH is higher than for information.

From the diagram in Figure 3.4, apart from the typical blocks namely, application, processing, transceiver, and battery, in addition, an EH node consists EH harvester to generate energy using sunlight and dedicated RF signals and power management module that instructs either to store the generated energy or to use it in immediate information transmissions [11]. Typically, the antenna used for RFEH can operate in multiple frequencies as the energy in the RF signals distributed over a range of frequencies. Figure 3.5 shows the architecture of RFEH over the nodes based on simultaneous and time switching of power and

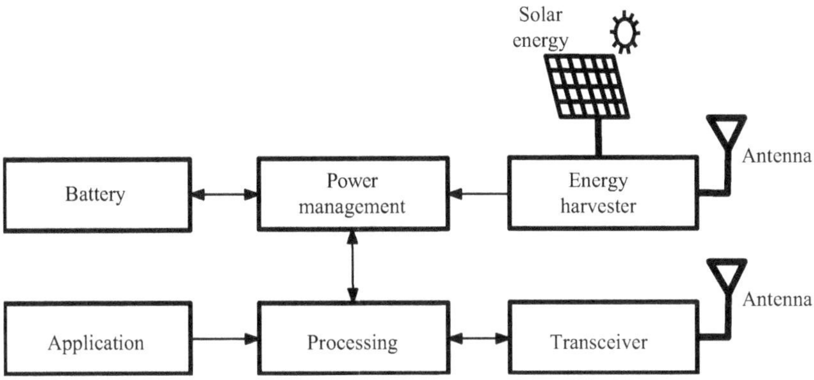

Figure 3.4 A general block diagram for an EH network node

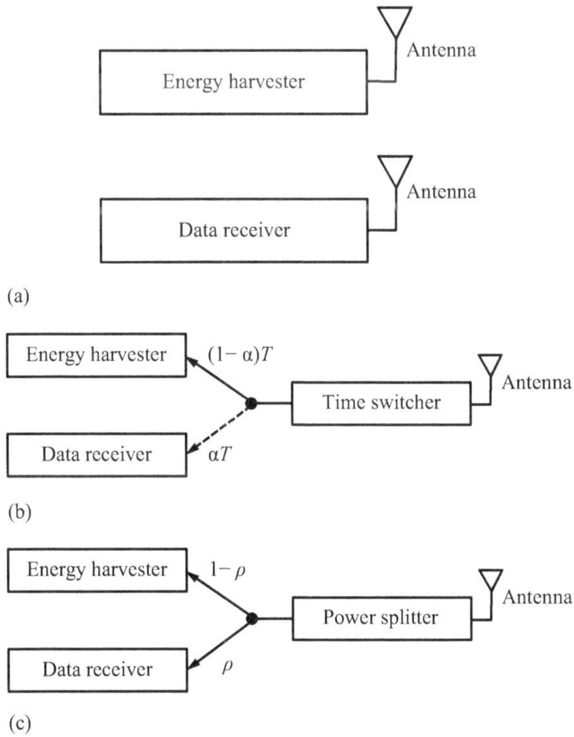

Figure 3.5 RFEH receiver based on (a) separate architecture, (b) time switching, and (c) power splitting

information transfers. A simple architecture of RFEH receiver is given by Figure 3.5(a) where energy harvester and data reception units are separate. The receiver is capable of EH using in-band or out-of-band RF signals. From in-band RFEH, the receiver harvest the energy from the same signal which has information, but in out-of-band RFEH, the energy is retrieved also from the signals operating at frequencies other than information signal [12]. As a signal contains both information as well as power, RFEH receiver can use the same signal for concurrent transmission of data and power which is also known as simultaneous wireless information and power transfer (SWIPT) [13]. In Figure 3.5(b), αT time slot is allocated using time switcher for information reception and in $(1-\alpha)T$, power is retrieved from the signal. In contrast, Figure 3.5(c) describes that power of the signal is split using power splitter, where its ρ times is used for decoding of information and $1-\rho$ times is inputted to energy harvester. Note that the parameters, $\rho, \alpha \in (0, 1)$.

Specifically, in the network architecture as shown in Figure 3.3, we have also shown that besides solar energy harvester, APs are also powered by smart grid that helps in information transfers among them using power line communication (PLC) [14]. Using it, the spectrum allocation for connectivity between APs can be avoided which improves the overall throughput of the network.

3.3 Energy harvesting models for intelligent wireless communications

The amount of energy to be harvested using ambient sources changes with location and time. For example, we get the solar power in a particular time slot of a day, besides it may also be affected by the weather conditions, diurnal cycle, and seasonal patterns [15]. On the other hand, the amount of available RF signal varies with time and frequency. For instance, in an urban area, the RF energy due to mobile signals is more in the daytime than at night. In contrast, over the spectrum, TV band has higher RF power than mobile band. In order to harvest the energy from an ambient source efficiently, we need to develop a prediction model for the source availability with high accuracy.

3.3.1 Solar energy predication model

To avoid the short term energy shortage in devices while performing the operations, accurate prediction of solar energy is essentially required in the short duration such as minutes or hours [16]. In current approaches, the solar energy is predicted using diurnal cycle which is divided into M equal time slots as shown in Figure 3.6, where the weather condition in subsequent days is expected to be the same. The length of a slot depends on the prediction required for an application and the prediction of energy for each slot is obtained at the onset of the slot. A day is split into slots to record the profile of generated energy of past days and using it,

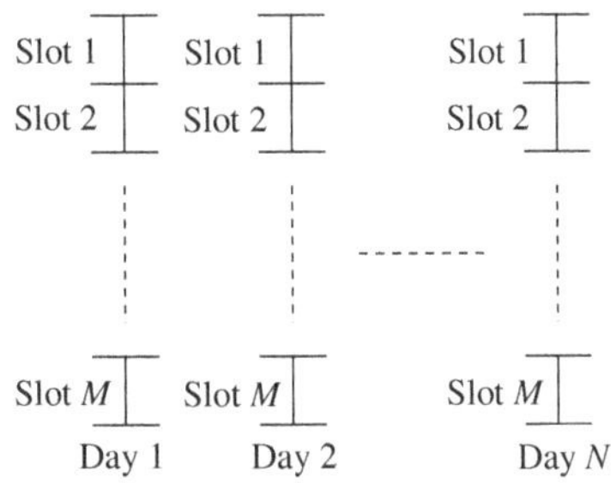

Figure 3.6 Dividing each day into M equal slots for predication of energy on the onset of a slot

the currently available solar energy is predicted. Based on it, different prediction models have been discoursed below.

3.3.1.1 Exponentially weighted moving average

The prediction model, exponentially weighted moving average (EWMA) is based on the assumption that the available energy profile at a slot of a day is almost analogous to the profile in the same slot of the preceding day [17]. The available energy in a slot is estimated by computing the weighted average of estimated energy and harvested energy (actual energy) of the preceding day in the same slot, can be expressed as:

$$\widehat{\mathscr{E}}_{\mathrm{E}}(n, m) = \omega \widehat{\mathscr{E}}_{\mathrm{E}}(n-1, m) + (1-\omega)\mathscr{E}(n-1, m) \tag{3.3}$$

where n is the present day, m is the slot of the day, $\omega \in (0, 1)$ is a weighting factor, $\widehat{\mathscr{E}}_{\mathrm{E}}$ is the estimated energy using EWMA, and $\mathscr{E}$ is the harvested energy in the given slot and day. In (3.3), the value of ω decides the dominance of $\widehat{\mathscr{E}}_{\mathrm{E}}$ or $\mathscr{E}$ of preceding day in the weighting average to obtain $\widehat{\mathscr{E}}_{\mathrm{E}}$ of present day. Further, ω can be optimized to adapt the seasonal variation that changes the energy profile with days. The disadvantage of EWMA is its high inaccurate estimated results in frequently changing weather conditions as the weighting factor ω used in the computation is fixed.

3.3.1.2 Modified EWMA

For optimal allocation of regularly harvested solar energy to network nodes, an accurate solar energy allocation (ASEA) scheme exploits the principle of EWMA to predict the available energy [18]. Its goal is to save energy for future use when the solar energy is inadequate or unavailable. In this regard, it is essential to predict (or estimate) the future available energy accurately that can be conceived by modifying EWMA by introducing a new parameter Υ. Using modified EWMA, the estimated energy $\widehat{\mathscr{E}}_{\mathrm{M}}(n, m)$ in slot m of day n is given by

$$\widehat{\mathscr{E}}_{\mathrm{M}}(n, m) = \widehat{\mathscr{E}}_{\mathrm{E}}(n, m) \cdot \Upsilon \tag{3.4}$$

where $\Upsilon \triangleq \frac{\mathscr{E}(n,m-1)}{\widehat{\mathscr{E}}_{\mathrm{E}}(n,m-1)}$ is the ratio of harvested energy to estimated energy in preceding slot using EWMA. Modified EWMA has a drawback as it gives significant error in the estimation if the predication error in the preceding slot occurs due to a temporary change in the weather condition.

3.3.1.3 Weather conditioned moving average

The demerits of EDMA and modified EDMA can be efficiently handled using weather conditioned moving average (WCMA) scheme that comprises present and preceding day weather conditions in the prediction of solar energy [19]. Using an energy generated matrix $\widehat{\mathscr{E}}$ of size $N \times M$ that consists of the energy values of M time slots for each of the past D days, where $\widehat{\mathscr{E}}_{\mathrm{W}}(n,m)$ denotes the energy generated in the mth time slot of the nth day. The prediction of available energy $\widehat{\mathscr{E}}_{\mathrm{W}}(n,m)$ in m^{th} slot of nth day as given by (3.5) depends on the generated energy $\widehat{\mathscr{E}}_{\mathrm{W}}(n,m-1)$ in the preceding slot $m-1$ of the same day n and average of the generated energies, $\mathscr{M}(n,m)$, of past D days in the mth time slot as expressed in (3.6):

$$\widehat{\mathscr{E}}_{\mathrm{W}}(n,m) = \omega\widehat{\mathscr{E}}_{\mathrm{W}}(n,m-1) + (1-\omega)\mathscr{M}(n,m)\mathrm{GAP} \tag{3.5}$$

$$\mathscr{M}(n,m) = \frac{\sum_{j=n-D}^{n-1} \widehat{\mathscr{E}}_{\mathrm{W}}(j,m)}{D} \tag{3.6}$$

The important factor in WCMA algorithm is GAP that measures the relative condition of solar energy on present day compared to preceding days. To find it, first, we define a vector $\mathscr{V} = \{v_1, v_2, \cdots, v_K\}$ of K elements which is the ratio of preceding K samples and mean of the available energy during preceding D days. An element $v_k \in \mathscr{V}$ is given by

$$v_k = \frac{\widehat{\mathscr{E}}_{\mathrm{W}}(n, m-K+k-1)}{\mathscr{M}(n, m-K+k-1)} \tag{3.7}$$

In case, $v_k > 1$ shows the available energy in time slot $(m-K+k-1)$ is greater than the mean value of energy over preceding D days in the same time slot that infers a sunny day. Otherwise, $v_k < 1$ implies a cloudy day. In order to assign more importance to the values nearer to slot m in time, we weight them using the vector $\mathscr{P} = \{p_1, p_2, \cdots, p_K\}$, where p_k is given as:

$$p_k = \frac{k}{K} \tag{3.8}$$

Using above parameters, the factor GAP is determined as:

$$\mathrm{GAP} = \frac{\mathscr{V} \cdot \mathscr{P}}{\sum \mathscr{P}}. \tag{3.9}$$

3.3.1.4 Proenergy prediction model

This prediction model also uses the harvested energy profile of preceding days to estimate the available energy profile on present day. But, it is distinct in computation of most similar day to the present day in terms of available energy which is obtained using the preceding energy profile. The estimated energy can be expressed as:

$$\widehat{\mathscr{E}}_{\mathrm{P}}(n,m) = \omega\mathscr{E}(n,m-1) + (1-\omega)\mathscr{E}_{\mathrm{MS}} \tag{3.10}$$

where $\mathscr{E}(n,m-1)$ is the harvested energy in the preceding slot and $\mathscr{E}_{\mathrm{MS}}$ is the harvested energy in the time slot m of the most similar (MS) day. In order to quantify the similarity of a day with present day, we calculate mean absolute error (MAE) using K preceding slots to present slot for every preceding D days and the day is MS day that has lowest MAE.

To further improve the performance, multiple energy profiles over the days are combined rather extracting it from MS day, it results in considering potential variation in weather instead of adopting single profile of a MS day in the estimated energy $\widehat{\mathscr{E}}_{\mathrm{P}}(n,m)$. To find it, the elements of $\mathscr{E}$ over the days are sorted according to MAE in the vector $\mathbf{E_{SP}} = \{e_1, e_2, \cdots, e_P\}$ having P profiles. The weighted profile for the available energy can be expressed as:

$$\mathscr{E}_{\mathrm{WP}} = \frac{\sum_{q=0}^{P} w_q e_q}{P-1}, \tag{3.11}$$

where w_q with respect present day profile $\mathscr{E}_C$ is given by

$$w_q = 1 - \frac{\mathrm{MAE}(e_q, \mathscr{E}_C)}{\sum_{q=1}^{P} \mathrm{MAE}(e_q, \widehat{\mathscr{E}}_C)} \tag{3.12}$$

Using (3.11), the energy prediction $\widehat{\mathscr{E}}_{\mathrm{P}}(n,m)$ is expressed as:

$$\widehat{\mathscr{E}}_{\mathrm{P}}(n,m) = \omega\mathscr{E}(n,m-1) + (1-\omega)\mathscr{E}_{WP} \tag{3.13}$$

3.3.1.5 QL-SEP algorithm

The Q-learning based solar energy prediction (QL-SEP) exploits the data, energy profile of past days and weather condition of present day to estimate the available energy. It uses the framework of EWMA which gives the status of solar energy in past days. Moreover, an additional feature is introduced to encounter present weather condition using a parameter, daily ratio (DR) φ. The energy prediction $\mathscr{E}_{\mathrm{Q}}$ using QL-SEP is expressed as:

$$\widehat{\mathscr{E}}_{\mathrm{Q}}(n,m) = \widehat{\mathscr{E}}_{\mathrm{E}}(n,m) \cdot (1 + \varphi(n,m)) \tag{3.14}$$

where $\widehat{\mathscr{E}}_{\mathrm{E}}$ is the predicted solar energy using (3.3). The DR φ in the current slot m of the present day n is given by

$$\varphi(n,m) = \frac{\sum_{k=1}^{K} p_{\mathrm{e}}(n, m-K+k-1) \cdot r(n, m-K+k-1) \cdot k}{K} \tag{3.15}$$

where p_{e} is the predication error and r is the reliability in the previous slot of the present day. Here, the reliability r is the quality of choice for the prediction and can be determined using Q-learning [20]. The Q-value for the desired action to obtain the solution can be determined as:

$$Q_{t+1}(\tau) = Q_t(\tau) + \vartheta(\varsigma - Q_t(\tau)) \tag{3.16}$$

where τ is the given slot, ς is the reward value, and ϑ is the learning rate. The reliability r takes $+1$ for positive feedback; otherwise, -1 for negative feedback. The Q value converges to $+1$ or -1 for the sequence of positive or negative feedback and the learning rate ϑ controls the Q-value.

3.3.2 RF energy prediction model

In radiofrequency energy harvesting (RFEH), the energy is generated by harvesting the ambient RF signals by collecting the signal using an antenna followed by converting it into DC power through a rectifier. In contrast, energy availability due to RF signal is more credible than the solar energy as it does not rely upon the nature. Besides, potential RF signal can be used in simultaneous wireless information and power transfer (SWIPT) for long-distance communications.

Statistically, the available RF energy changes with time and frequency. Usually, the amount of RF signals is higher during the daytime than at night in urban areas. While in frequency, the RF energy in TV band is higher than in mobile band. So, it is desired to choose appropriate time and frequency to maximize the generated energy by harvesting.

Most basic model for available RF energy is the Bernoulli model that assumes probability t for available RF energy, otherwise, it is absent with probability $1-t$ [21,22]. In [23], a simple model is considered where three levels 0, ξ, and 2ξ of available energy are assumed to occur with equal probability. However, these theoretical models are not tractable in realistic scenarios. The algorithms based on machine learning (ML) estimate the available RF energy efficiently in time and frequency [24]. In supervised learning, the feature vector is the harvested energy of the q frequency bins in time slot m represented as:

$$\mathscr{E}^{m,l-1:l-q} = \left[e^{m,l-1}, e^{m,l-2}, \cdots, e^{m,l-q}\right]^T \tag{3.17}$$

and target label is harvested energy of frequency bin l in time slot m denoted as $e^{m,l}$. Here $q \leq L$, where q is the length of the feature vector and L is the total number of frequency bins over a frequency band.

In order to get a trained model, feature vectors are trained for their defined target labels. For the given target label $e_{trn}^{m,l}$, the training phase using linear regression (LR) is given by

$$\begin{aligned} e_{trn}^{m,l} &= \alpha_0 + \alpha_{l-1} e_{trn}^{m,l-1} + \alpha_{l-2} e_{trn}^{m,l-2} + \cdots + \alpha_{l-q} e_{trn}^{m,l-q} \\ &= \alpha_0 + \sum_{j=1}^{q} \alpha_{l-j} e_{trn}^{m,l-j} \\ &= \alpha_0 + \left[\alpha_{l-1}, \alpha_{l-2}, \cdots, \alpha_{l-q}\right] \mathscr{E}_{trn}^{m,l-1:l-q} \end{aligned} \tag{3.18}$$

where $\mathscr{E}_{trn}^{m,l-1:l-q} = \left[e_{trn}^{m,l-1}, e_{trn}^{m,l-2}, \cdots, e_{trn}^{m,l-q}\right]^T$. From the given feature vectors and their defined target labels, the model parameters are trained represented by $\mathscr{A}_{trn} = [\alpha_{l-1}^t, \alpha_{l-2}^t, \cdots, \alpha_{l-q}^t]$. Using it, for a given feature vector $\mathscr{E}^{m,l-1:l-q}$, target label is estimated as:

$$\widehat{\mathscr{E}}^{m,l} = \alpha_0^t + \mathscr{A}_{trn} \mathscr{E}^{m,l-1:l-q} \tag{3.19}$$

Using decision tree (DT), the data related to energy harvesting is divided into subsets and a tree is made for each of them. Thereafter, the subtrees are combined to form a single tree [25]. The data in DT is represented as:

$$\left(\widehat{\mathscr{E}}_{trn}^{m,l-1:l-q}, e_{trn}^{m,l}\right) = \left(e_{trn}^{m,l-1}, e_{trn}^{m,l-2}, \cdots, e_{trn}^{m,l-q}, e_{trn}^{m,l}\right) \tag{3.20}$$

To manage the data for analysis, $(\varepsilon_{trn}^{m,l-1:l-q})$ is split into smaller regions (leaves). The partition of tree is determined by a least-square error criterion. In a binary tree, the optimal split s at node z maximizes $\Delta\theta(s, z)$ as given by

$$\Delta\theta(s, z) = \theta(z) - p_l \theta(z_l) - p_r \theta(z_r) \tag{3.21}$$

where $\theta(z)$ is mean squared error at node z, while p_l and p_r are the fraction of data that lies at left and right branch of node z with corresponding errors $\theta(z_l)$ and $\theta(z_r)$, respectively [26]. Apart from the energy prediction, the efficiency of the generated energy can be maximized by accurately modeling the channel at the receiver [27]. Using the predicted energy states, next, we describe the optimal policies for efficient power management in the self-sustainable network.

3.4 AI-assisted online algorithms for optimal energy harvesting communications

Transmission policy for battery-powered and energy harvesting network is different, because in the former, the amount of source energy is fixed, while the later one harvests random available energy. An optimal transmission policy is procured to efficiently utilize the harvested energy, to maximize the throughput, and to minimize end-to-end transmission delay. There are two different policies: (1) online and (2) offline. In online, the network devices have the statistical knowledge of the

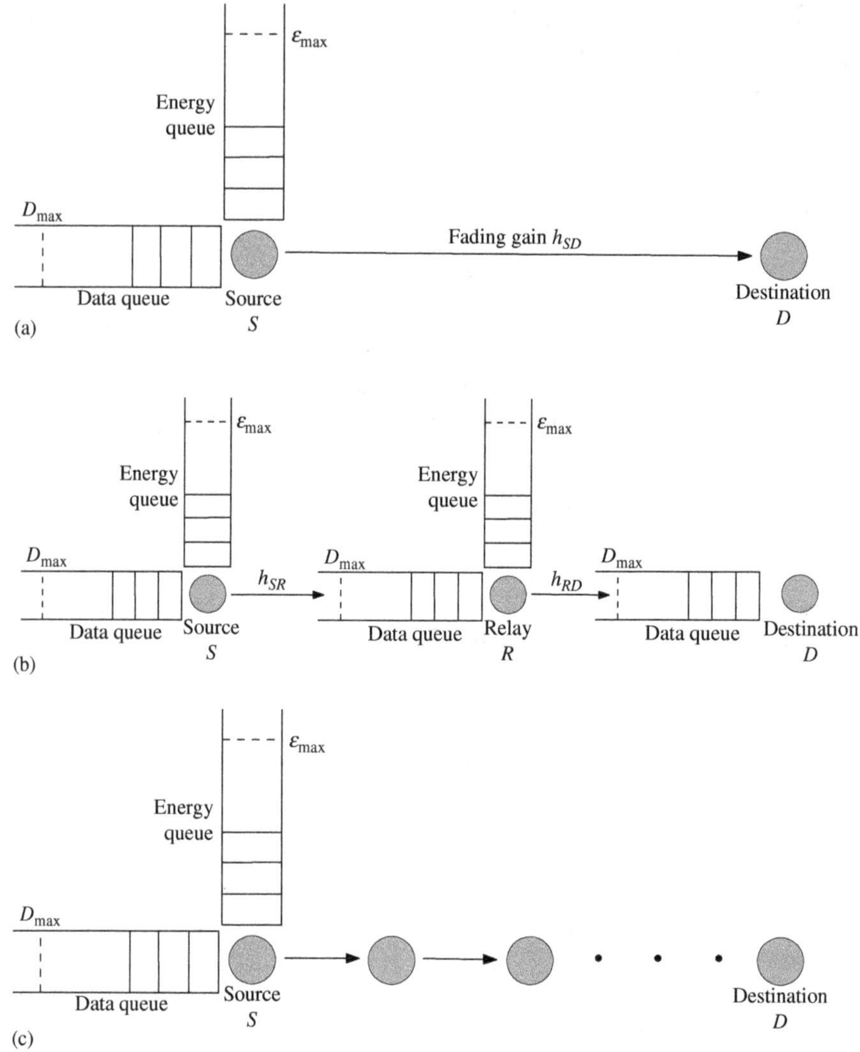

Figure 3.7 EH wireless networks: (a) single-hop communication; (b) dual-hop communication; and (c) multi-hop communication

available energy to be harvested, while under offline policy, the devices have full knowledge of available energy along with their arrival time. Although offline policy works in ideal scenario, using it, we obtain the heuristic solutions to find optimal transmission policy. From the existing works, the EH communications can be categorized into single-hop, dual-hop, and multihop as depicted in Figure 3.7. In a single-hop EH communication (cf. Figure 3.7(a)), the source (or transmitter) equipped with EH unit has a data and energy queue, where respective data and energy are stored in packets. $\mathcal{E}_{max}$ and D_{max} in the energy and data queue respectively denotes the

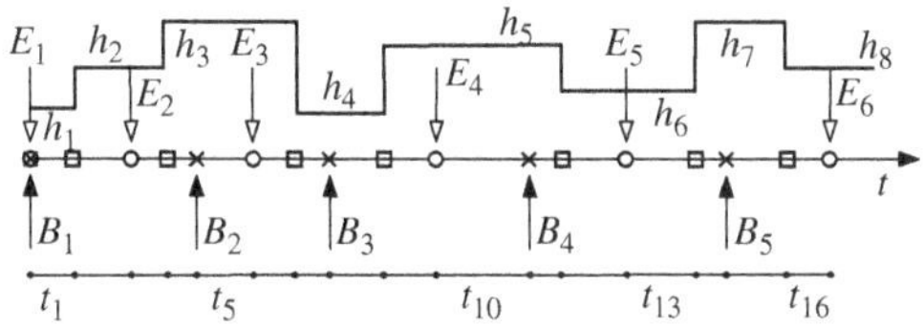

Figure 3.8 Epoch due to events, channel state change, harvested energy, data rate arrival at the transmitter (source)

maximum capacity of the queue and arrival of energy or data, exceeding the maximum capacity, are discarded. In the given architecture, to investigate the underlying transmission policy, communication in general consists of three random events at the transmitter, such as channel state change, harvested energy, and arrival of data as depicted in Figure 3.8. The time interval between two consecutive events is termed as epoch.

Although, the direct transmission from the source to destination is simple, but in many instances, due to obstacles or high fading and lossy channels, the direct transmission is not feasible. Here a relay is essentially required to forward the data from the source to destination using two-hop communication as shown in Figure 3.7(b). Further, more than one intermediate relays can be introduced based on requirement for multihop communication that is more realistic and complicated (cf. Figure 3.7(c)).

Different from the battery-powered networks, transmission policies in energy harvesting wireless communications have the following causality constraints: (1) the generated energy cannot be consumed before it is harvested, (2) data cannot be transmitted from the source before it is arrived on it, (3) the total dissipated energy cannot be greater than entire harvested energy, and (4) the total transmitted data cannot be greater than entire arrived data. In case of finite energy and data queue capacity, besides above constraints, it is also ensured that instantaneous generated or dissipated energy and arrived or transmit data cannot be more than the respective capacities, otherwise, they may be lost after exceeding the queue capacity.

The system performance of an EH wireless communication relies on the generated energy profile with time (EH profile) [28]. For example, the policy P0 represented by the dashed line in Figure 3.9 provides best performance as it supplies transmit power at a constant rate. However, it violets one of the causality constraint at the intersection of the dashed line with EH profile where total dissipated energy is greater than the entire harvested energy.

In contrast, P1 and P2 are always below the EH profile and feasible policies, where P1 is better than P2 in the power supply.

3.4.1 Offline policies

In a single-hop model as shown in Figure 3.7(a), the objective of the network is to minimize the delivery time T of the entire data while transmission from source to destination. In this context, one of the optimal offline policy has been described in [29] which is investigated in two scenarios: (1) entire data are available at the

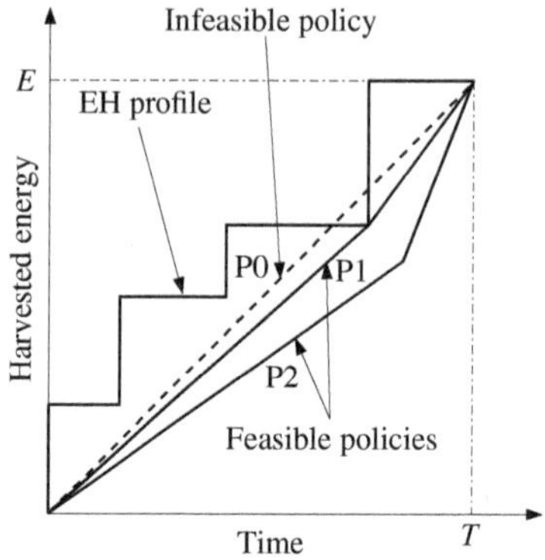

Figure 3.9 EH profile with insights on feasible policies. Adapted from [28]

source before the beginning of the transmission and (2) data may arrive while transmission of the available data. Here, it is assumed that the energy and data queues have infinite capacity, perfect channel state information (CSI) is available at the transmitter and receiver, and entire harvested energy is used at the end of the transmission time T. In this regard, the authors have described three requisite lemmas. The first lemma describes that the transmission power and data rate increase monotonically under the optimal policy. The second lemma illustrates, the two remain constant in the interval of arrival of two consecutive energy packets. Whereas, the third lemma describes that the total harvested energy in time T is completely used in transmission of entire data. For the two scenarios (as described above), two algorithms have been discoursed to minimize the transmission time T.

To investigate a more realistic scenario, authors in [30] consider the point-to-point communication in a fading channel where the CSI changes with time. Here, the epoch (cf. Figure 3.8) depends on the two events, harvested energy and channel state change. It is assumed that the entire data are available before the transmission, perfect CSI is known at the transmitter in each epoch, and battery storage has a finite capacity. In the framework, two objectives have been set to maximize the throughput for a given deadline time T and minimize the time T to transmit the entire available data. Offline approaches have been adopted for both the objectives where all the events are deterministically known. For the offline policy, the directional water-filling algorithm used to maximize the throughput and using the obtained result, the optimal policy to minimize T is determined by mapping the problems of the two objectives using maximum departure curve. In contrast, the framework in [31] also considers the arrival of data apart from channel state change and harvested energy to determine the epoch. To minimize the transmission completion time T, an equivalent convex problem is formulated which minimizes the power dissipation in transmission of data. Further, the problem was solved by formulating a sequential unconstrained minimization technique (SUMT) [32] where the objective of the problem is added with penalty terms corresponding to the constraints.

The frameworks in [30,31] assumed that the perfect CSI is known at the transmitter which is not realistic. In [33], a channel training optimization is discoursed to obtain accurate CSI at the transmitter. Furthermore, compared to conventional communication system, it has been shown that the optimization of training of power allocation and training period is a coupled problem. In this regard, two solution methodologies have been described that performed the optimization based on either instantaneous or average EH profile.

Different from the above works, authors in [34] consider an EH source and a non-EH relay node. As the non-EH relay is half-duplex, it can neither harvest energy nor transmit and receive data simultaneously. Here, throughput maximum policy is adopted using an improved directional water-filling algorithm. On the other hand, in [35], source and relay, both are equipped with EH transmitter and optimal offline policy is investigated to maximize the throughput within a transmission deadline T.

3.4.2 Online policies

Online policies based on AI-assisted algorithms for EH wireless networks are more realistic than offline as the events, harvested energy, data arrival, and channel state change are statistically known at the source. In [36], based on the channel state and energy queue length, the source decides whether to transmit it instantly or to delay it for its transmission in the later time slot. The goal of the framework is to maximize the throughput of the network. The problem is formulated as Markov decision process (MDP). Further, the authors in [37] improve the decision process by investigating the nonlinearity in power-rate relation which is one of the optimal policy for a general discrete transmission set. In [38], the MDP to maximize throughput is efficiently solved by defining piecewise linear fit function to the battery state. It makes the problem convex in each iteration and closed-form of the optimal solution is obtained.

For online power control schemes, using statistical knowledge of harvested energy, the problem can be designed as an MDP in a point-to-point communication. The important factors of MDP are states $s \in \mathscr{S}$, actions $a \in \mathscr{A}$, rewards $r_a(s) = \mathbb{R}$, and state transition probabilities $p_a(s'|s)$. The states are regarded by the channel condition, harvested energy, and data arrival at the source. Whereas, the power dissipation and transmission of data can be considered as actions of the system. The reward of the framework is the performance metrics namely, outage probability [39], throughput [39], and error rate [40]. For a given action a, the state transition probability describes the transition from the present state s to the next state s'. The objective of the designed problem is to find the optimal policy $\pi(s)$ that maximizes the expected infinite-horizon reward $\mathscr{V}_\pi(s_0)$ starting from the state s_0 as:

$$\mathscr{V}_\pi(s_0) = \mathbb{E}_\pi\left[\sum_{j=0}^{\infty} \beta^j R_{\pi(s_j)}(s_j)\right]; s_j \in \mathscr{S}, \pi(s_j) \in \mathscr{A} \tag{3.22}$$

where $\beta \in [0, 1)$ is a discount factor. In the case of recurrent Markov chain, the optimal solution follows the Bellman's equation [98]:

$$\mathcal{V}_{\pi^*}(s) = \max_{a \in \mathcal{A}} \left[R_a(s) + \beta \sum_{s' \in \mathcal{S}} p_a(s'|s) \mathcal{V}_{\pi^*}(s') \right] \tag{3.23}$$

where (3.23) is independent of the initial state s_0 and the algorithms involved to solve it are value iteration, policy iteration, and linear programming that operate with high complexity.

In [30], an optimal online policy under MDP is evaluated using dynamic programming that maximizes the throughput within a given deadline. Here, many suboptimal policies are investigated with respect to the different channel state and energy harvesting. Authors in [36] designed the MDP for EH source node having infinite energy queue capacity. Its objective is to decide whether to transmit the data or defer it in each time slot to maximize the average throughput. In this scenario, the optimal policy is threshold type which relies on channel and energy storage capacity. In [37], the policy constitutes discrete power levels where a particular power level is preferred for a given battery level.

Principally, the offline algorithms provide significant performance enhancement than online algorithms due to availability of noncausal information of harvested energy, data arrival, and channel state. In [41], the performance of online algorithm is determined by finding the competitive ratio defined as the maximum ratio of the gain between offline and online policies with respect to sequences of harvested energy and channel state.

Instead of performance metrics based on throughput, authors in [42] find the optimal threshold in threshold-based algorithm to maximize reward rate as a function of message values. An energy-efficient transmission policy is adopted in [43] to maximize the coverage in EH body sensor network. Using dynamic programming, optimal and suboptimal transmission policies are described in [44] to minimize the outage probability, whereas outage based on data rate is minimized in [39].

As the EH process varies slowly than the channel state, in [45], a dual-stage power management is discoursed where outer stage manages the power for the inner stage to maximize the long-term utility while keeping the energy neutrality. Whereas, the inner stage optimizes the transmissions to maximize the short-term utility. A string tautening method is proposed in [46] that comprises on, off, and first-on-then-off policies for optimal scheduling under the constraint of delivery of data within a deadline (Figure 3.9).

Without modeling of the environment to predict the available energy resources, Q-learning can provide an optimal policy for a given MDP. It learns with time that for an action at a given state what is the long-term reward instead of evaluating it using state transition probability. Authors in [47] discourse Q-learning and speedy Q-learning that provide the online transmission policies by learning the joint variation of harvested energy and data arrivals. A learning theoretic approach is proposed in [48] where the priori information about the random events at the source is not known. Using it, the transmission policy is optimized to maximize the expected total transmitted data in the point-to-point communication.

3.5 Optimal resource allocation in EH self-sustainable communication systems

In this section, we discourse the optimizations for the resource allocation to enhance the system performances over the distinct self-sustainable wireless networks.

3.5.1 Self-sustainable D2D communication

In device-to-device (D2D) communications as depicted in Figure 3.10, there are two modes: (1) cellular mode and (2) D2D mode. In the former case, the user devices lie under the cellular network and communicate via the base stations (BSs). Whereas, in the later, the communication is established between user devices without BSs where the spectrum is reused that is already used by the cellular devices. With the exponential growth of devices and data traffic, D2D communication also provides active user cooperation which decreases the power dissipation and extends the lifetime of the devices. The D2D communication is also energy-efficient because a device can act as a relay between an end-to-end communication. As for the distant communication, the channel condition deteriorates that requires high transmit power for error-free transmission. Therefore, the cooperative communication using relays [49–52] decreases the power dissipation by reducing the transmission range. Further, D2D communication enables the devices to communicate directly without any infrastructure as well as it helps in the improvement of throughput, spectral efficiency, and energy efficiency.

In particular, the framework described in [53] consists a hybrid access point (HAP) and N RF energy transferred (RF-ET) users powered by the HAP as shown in Figure 3.10. While performing, each user either selects cellular mode or D2D mode. In order to increase the throughput, the transmission time is to be optimized for data transfer (DT) and energy transfer (ET) in both the modes. Therefore, there exist two tradeoffs: (1) selection of D2D and cellular modes and (2) time switching between DT and ET. Based on it, a combinatorial joint problem is designed to maximize the sum-throughput which is decoupled to obtain its optimal solution.

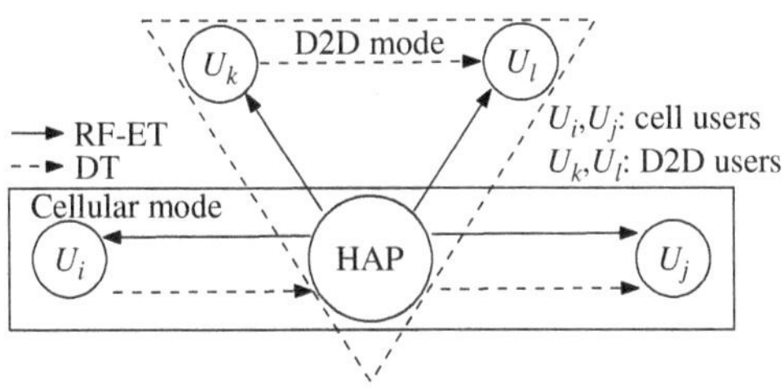

Figure 3.10 Cellular and D2D modes in self-sustainable D2D communications

3.5.2 MIMO simultaneous wireless information and power transfer (SWIPT)

In recent developments, the proposed scheme, simultaneous wireless information and power transfer (SWIPT) is capable to transmit the power as well as data information to the receiver at the same time [54]. However, the trade-off between the data transmission rate and harvested energy to be optimized to enhance the system performance [55]. Generally, the data and power reception at the same receiver is not reliable as the information content is lost while performing the EH using the same signal. Therefore, at the SWIPT receiver, either the two antennas are used separately or the received signal is spitted for the two operations. Based on it, three types of SWIPT receivers, separate receiver, time switching, and power splitting as shown in Figure 3.5 are used.

As multiple-input multiple-output (MIMO) encounter the fading channels efficiently thorough its diversity to receive multiple component of the signals at the receiver. These characteristics are exploited to maximize the harvested energy in an MIMO SWIPT system. High beamforming gain using passive intelligent surface (PIS) assisted MISO is investigated in [56]. Further, for efficient PIS-assisted energy transfer to EH user, a novel channel estimation (CE) protocol is discoursed. In [57], a multiple-input single-output (MISO) wireless network is considered where energy-constrained pilot carrier is used to estimate the channel. Using it, the sum harvested energy of the users is maximized by applying the transmit precoding. Whereas, in [58], a nonconvex joint problem is formulated to maximize the harvested energy by optimizing the transmit precoding and power splitting for SWIPT at the receiver. The global solution is obtained by formulating an equivalent generalized convex problem. On the other hand, in [59,60], the precoding and power splitting are jointly optimized in MISO SWIPT masticating Internet of things (IoT). Here, the problem is designed to maximize the minimum of the harvested energy in IoT. In order to show the global optimality of the solution, an equivalent problem, semi-definite relaxation is formulated. Using it, upper and lower bounds of the solution are obtained; thereafter, the optimal solution is determined by the discoursed global optimization algorithm.

3.5.3 Resource allocation in distinct EH wireless networks

There have been various energy-efficient resource allocation schemes for different radio networks as discoursed in [61–82]. Authors in [83] and [84] consider a simultaneous transmission of power and information in an orthogonal frequency division multiplexing (OFDM) system where the spectral efficiency for data transfer is maximized through optimal power allocation (PA) and the system achieves the self-sustainability through EH using power transfer. In [85], the trade-off between self-sustainability and data rate transfer is studied and optimized for all the considered configurations. In [86], the closed-form of end-to-end throughput is obtained in a relay powered cooperative network; thereafter, the performance enhancement is quantified numerically with respect to different system parameters. Peak-to-average power ratio (PAPR) of the OFDM system is minimized in [87] while keeping the same self-sustainability performance. In [88], the capacity is maximized under the constraint of maximum transmit power and minimum self-sustainability using EH. The network delay is minimized in [89] through Markov decision process by joint optimization of PA and routing of information. Here, the APs harvest the energy of users and using it, the users communicate to APs. For maximizing the conversion efficiency in EH, authors used beam switching in [90] that exploits beamforming to charge the user devices. In order to minimize overall power consumption through joint optimization of PA and EH, they obtain an optimal solution by Markov decision processes. The definition and corresponding expression for self-sustainability of an EH system is obtained in [91]. Using it, the bounds on the eventual energy outage of the network are evaluated.

3.6 Conclusion and future directions

This chapter investigates the essential aspects of efficient self-sustainable green EH wireless networks. In a practical framework, the existing bell-shaped tradeoff between energy efficiency and spectral efficiency is optimized to reduce the power dissipation while providing the services. However, the saved energy is not sufficient for self-sustainable wireless networks; it relies on the EH that generates the energy using ambient sources. From this perspective, various energy prediction models as well as intelligent learning-based predictions are discoursed to efficiently manage the generated energy in the self-sustainable network. Furthermore, the offline and online optimal policies are used for proper power management. Lastly, this chapter describes the optimal resource allocations such as transmit power, EH, time scheduling, and routing of information for the enhancement of the system performances.

Using the prediction models, an intelligent MAC protocol can be designed to predict the required energy with minimal error in a particular slot that leads to the efficient management of available energy without any interruption in performance. Also, the prediction of RF energy over frequencies and time would increase the efficiency of harvesters to generate maximum energy. Further, we need to develop a model to integrate the energy generated from the heterogeneous sources such as solar, RF, and vibrations to improve the continuous recharging aspects. While performing the power management, most of the existing approaches use the infinite capacity of battery storage which opens the challenge for a next-generation realistic network where the size of devices is small with limited battery capacity. From the state-of-the-art, the cost of power dissipation at the receiver without EH capability is not considered as user devices are desired to be self-sustainable while downlink transmission in a cellular system. Most of the frameworks have assumed the energy state information (ESI) is perfectly known; however, it is time-varying and depends on the energy availability over the ambient sources. To improve the transmission efficiency of self-sustainable devices having limited energy, multiple antenna techniques such as beamforming and MIMO can be exploited. In SWIPT, the energy receiver is vulnerable in secured communication as it may as an eavesdropper overhears the information. Therefore, we need to address the physical-level security in the EH system. For complete self-sustainability of the network, each of the network nodes needs to be incorporated with the EH module. But, it gives many challenges in routing, relaying, energy, spectrum, information sharing, security, and cross-layer optimization. While allocating resources optimally, most of the frameworks have investigated the optimizations without considering other power consumption due to processing and circuitry in the devices. Also, the optimization of the location of network nodes and their number has been ignored in the literature which significantly affects the performance of the network [92,93].

References

[1] Fettweis, G. and Zimmermann, E. "ICT energy consumption-trends and challenges." In: *Proceeding of International Symposium on Wireless Personal Multimedia Communications*. Lapland, Finland; 2008. pp. 1–6.

[2] Chen, Y., Zhang, S., Xu, S., *et al.* "Fundamental trade-offs on green wireless networks." *IEEE Communications Magazine.* 2011;49(6): 30–37.

[3] Miao, G., Himayat, N., Li, G. Y., *et al.* "Distributed interference-aware energy-efficient power optimization." *IEEE Transactions on Wireless Communications.* 2011; 10(4): 1323–1333.

[4] Perera, T. D. P., Jayakody, D. N. K., Sharma, S. K., *et al.* "Simultaneous wireless information and power transfer (SWIPT): Recent advances and future challenges." *IEEE Communications Surveys and Tutorials.* 2017; 20(1): 264–302.

[5] Zhang, Y., Zhang, J., Sun Y., *et al.* "Energy-efficient transmission for wireless powerec D2D communication networks." *Proceedings of IEEE ICC.* 2017. Paris, France. pp. 1–7.

[6] Gelenbe, E., Gesbert, D., Gunduz, D., *et al.* "Energy harvesting communication networks: Optimization and demonstration (the E-CROPS project)." *Proceedings of Tyrrhenian International Workshop on Digital Communications-Green ICT (TIWDC)*; 2013. pp. 1–6.

[7] Cammarano, A., Petrioli, C., and Spenza, D. "Pro-energy: A novel energy prediction model for solar and wind energy-harvesting wireless sensor networks." *Proceedings of IEEE International Conference on Mobile Ad-Hoc and Sensor Systems (MASS).* 2012. Las Vegas, NV, USA; pp. 75–83.

[8] Sharma, N., Sharma, P., Irwin, D., *et al.* "Predicting solar generation from weather forecasts using machine learning." *2011 IEEE international conference on smart grid communications (SmartGridComm).* Brussels, Belgium; 2011. pp. 528–533.

[9] Azmat, F., Chen, Y., and Stocks, N. "Predictive modelling of RF energy for wireless powered communications." *IEEE Communications Letters.* 2015; 20(1): 173–176.

[10] Ahmed, I., Butt, M. M., Psomas, C., *et al.* "Survey on energy harvesting wireless communications: Challenges and opportunities for radio resource allocation." *Computer Networks.* 2015; 88: 234–248.

[11] Sudevalayam S., and Kulkarni P. "Energy harvesting sensor nodes: Survey and implications." *IEEE Communications Surveys and Tutorials.* 2010; 13(3): 443–461.

[12] Lu, X., Wang, P., Niyato, D., *et al.* "Wireless networks with RF energy harvesting: A contemporary survey." *IEEE Communications Surveys and Tutorials*. 2014; 17(2): 757–789.

[13] Perera, T. D. P., Jayakody, D. N. K., Sharma, S. K., *et al.* "Simultaneous wireless information and power transfer (SWIPT): Recent advances and future challenges." *IEEE Communications Surveys and Tutorials*. 2017; 20(1): 264–302.

[14] da Silva Costa, L. G., de Queiroz, A. C. M., Adebisi, B., *et al.* "Coupling for power line communications: A survey." *Journal of Communications and Inform System*. 2017; 32(1): 8–22.

[15] Cammarano, A., Petrioli, C., and Spenza, D. "Pro-energy: A novel energy prediction model for solar and wind energy-harvesting wireless sensor networks." *Proceedings of International Conference on Mobile Ad-Hoc and Sensor Systems (MASS)*. 2012. Las Vegas, NV, USA. pp. 75–83.

[16] Mulder, F. "Implications of diurnal and seasonal variations in renewable energy generation for large scale energy storage." *Journal of Renewable and Sustainable Energy*. 2014; 6(3): 033105.

[17] Kansal, AA., Hsu, J., Zahedi, S., *et al.* "Power management in energy harvesting sensor networks." *ACM Transactions on Embedded Computing Systems (TECS)*. 2007; 6(4): 32–es.

[18] Noh, D. K. and Kang, K. "Balanced energy allocation scheme for a solar-powered sensor system and its effects on network-wide performance." *Journal of Computer and System Sciences*. 2011; 77(5): 917–932.

[19] Piorno, J. R., Bergonzini, C., Atienza, D., *et al.* "Prediction and management in energy harvested wireless sensor nodes." *Proceedings of IEEE International Conference on Wireless Communication, Vehicular Technology, Information Theory and Aerospace and Electronic Systems Technology*. 2009. Aalborg, Denmark. pp. 6–10.

[20] Sutton, R.S. and Barto, A.G. *Reinforcement learning: An introduction*. MIT Press, 1998.

[21] Medepally, B., Mehta, N.B., and Murthy, C.R. "Implications of energy profile and storage on energy harvesting sensor link performance." *Proceedings of IEEE GLOBECOM*. 2009. Honolulu, HI, USA. pp. 1–6.

[22] Lei, J., Yates, R., and Greenstein, L. "A generic model for optimizing single-hop transmission policy of replenishable sensors." *IEEE Transactions Wireless Communications*. 2009; 8(2): 547–551.

[23] Minasian, A., ShahbazPanahi, S., and Adve, R.S. "Energy harvesting cooperative communication systems." *IEEE Transactions on Wireless Communications*. 2014; 13(11): 6118–6131.

[24] Azmat, F., Chen, Y., and Stocks, N. "Predictive modelling of RF energy for wireless powered communications." *IEEE Communications Letters*. 2015; 20(1): 173–176.

[25] Flach, P. *Machine learning: The art and science of algorithms that make sense of data*. Cambridge University Press; 2012.

[26] Torgo, L. "Functional models for regression tree leaves." *Proceedings of International Conference on Machine Learning (ICML)*; 1997. pp. 385–393.
[27] Kumar, S., Mishra, D., and De, S. "An accurate channel model for optimizing effect of non-LOS component in RF energy transfer." *Proceedings of IEEE National Conference on Communication (NCC)*. Chennai, India; 2017. pp. 1–6.
[28] Luo, Y., Zhang, J., and Letaief, K. B. "Training optimization for energy harvesting communication systems." *Proceedings of IEEE GLOBECOM*. Anaheim, CA, USA; 2012. pp. 3365–3370.
[29] Yang, J. and Ulukus, S. "Optimal packet scheduling in an energy harvesting communication system." *IEEE Transactions Communications*. 2011; 60(1): 220–230.
[30] Ozel, O., Tutuncuoglu, K., Yang, J., *et al.* "Transmission with energy harvesting nodes in fading wireless channels: Optimal policies." *IEEE Journal on Selected Areas in Communications*. 2011; 29(8): 1732–1743.
[31] Ozcelik, F. M., Uctu, G., and Uysal-Biyikoglu, E. "Minimization of transmission duration of data packets over an energy harvesting fading channel." *IEEE Communications Letters*. 2012; 16(12): 1968–1971.
[32] Luenberger, D. G., *Linear and Nonlinear Programming*. 2nd ed. Reading, MA: Addison-Wesley, 1984.
[33] Luo, Y., Zhang, J., and Letaief, K. B. "Training optimization for energy harvesting communication systems." *Proceedings of IEEE GLOBECOM*. 2012. Anaheim, CA, USA; pp. 3365–3370.
[34] Luo, Y., Zhang, J., and Letaief, K. B. "Throughput maximization for two-hop energy harvesting communication systems." *Proceedings of IEEE ICC*. 2013. Budapest, Hungary. pp. 4180–4184.
[35] Orhan, O., and Erkip, E. "Energy harvesting two-hop networks: Optimal policies for the multi-energy arrival case." *Proceedings IEEE Sarnoff Symposium*. 2012. Newark, NJ, USA; pp. 1–6.
[36] Kashef, M. and Ephremides, A. "Optimal packet scheduling for energy harvesting sources on time varying wireless channels." *Journals of Communications and Networks*. 2012; 14(2): 121–129.
[37] Bacinoglu, B. T. and Uysal-Biyikoglu, E. "Finite-horizon online transmission scheduling on an energy harvesting communication link with a discrete set of rates." *Journal of Communications and Networks*. 2014; 16(3): 393–300.
[38] Wang, Z., Aggarwal, V., and Wang, X. "Power allocation for energy harvesting transmitter with causal information." *IEEE Transactions on Communications*. 2014; 62(11): 4080–4093.
[39] Li, W., Ku, M. L., Chen, Y., *et al.* "On outage probability for stochastic energy harvesting communications in fading channels." *IEEE Signal Process Letters*. 2015; 22(11): 1893–1897.
[40] Ku, M. L., Li, W., Chen, Y., *et al.* "On energy harvesting gain and diversity analysis in cooperative communications." *IEEE Journal on Selected Areas Communications*. 2015; 33(12): 2641–2657.

[41] Vaze, R., Garg, R., and Pathak, N. "Dynamic power allocation for maximizing throughput in energy-harvesting communication system." *IEEE/ACM Transactions on Networking.* 2013; 22(5): 1621–1630.

[42] Lei, J., Yates, R., and Greenstein, L. "A generic model for optimizing single-hop transmission policy of replenishable sensors." *IEEE Transactions on Wireless Communications.* 2009; 8(2): 547–551.

[43] Seyedi, A. and Sikdar, B. "Energy efficient transmission strategies for body sensor networks with energy harvesting." *IEEE Transactions on Communications.* 2010; 58(7): 2116–2126.

[44] Huang, C., Zhang, R., and Cui, S. "Optimal power allocation for outage probability minimization in fading channels with energy harvesting constraints." *IEEE Transactions on Wireless Communications.* 2014; 13(2): 1074–1087.

[45] Reddy, S., and Murthy, C. R. "Dual-stage power management algorithms for energy harvesting sensors." *IEEE Transactions on Wireless Communications.* 2012; 11(4): 1434–1445.

[46] Chen, X., Ni, W., Wang, X., *et al.* "Provisioning quality-of-service to energy harvesting wireless communications." *IEEE Communications Magazines.* 2015; 53(4): 102–109.

[47] Prabuchandran, K., Meena. S. K., and Bhatnagar, S. "Q-learning based energy management policies for a single sensor node with finite buffer." *IEEE Wireless Communications Letters.* 2012; 2(1): 82–85.

[48] Blasco, P., Gunduz, D., and Dohler, M. "A learning theoretic approach to energy harvesting communication system optimization." *IEEE Transactions on Wireless Communications.* 2013; 12(4): 1872–1882.

[49] Prasad, G., Mishra, D., and Hossain, A. "QoS-aware green communication strategies for optimal utilization of resources in 5G networks." *Paving the way for 5G Through the convergence of wireless systems*. IGI Global. 2019. pp. 186–208.

[50] Prasad, G., Mishra, D., and Hossain, A. "Joint optimal design for outage minimization in DF relay-assisted underwater acoustic networks." *IEEE Communications Letters.* 2018; 22(8): 1724–1727.

[51] Prasad, G., Mishra, D., Tourki, K., *et al.* "QoS and energy aware optimal resource allocations in DF relay-assisted FSO networks." *IEEE Transactions on Green Communication Networking (Early Access).* 2020.

[52] Prasad, G., Mishra, D., Tourki, K., *et al.* "QoS-aware power allocation and relay placement in green cooperative FSO communications." *Proceedings of IEEE Wireless Communications Networking Conferences (WCNC).* Marrakesh, Morocco. 2019. pp. 1–6.

[53] Mishra, D., De, S., Alexandropoulos, G. C., *et al.* "Energy-aware mode selection for throughput maximization in RF-powered D2D communications." *Proceedings on IEEE GLOBECOM.* Singapore. 2017. pp. 1–6.

[54] Mishra, D., De, S., Jana, S., *et al.* "Smart RF energy harvesting communications: Challenges and opportunities." *IEEE Communications Magazine.* 2015; 53(4): 70–78.

[55] Varshney, L. R. "Transporting information and energy simultaneously." *Proceedings of IEEE International Symposium on Information Theory.* Toronto, ON, Canada. 2008. pp. 1612–1616.

[56] Mishra, D. and Johansson, H. "Channel estimation and low-complexity beamforming design for passive intelligent surface assisted MISO wireless energy transfer." *Proceedings of IEEE International Conferences on Acoustics, Speech and Signal Processing (ICASSP).* Brighton, UK. 2019. pp. 4659–4663.

[57] Mishra, D. and Johansson, H. "Efficacy of multiuser massive MISO wireless energy transfer under IQ imbalance and channel estimation errors over rician fading." *Proceedings of IEEE International Conferences on Acoustics, Speech and Signal Processing (ICASSP).* Calgary, AB, Canada. 2018. pp. 3844–3848.

[58] Mishra, D. and Alexandropoulos, G. C. "Transmit precoding and receive power splitting for harvested power maximization in MIMO SWIPT systems." *IEEE Transactions on Green Communications and Networking.* 2018; 2(3): 774–786.

[59] Mishra, D., Alexandropoulos, G. C., and De, S. "Harvested power fairness optimization in MISO SWIPT multicasting IoT with individual constraints." *Proceedings of IEEE ICC.* Kansas City, MO, USA. 2018. pp. 1–6.

[60] Mishra, D., Alexandropoulos, G. C., and De, S. "Energy sustainable IoT with individual QoS constraints through MISO SWIPT multicasting." *IEEE Internet of Things Journal.* 2018; 5(4): 2856–2867.

[61] Bourdena, A., Mavromoustakis, C. X., Kormentzas, G., *et al.* "A resource intensive traffic-aware scheme using energy-aware routing in cognitive radio networks." *Future Generation Computer Systems.* 2014;39: 16–28.

[62] Dimitriou, C. D., Mavromoustakis, C. X., Mastorakis, G., *et al.* "On the performance response of delay-bounded energy-aware bandwidth allocation scheme in wireless networks." *Proceedings of IEEE ICC Workshops.* Budapest, Hungary. 2013. pp. 631–636.

[63] Mavromoustakis, C.X., Bourdena, A., Mastorakis, G., *et al.* "An energy-aware scheme for efficient spectrum utilization in a 5G mobile cognitive radio network architecture." *Telecommunication Systems.* 2015; 59(1): 63–75.

[64] Mavromoustakis, C. X., Mastorakis, G., Bourdena, A., *et al.* "Energy efficient resource sharing using a traffic-oriented routing scheme for cognitive radio networks." *IET Networks.* 2014; 3(1): 54–63.

[65] Papanikolaou, K., Mavromoustakis, C. X., Mastorakis, G., *et al.* "Energy consumption optimization using social interaction in the mobile cloud." *Proceedings of International Conferences Mobile Networks Manage.* Cham, Switzerland. 2014. pp. 431–445.

[66] Mastorakis, G., Bourdena, A., Mavromoustakis, C. X., *et al.* "An energy-efficient routing protocol for ad-hoc cognitive radio networks." *Proceedings of Future Network and Mobile Summit.* Lisboa, Portugal; 2013. pp. 1–10.

[67] Mavromoustakis, C. X., Kormentzas, G., Mastorakis, G., *et al.* "Joint energy and delay-aware scheme for 5G mobile cognitive radio networks." *Proceedings of IEEE GLOBECOM.* Austin, TX, USA. 2014. pp. 2624–2630.

[68] Mavromoustakis, C. X., Dimitriou, C. D., Mastorakis, G., *et al.* "Real-time performance evaluation of F-BTD scheme for optimized QoS energy conservation in wireless devices." *Proceedings of IEEE GLOBECOM Workshops*. Atlanta, GA, USA. 2013. pp. 1151–1156.

[69] Mastorakis, G., Mavromoustakis, C. X., Bourdena, A., *et al.* "An energy-efficient routing scheme using backward traffic difference estimation in cognitive radio networks." *Proceedings of IEEE International Symposium on a World of Wireless, Mobile and Multimedia Networks (WoWMoM)*. Madrid, Spain. 2013. pp. 1–6.

[70] Mastorakis, G., Mavromoustakis, C. X., Bourdena, A., *et al.* "Optimizing radio resource management in energy-efficient cognitive radio networks." *Proceedings of ACM Workshop on High Performance Mobile Opportunistic Systems*. Barcelona, Spain. 2013. pp. 75–82.

[71] Sundarasekar, R., Shakeel, P. M., Baskar, S., *et al.* "Adaptive energy aware quality of service for reliable data transfer in under water acoustic sensor networks." *IEEE Access*. 2019; 7: 80093–80103.

[72] Mavromoustakis, C. X., Mastorakis, G., Batalla, J. M., *et al.* "Social-oriented mobile cloud offload processing with delay constraints for efficient energy conservation." *Proceedings of IEEE ICC*. Paris, France. 2017. pp. 1–7.

[73] Skourletopoulos, G., Mavromoustakis, C. X., Mastorakis, G., *et al.* "Elasticity debt analytics exploitation for green mobile cloud computing: An equilibrium model." *IEEE Transactions Green Communications Networking*. 2018; 3(1): 122–131.

[74] Mavromoustakis, C. X., Kormentzas, G., Mastorakis, G., *et al.* "Context-oriented opportunistic cloud offload processing for energy conservation in wireless devices." *Proceedings of IEEE GLOBECOM Workshops*. Austin, TX, USA. 2014. pp. 24–30.

[75] Mavromoustakis, C. X., Dimitriou, C. D., and Mastorakis, G. "Using real-time backward traffic difference estimation for energy conservation in wireless devices." *Proceedings of International Conferences on Advances in P2P Systems (AP2PS)*. 2012. pp. 18–23.

[76] Bourdena, A., Mavromoustakis, C., Mastorakis, G., *et al.* "Using socio-spatial context in mobile cloud offload process for energy conservation in wireless devices." *IEEE Transactions Cloud Computer*. 2015; 7(2): 392–402.

[77] Mavromoustakis, C. X., Dimitriou, C., and Mastorakis, G. "On the real-time evaluation of two-level BTD scheme for energy conservation the presence of delay sensitive transmissions and intermittent connectivity in wireless devices." *Journal on Advances in Networks and Services*. 2013; 6(3): 148–161.

[78] Mavromoustakis, C. X., Mastorakis, G., Papadakis, S., *et al.* "Energy consumption optimization through pre-scheduled opportunistic offloading in wireless devices." *Proceedings of International Conferences on Emerging Network Intelligence*, EMERGING. 2014. pp. 24–28.

[79] Mavromoustakis, C. X., Andreou, A., Mastorakis, G., *et al.* "On the performance evaluation of a novel offloading-based energy conservation mechanism

for wireless devices." *International Conferences on Mobile Networks and Manage*. Wuerzburg, Germany. 2014. pp. 179–191.

[80] Mavromoustakis, C. X., Batalla, J. M., Mastorakis, G., *et al.* "Socially oriented edge computing for energy awareness in IoT architectures." *IEEE Communications Magazine.* 2018; 56(7): 139–145.

[81] Marin, R. C., Ciobanu, R. I., Dobre, C., *et al.* "A context-aware collaborative model for smartphone energy efficiency over 5G wireless networks." *Computer Networks.* 2017; 129: 352–362.

[82] Mastorakis, G., Mavromoustakis, C. X., Bourdena, A., *et al.* "Maximizing energy conservation in a centralized cognitive radio network architecture." *Proceedings of International Workshop on Computer Aided Modeling and Design of Communications Links and Networks (CAMAD)*. Berlin, Germany. 2013. pp. 175–179.

[83] Ng, D. W. K. and Schober, R. "Spectral efficient optimization in OFDM systems with wireless information and power transfer." *Proceedings of IEEE European Signal Processing Conferences (EUSIPCO).* Marrakech, Morocco. 2013. pp. 1–5.

[84] Maso, M., Lakshminarayana, S., Quek, T. Q. S., *et al.* "A composite approach to self-sustainable transmissions: Rethinking OFDM." *IEEE Transactions on Communications.* 2014; 62(11): 3904–3917.

[85] Maso, M., Lakshminarayana, S., Quek, T. Q. S., *et al.* "The price of self-sustainability for block transmission systems." *IEEE Journal on Select Areas in Communications.* 2015; 33(8): 1549–1562.

[86] Mishra, D. and De, S. "Achievable throughput in relay-powered RF harvesting cooperative sensor networks." *Proceedings of IEEE National Conferences on Communications (NCC)*. Guwahati, India. 2016. pp. 1–6.

[87] Yang, C. and Varshney, L. R. "Self-sustainable OFDM transmissions with smooth energy delivery." *Proceedings of IEEE International Symposiums on Next-Generation Electronics (ISNE)*. Hsinchu, Taiwan. 2016. pp. 1–2.

[88] Xu, W., Li, S., Lee, C., *et al.* "Optimal secure multicast with simultaneous wireless information and power transfer in presence of multiparty eaves-dropper collusion." *IEEE Transactions on Vehicular Technology.* 2016; 65 (11): 9123–9137.

[89] Yao, Q., Huang, A., Shan, H., *et al.* "Adaptive harvest-then-cooperate: Delay-aware wireless powered communication networks." *Proceedings of IEEE International Workshop on Signal Process*. Advances in Wireless Communications. (SPAWC). Edinburgh, UK. 2016. pp. 1–5.

[90] Yao, Q., Quek, T. Q. S., Huang, A., *et al.* "Joint downlink and uplink energy minimization in WET-enabled networks." *IEEE Transactions on Wireless Communications.* 2017; 16(10): 6751–6765.

[91] Guruacharya, S. and Hossain, E. "Self-sustainability of energy harvesting systems: concept, analysis, and design." *IEEE Transactions on Green Communications and Networking.* 2018; 2(1): 175–192.

[92] Prasad, G., Mishra, D., and Hossain, A. "Joint optimization framework for operational cost minimization in green coverage-constrained wireless

networks." *IEEE Transactions on Green Communications and Networking*. 2018; 2(3): 693–706.

[93] Prasad, G., Mishra, D., and Hossain, A. "Coverage-constrained base station deployment and power allocation for operational cost minimization." *Proceedings of IEEE International Symposium on Personal, Indoor, and Mobile Radio Communications (PIMRC)*. Montreal, QC, Canada. 2017. pp. 1–5.

Chapter 4

Discrete wavelet transform applications in the IoMT

Tamara K. Al-Shayea[1], Constandinos X. Mavromoustakis[1], George Mastorakis[2], Jordi Mongay Batalla[3], Evangelos Pallis[4], Evangelos K. Markakis[4], Imran Khan[5] and Dinh-Thuan Do[6]

Recent research efforts have explored methods to achieve the wavelet transform as the most significant tool in medical image enhancement and processing. The problems of image fusion, compression, edge detection, denoising, and contrast enhancement can be handled by discrete wavelet transform (DWT) in the Internet of medical things (IoMT) framework. In this chapter, we present the novel DWT with orthogonal and biorthogonal wavelets application. Multiple applications of the wavelet transform in medical images have been submitted. These applications demonstrate the successful impact of applying DWT.

The DWT has the ability to enhance the medical image and remove noise. The DWT in image compression can separately reduce the computational complexity into high and low frequency. This process reduces the image data in order to be able to store or transmit data in an efficient form. There are some advantages in using fusion based on DWT during other traditional methods, for example, reduced features and energy compaction. In digital watermarking, DWT technique is used for embedding and extraction of watermark in the original image.

[1]Department of Computer Science, University of Nicosia, Nicosia, Cyprus
[2]Department of Management Science and Technology, Hellenic Mediterranean University, Agios Nikolaos, Crete, Greece
[3]Internet Architectures and Applications Department, National Institute of Telecommunications and Warsaw University of Technology, Warsaw, Poland
[4]Department of Electrical and Computer Engineering, Hellenic Mediterranean University, Heraklion, Crete, Greece
[5]Department of Electrical Engineering, University of Engineering and Technology Peshawar, Khyber Pakhtunkhwa, Pakistan
[6]Department of Computer Science and Information Engineering, Faculty of Electronics Technology, Industrial University of Ho Chi Minh City (IUH), Ho Chi Minh City, Vietnam

4.1 Introduction

The Internet of medical things (IoMT) is a relatively novel technology that has significantly been very effective in the health care sector throughout the previous decade [1]. The progress of 5G technology, cloud storage, and big data has actively promoted the rapid evolution of the IoMT. Several applications have been formed with regards to the rigid security requirements and precision needed for diagnosis of disease and medical analysis that is supplied by IoMT. The fast transfer of medical information can be facilitated by IoMT that performs a prominent role in protecting pathological information and private information of patients. These applications assist in the process of diagnosing and treatment of doctors [2,3]. The IoMT gathers the safety and reliability, conventional medical devices, and scalability abilities of conventional Internet of things (IoT). It is capable of finding the solution to several diseases through its capability of managing various devices, widespread for various patients [4]. The IoMT develops computational and mathematical methods for solving problems pertaining to medical images and their use for biomedical research and clinical care. The IoMT is a set of biomedical equipment and similar implementations, which connect the health care information technology schemes by virtual computer systems. Particularly in the diagnostics area, medical image categorization has a major part to do in forecasting and initial diagnosis of hard illnesses by improving access to emerging technology for any diagnosis. However, categorization of medical images is deemed as a challenging task in computer-based diagnostic [5]. Medical image has always contained critical information, and protecting the privacy and security of the patient has remained a major challenge [3]. Wearable devices became common with broad implementations in the health surveillance system that has developed the rapid enhancement for IoMT [6]. The secure IoMT-based healthcare system is a key issue in the face of ever-growing challenges, especially with regard to medical images [7]. The IoMT has been applied successfully in medical applications involving medical image processing algorithms [8]. The IoMT has earned massive attraction from the medical image processing. In the future, discrete wavelet transform (DWT) applications will support IoMT concept, in which most of the medical devices used by doctors on a daily basis will be connected to each other via the Internet, hence, will result in increased coherence between both authorized users and medical devices. In this chapter, we introduce an important theory of DWTs and wavelets families. We also demonstrate several implementations of the wavelet transform in medical images. Moreover, we also highlight the importance of DWT applications.

4.2 The discrete wavelet transform

The diversity of resolution decomposition of a signal is regarded to be the major concept of the wavelet transforms. The major idea of the wavelet transform is the multiresolution decomposition of signals and images. It is applicable to be used for small items requiring high resolution; on the other hand, the lower resolution is sufficient for larger items. There is a common way in multiresolution decomposition to generate an approximated object through the use of a scaling function, specified object, and also wavelet functions (low and high pass filters), and specified objects include the dissimilarity adjacent. The process of analyzing the wavelet consists of various forms: the DWT, transform, and the continuous wavelet series. Each wavelet has a different basis procedure and it is dissimilar either in being discrete or continuous. The filter coefficients are considered as basic procedures for the DWT. The DWT is applied in practical implementations on large scale.

The DWT is employed to implement the wavelet transform utilizing a discrete set of the wavelet scales and translations obeying some specific principles. More to the point, the signal into mutually orthogonal set of wavelet is decomposed by this transform and it differs from the continuous wavelet transform (CWT), or its implementation for the discrete-time series known as discrete-time continuous wavelet transform (DT-CWT). The wavelet can be established from a scaling function that demonstrates its scaling features. The scaling functions should be orthogonal to its discrete translations and should include mathematical conditions on them, which have been mentioned everywhere such as the dilation equation:

$$\phi(x) = \sum\nolimits_{k=-\infty}^{\infty} a_k \phi(Sx - k) \tag{4.1}$$

where S is a scaling factor.

$$\psi = \sum\nolimits_{k=-\infty}^{\infty} (-1)^k a_{N-1-k} \psi(2x - k) \tag{4.2}$$

where N is an even integer. The group of wavelets then forms an orthonormal basis which is used to decompose signal.

The Haar DWT filters are deemed orthogonal and symmetric. The first row of the transform matrix constitutes the basic procedure that divides the low-frequency component of the signal, while the second row constitutes the basic procedure that divides the high-frequency component:

$$\begin{bmatrix} Y_{\varnothing}(0,\ 0) \\ \mathrm{Y}_{\psi}(0,0) \end{bmatrix} = \frac{1}{\sqrt{2}} \begin{bmatrix} 1 & 1 \\ 1 & -1 \end{bmatrix} \begin{bmatrix} \mathrm{y}(0) \\ \mathrm{y}(1) \end{bmatrix} \text{or } \mathrm{Y} = \mathrm{Z}_{2,0}\mathrm{y} \tag{4.3}$$

where Y represents coefficient, $Z_{2,0}$ represents transform, and y represents input matrices. The input data length is the subscript of 2, and 0 refers to the measurement of decomposition. For S equals 2, it has one measurement. It is expected that the length of S of a column sequence $y(z)$ is a power of 2.

$$\{\mathrm{y}(0), \mathrm{y}(1), \mathrm{y}(2), \cdots, \mathrm{y}(\mathrm{Z}-1)\} \tag{4.4}$$

Thus $Z = 2^I$, where I is a positive integer. The coefficients are dedicated as $Y_{\varnothing}(i_0, k)$, where i_0 is a given beginning measurement in the range from 0 to $I - 1$. The index range i identifies the measurement of decomposition, is

$$i = i_0,\ i_0 + 1,\ \cdots,\ I - 1 \tag{4.5}$$

It is an index of the sub-band of the spectrum regarding the basic procedure. The degree of indicator k, which specifies the time scale of the basic procedure, is

$$k = 0,\ 1,\ \cdots,\ 2^i - 1 \tag{4.6}$$

Altogether, indicators i and k refer to a specific set of frequency contents at specific times.

For $S = 2, I = 1, i = 0$, and $k = 0$. The symbol $\varnothing$ $X_{\varnothing}(i_0, k)$ clarifies that it is an approximated coefficient acquired by particular averaging of the input signal. The symbol ψ in $X_{\psi}(i, k)$ clarifies that it is a coefficient acquired by distinguishing the input signal.

The two-point one level inverse DWT of the coefficients $\{Y_{\varnothing}(0,0), Y_{\psi}(0,0)\}$ is known as

$$\begin{bmatrix} y\,(0) \\ y\,(1) \end{bmatrix} = \frac{1}{\sqrt{2}} \begin{bmatrix} 1 & 1 \\ 1 & -1 \end{bmatrix} \begin{bmatrix} Y_{\varnothing}\,(0,\ 0) \\ Y_{\psi}\,(0,\ 0) \end{bmatrix} \text{ or } y = Z_{2,\,0}^{-1} Y \tag{4.7}$$

The inverse transform matrix $Z_{2,0}^{-1} = Z_{2,0}^{T}$ is the inverse and also the transpose of the transform matrix $Z_{2,0}$. With random real values:

$$\begin{bmatrix} x \\ y \end{bmatrix} = \begin{bmatrix} a\,c \\ b\,d \end{bmatrix} \begin{bmatrix} a\,c \\ b\,d \end{bmatrix} \begin{bmatrix} x \\ y \end{bmatrix} = \begin{bmatrix} (a^2 + c^2)x + (ab + cd)y \\ (ba + dc)x + (b^2 + d^2)y \end{bmatrix} = \begin{bmatrix} x \\ y \end{bmatrix} \tag{4.8}$$

With the orthogonality terms:

$$a^2 + b^2 = 1 = c^2 + d^2 \text{ and } ac + bd = 0 \tag{4.9}$$

$$a^2 + c^2 = 1 = b^2 + d^2 \text{ and } ab + cd = 0 \tag{4.10}$$

The equation is satisfactory. In case of alternating orthogonal rows of a matrix, the columns will fulfill a similar condition. An orthogonal matrix of order Z is a group of Z alternately orthogonal unit vectors [9,10].

In Figure 4.1, we illustrate that an image can be decomposed employing sub-band decomposition. We start with an $N \times M$ image. Each row is filtered, then down sample to have two $N \times M/2$ images. Each column is filtered and subsample the filter output to have four $N/2 \times M/2$ images [11]. Figure 4.2 illustrates the three levels wavelet decomposition tree. Discrete wavelet transform decomposition at level three is illustrated in Figure 4.3.

4.3 DWT applications

The applications of DWT are extremely important, which have been chosen empirically in previous works. The objective of this section aims to develop a framework for the optimal DWT applications to find the best solutions.

4.3.1 Image denoising

Noise is an inadmissible signal that alters the property and performance of the signal. Image can be corrupted with noise such as salt and pepper, Gaussian,

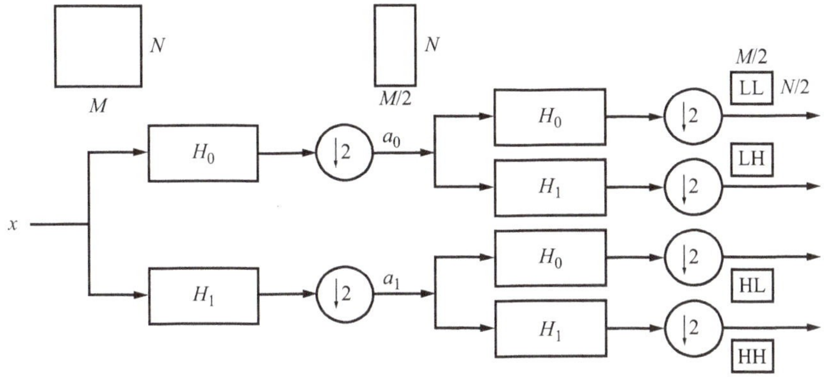

Figure 4.1 Sub-band decomposition of an NXM image. Adapted from [11]

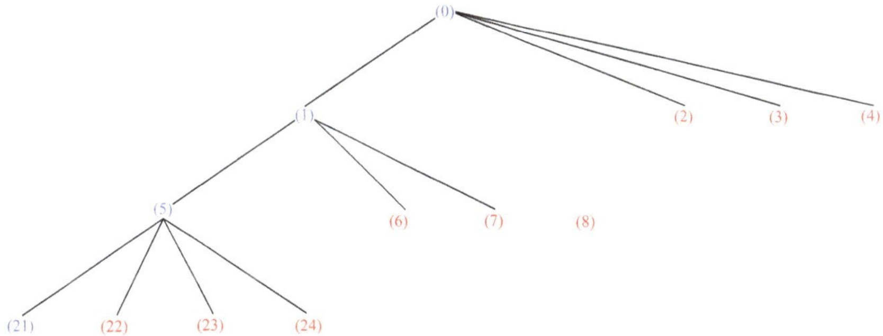

Figure 4.2 Three levels wavelet decomposition tree

and speckle distribution [12]. Image denoising has been employed to overcome the noise in order to retain all details features and vital signal features in the image. Wavelet denoising is applied to remove the noise in the signal and maintain the features of the signal wavelet while denoising is applied, without regard to its frequency content. In modern medical imaging systems, denoising of tomographic image is a major issue. However, this problem has been technically solved by using DWT [13,14]. The objective of denoising reconstructs the host image from its noisy observation as precise as possible while keeping significant detail features, for example, edge and texture in the denoised image. We seek to achieve this goal. Image denoising has been widely used toward that goal during the past decades. Denoising schemes can generally be categorized into three types of techniques: spatial domain, transform domain, and hybrid techniques [15]. In medical image processing, denoising is a crucial step to address them [16]. Due to its popularity, the wavelet transforms constitute an important tool in massive research and implementation domains. During the use of wavelet transforms, in wavelet domain, different algorithms for denoising have been implemented. Wavelets have been performed extremely well in image denoising because the possibility of multi-resolution analysis. Wavelet thresholding is a method for estimating signal that

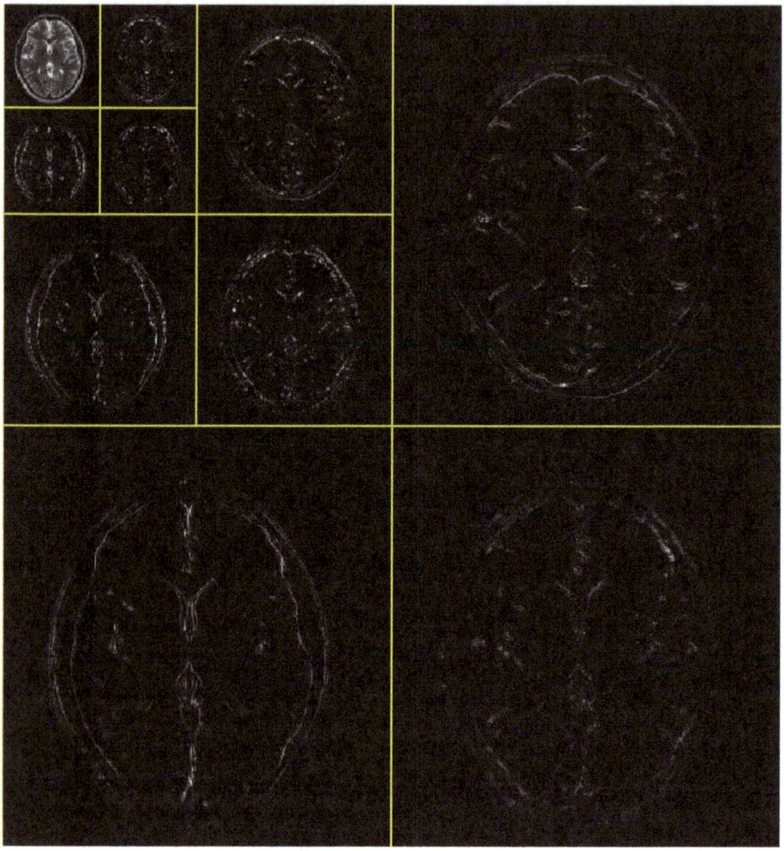

Figure 4.3 Discrete wavelet transform decomposition at level three

takes advantage of the abilities of wavelet transform for signal denoising [17]. Image noise changes information of an image where it may be color information in images or the random variation of brightness. Image denoising is designed to remove this change and recover the original image [18]. In general, noise has damaged the image during transmission, compression, storage, and retrieval operations. These impacts are related to distortion and loss of image information. Medical image plays a significant part in diagnosing and treatment of different diseases. The medical image must be noise-free for optimal decision making in diagnosis, but image quality may be affected by the noise and this results in degradation in several types of medical images such as magnetic resonance imaging (MRI) and ultrasound imaging. The main objective is to remove the noise from the medical image without impacting the important information [19–21]. Discrete wavelet transform has proved to be an effective and sufficient tool for reducing noise and it has been implanted in the field of digital image processing [22,23]. Figure 4.4 illustrates the original medical image with Gaussian noise

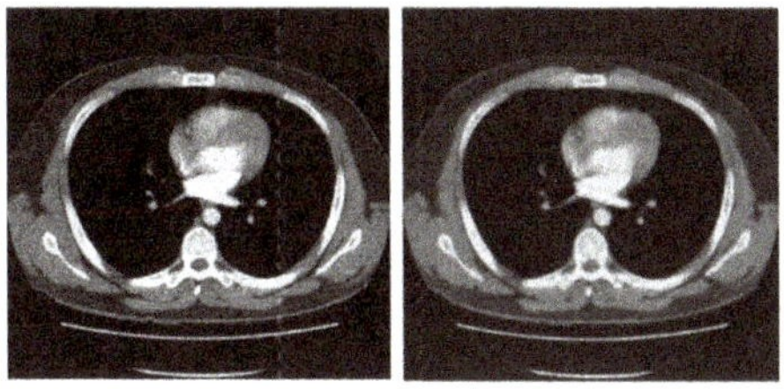

Figure 4.4 (a) Magnetic resonance imaging (MRI) original medical image with Gaussian noise and (b) MRI medical image after denoising

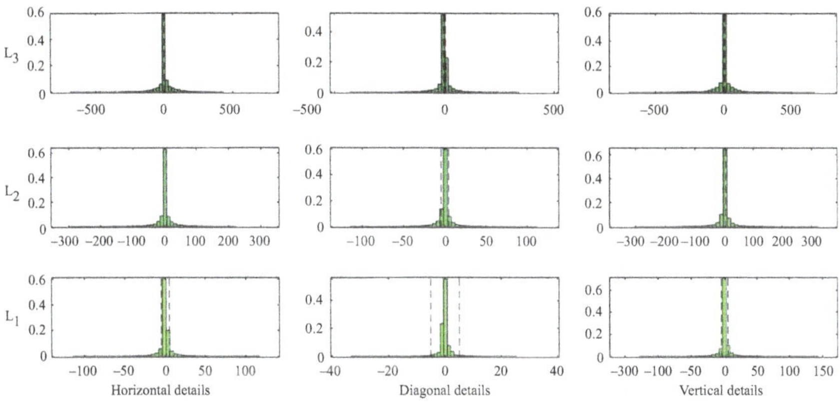

Figure 4.5 Magnetic resonance imaging medical image histogram for level one, level two, and level three

and the same image after denoising. The histogram for the medical image for three levels is shown in Figure 4.5.

4.3.2 Image fusion

The process of the image fusion can be defined by combining all the important information from multiple images, and their incorporation into fewer images to be as one image. Fused images may be established from multiple images from the same imaging modality or by bringing together information from multiple modalities. Usually, fusion can be accomplished fundamentally by employing spatial domain and transform domain fusion approach [24]. The image fusion is the process by which two or more images are infused into a single image with substantial content. In several fields, fusion is considered as a significant technology, comprising robotics, remote sensing, and medical implementations. The image fusion produces a composite image that is more appropriate and understandable for the human and machine perception, and the main feature of fusion image is to enhance reliability and capability by getting better or more reliable image. Accordingly, the reduction of the amount of data is not only the main purpose of image fusion [25]. Medical image fusion plays a central role in the diagnosis of patients, which can be utilized to multiple images of a patient, and these images are registered or merged to offer further information. Magnetic resonance imaging (MRI), computed tomography (CT), positron emission tomography (PET), and single-photon emission computed tomography (SPECT) are the common imaging modalities for the detection of different diseases. They also provide information about a human body in several methods. Discrete wavelet transform is a widely used wavelet transform for image fusion which can offer spectral as well as more directional information with three spatial orientations that are horizontal, vertical, and diagonal [26]. Discrete wavelet transform and inverse DWT can be applied to obtain the fused image that has more accurate information in comparison to the source images [27]. The wavelet transform-based fusion overcomes other techniques in terms of signal representation, complementary information, and redundancy. Such features aid in making transforms appropriate for multisensory image fusion [28]. Good quality fused image and better signal to noise ratio are provided by DWT. The usage of DWT in image fusion is to reduce spatial distortion and spectral degradation produced in the fused image [24,29]. Original medical CT image one and two are shown in Figure 4.6. Figures 4.7, 4.8, 4.9, 4.10, 4.11, 4.12, and 4.13 show fused image using Haar, Sym6, Coif3, Bior3.9, rbio1.1, rbio4.4, and Fk22 wavelets respectively, with maximum, minimum, and mean fusion methods.

4.3.3 Image compression

The image compression is an essential technique for dealing with the medical images. Medical images are widely used in our present-day world. Medical image is highly significant in the area of medical science for the future reference of the patient, and therefore should be stored. This image needs a process of compression before storage [30,31]. The use of medical image has become essential for the diagnosis of the patient

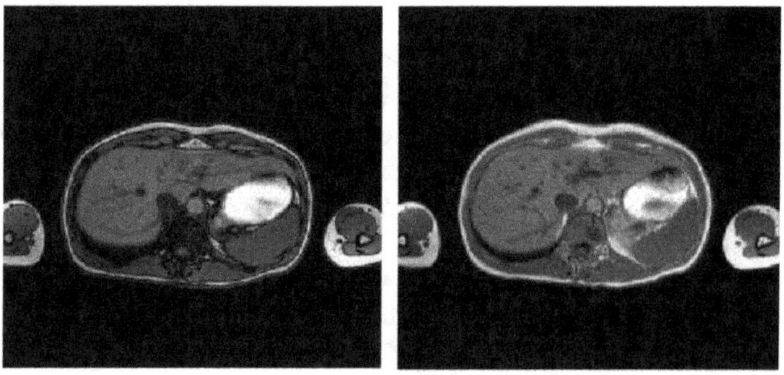

(a) (b)

Figure 4.6 (a) Original computed tomography (CT) image1 and (b) original CT image2

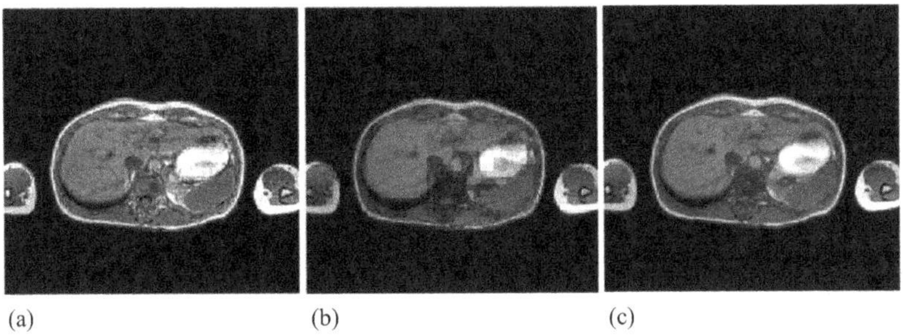

Figure 4.7 Fused image using discrete wavelet transform (Haar): (a) maximum method, (b) minimum method, and (c) mean method

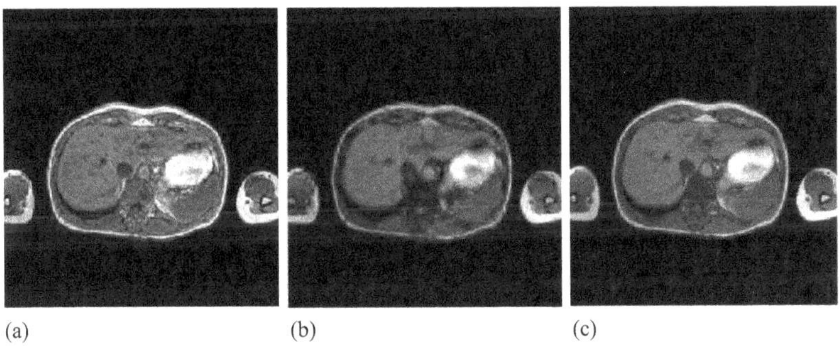

Figure 4.8 Fused image using discrete wavelet transform (Sym6): (a) maximum method, (b) minimum method, and (c) mean method

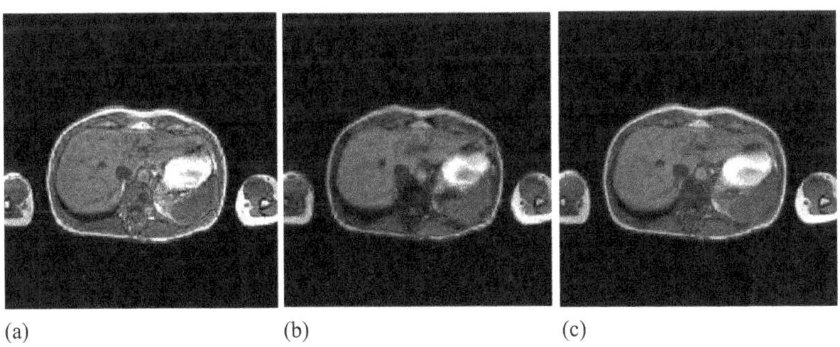

Figure 4.9 Fused image using discrete wavelet transform (Coif3): (a) maximum method, (b) minimum method, and (c) mean method

and hence the massive number of medical images are utilized on a large scale. There are various compression techniques that use different medical images such as magnetic resonance images (MRI) and X-ray angiograms (XA), etc. [30]. Image compression is a sort of data compression implemented to digital images, and it can significantly help

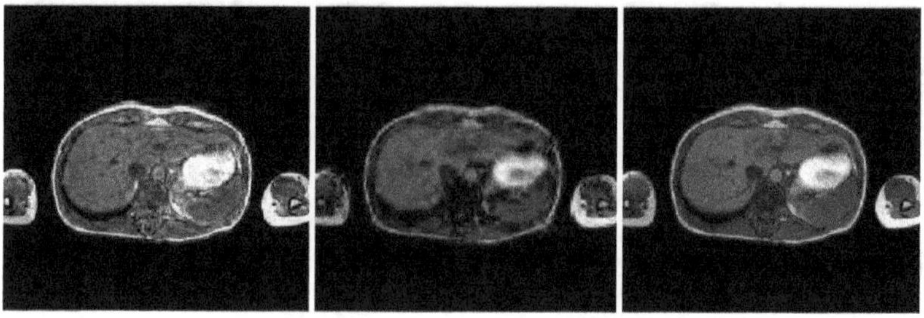

(a) (b) (c)

Figure 4.10 Fused image using discrete wavelet transform (Bior3.9): (a) maximum method, (b) minimum method, and (c) mean method

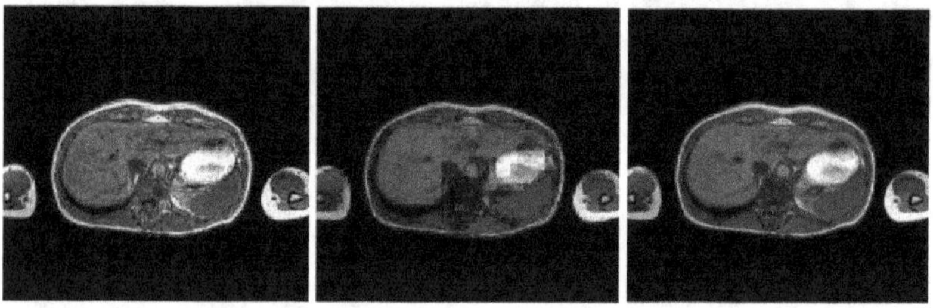

(a) (b) (c)

Figure 4.11 Fused image using discrete wavelet transform (Rbio1.1): (a) maximum method, (b) minimum method, and (c) mean method

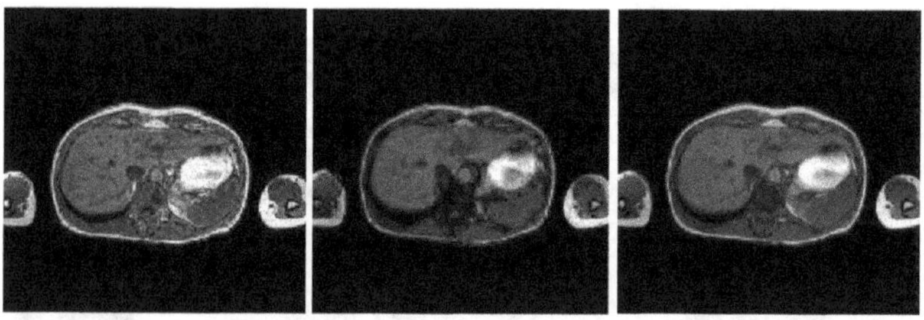

(a) (b) (c)

Figure 4.12 Fused image using discrete wavelet transform (Rbio4.4): (a) maximum method, (b) minimum method, and (c) mean method

reduce the size of image, transmission, and storing without any impact on picture quality. Discrete wavelet transform is achieved for the effective image compression [32]. Compression of medical image plays a significant role in storage and transmission. Therefore, there is a requirement for efficient image compression for managing and storing the medical images [30,33]. The transform-based techniques are the most

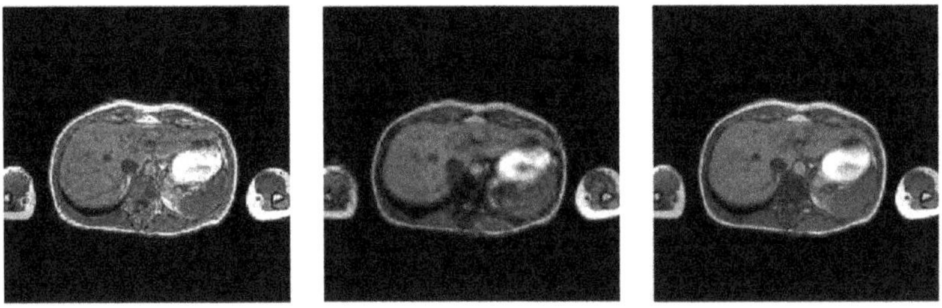

(a) (b) (c)

Figure 4.13 Fused image using discrete wavelet transform (FK22): (a) maximum method, (b) minimum method, and (c) mean method

commonly lost image compression techniques which include DWT, and DWT is distinguished by its ability to follow up the time and frequency simultaneously and constitute local features better [34]. Massive storage capacity and transmission bandwidth should be provided for uncompressed digital images. Effective image compression solutions have become crucially important with the intensively current growth of data by multimedia-based web implementations [35], in the areas of image processing, interpretation, analysis, and archiving. Researchers in the entire world are tending to design schemes which have the feature of easing various advantages in one scheme, for instance, compression, fusion, enhancement, security, etc. The image databases are major and contain plenty of information. It is therefore imperative to ensure that the compression of these high-quality images and securing these databases. Accordingly, DWT is considered as a guide to image compression and also an image security technique [36,37]. The most notable feature of employing DWT is that it will not only be able to compress an image but also will aid to maintain the quality of the image as previously in its original form, which has not been possible in advance in other image compression methods [38]. Discrete wavelet transform is deemed to be one of the advantageous techniques, which can be able to improve the quality of the compressed image and provide a mathematical method of encoding information in a way that it is layered according to the level of detail [39]. The decrease of the redundancy of the image that leads to increase the capacity of storage and efficient transmission is the main aim of image compression. Image compression techniques are mainly categorized by two types and these techniques are Lossless and Lossy image compression.

1. Lossless image compression

   Lossless compression techniques, as its name suggests, include information that cannot be lost. In lossless compression, the original data can be reconstructed completely from the compressed data. It benefits from all the information in the original image while being compressed, so when the image is decompressed, it will be completely corresponding to the original image. It is necessary that the reconstruction is identical to the original image. Lossless compression is generally employed for implementations that afford any difference between the original and reconstructed data. It is designed for lower ratios but keeps all the pixels of the

Table 4.1 Comparison between different discrete wavelet transform filters

Wavelets	Size of the original image	Size of the compressed image	Retained energy	No. of zeros	Compression ratio
Haar	100 KB	24 KB	97.82%	97.82%	4.166
Sym6	100 KB	32 KB	98.04%	98.04%	3.125
Sym6	100 KB	28 KB	96.95%	98.14%	3.571
Coif3	100 KB	32 KB	97.92%	97.92%	3.125
Coif3	100 KB	28 KB	96.30%	98.01%	3.571
Bior3.9	100 KB	32 KB	97.66%	97.66%	3.125
Bior3.9	100 KB	28 KB	91.01%	97.94%	3.571
Rbio1.1	100 KB	36 KB	99.36%	95.31%	2.777
Rbio1.1	100 KB	12 KB	94.24%	98.42%	8.333
Rbio4.4	100 KB	32 KB	97.96%	97.96%	3.125
Fk22	100 KB	28 KB	96.50%	97.88%	3.571
Fk22	100 KB	32 KB	97.80%	97.80%	3.125

original image. Images in which geometric forms are comparatively simple can be deemed for lossless image compression. In lossless image compression, there are common techniques.

- Huffman encoding
- Run length encoding
- Arithmetic coding
- Entropy encoding
- Lempel–Ziv–Welch coding

2. Lossy image compression

Lossy image compression can be used in digital images to increase storage capacities with minimal degradation of picture quality. It is designed for higher compression rates but suffers from a less resolution in the compressed image. In lossy compression techniques, the reconstructed image has lost some information and data. There are two methods of lossy compression predictive and transform coding [11,40]. Table 4.1 shows a comparison between multiple wavelet filters. Figure 4.14 demonstrates compressed image using multiple wavelet filters.

4.3.4 Image watermarking

Digital watermarking is used to guarantee and facilitate data security, authentication, and copyright protection of digital data by protecting intellectual property from illegal copying. Digital watermarking technique is employed to embed data covertly in the original image [41]. Watermarking is considered as the most important process of hiding the information, and this process can embed and extract digital data without destroying its value. Watermarking information can be implemented in audio, video, text, or image. Digital watermarking is accomplished in spatial and frequency domain, where the robustness of digital data is used in the frequency domain more than spatial domain. The watermark has been hidden in the original data in such a way that it cannot

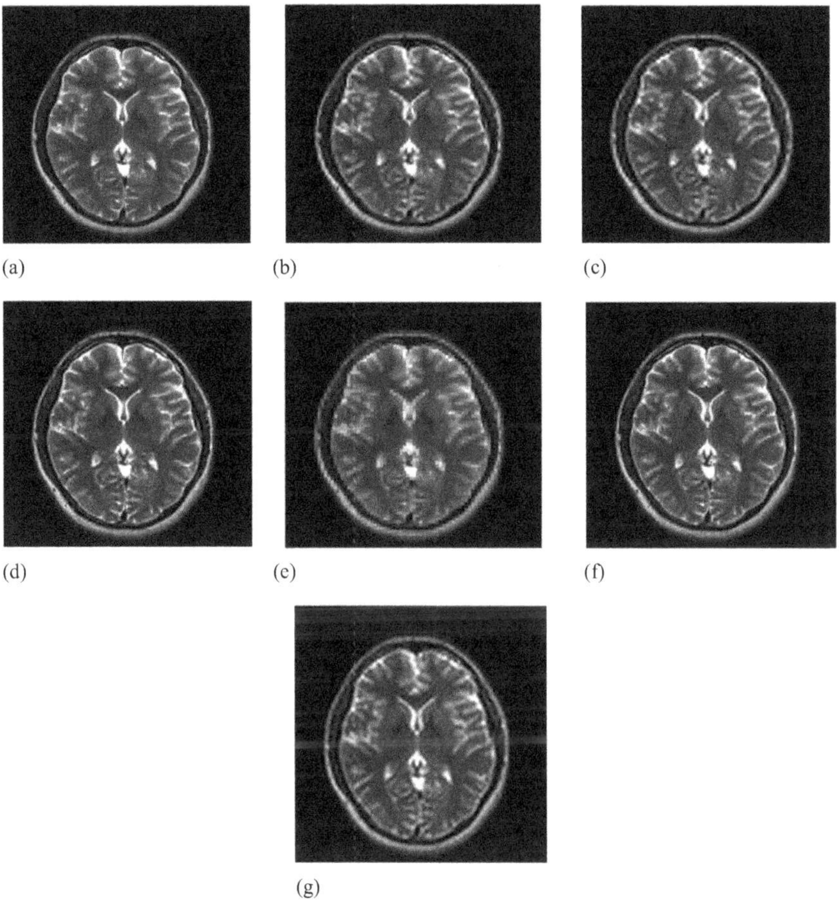

Figure 4.14 *(a) Compressed image using discrete wavelet transform (Haar), (b) compressed image using discrete wavelet transform (Sym6), (c) compressed image using discrete wavelet transform (Coif3), (d) compressed image using discrete wavelet transform (Bior3.9), (e) compressed image using discrete wavelet transform (Rbio1.1), (f) compressed image using discrete wavelet transform (Rbio4.4), and (g) compressed image using discrete wavelet transform (Fk22)*

be separated from the data and it has been designed so that it can be resistant to many operations without degrading the original document. Several researchers had proposed techniques in both domains and they had hybrid between domains. The watermark remains accessible but permanently marked. In watermarking, the important information is the external data. The watermark is the internal data and further data are responsible for protecting the external data and to prove ownership. Digital watermarks can be categorized into three sorts which are robust, semifragile, and fragile. The DWT, discrete cosine transformation (DCT), and discrete Fourier transform (DFT) are the most-used methods in the frequency domain. Frequency domain watermarking techniques have been used such as DWT and DCT for increasing higher efficiency. The DWT watermarking has been applied to the filtered image. These methods can be used in many types of digital data. Digital watermarking is of key importance to the future development area of research, and commercialization of watermarking techniques is considered essential support in addressing some of the challenges faced by the rapid generation of digital content [41,42]. Image watermarking has been generally proved as a relevant technique for enhancing integrity, authenticity, and data security where medical image is stored, transmitted, and retrieved via networks. Medical image watermarking is paramount. It should ensure the confidentiality of information. Private medical data that have been embedded into the original image must be imperceptible and could only be restored by authorized users, preventing any potential access, change, or destruction. The enhanced quality of medical services precisely relies on the capability of the techniques employed to assure ethics and protect the transmitted medical confidences [43,44]. In an era of new threats and challenges for medical image that requires retention of its information, there is an increased importance to ensure reliable protection of medical image. The safety of medical image is of critical importance before its submission to the initial medical diagnose [45]. Medical images such as X-rays, USG, and MRI are totally important through the provision enhanced diagnosis and treatment of patients [46]. The process of the integrated digital watermarking technique into medical image must be evaluated, validated, and verified for its applicability and appropriateness to medical domains. This is important to guarantee the ability of any approach to tackle security threats that may face medical image through routine medical practices [47]. Discrete wavelet transform is the most appropriate process for digital watermarking in the frequency domain that leads to a promoted robust double watermarking technique for medical image [48]. The medical image in the watermarking technique splits into regions and the watermark information can be embedded in the spatial domain and transform domain. Extreme care is required before embedding watermarking information in medical images to protect the image quality and avoid the wrong diagnosis [49]. Figure 4.15 illustrates the X-ray medical image, the watermark image, and the watermarked medical image.

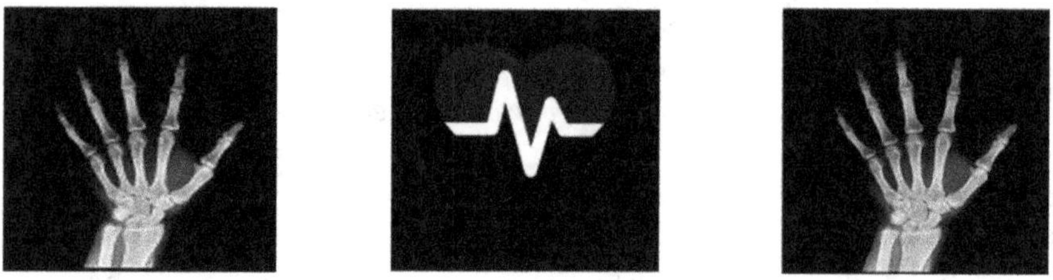

(a) (b) (c)

Figure 4.15 (a) X-ray original image, (b) watermark image, and (c) the watermarked image

4.4 Conclusions

We conclude, based on DWT, that the DWT has been decomposed an image regarding its features analogically to several spectral sub-bands, as familiarly known sub-bands of the spectrum. The DWT is fundamentally composed of a group of bandpass filters. The basic procedures of the DWT can be bounded in terms of time and frequency domain. Through the use of low pass and high pass filters frequently and effectively, we can achieve the decomposition of an image by the DWT and its reformation. The Haar wavelet is deemed to be the earliest wavelet of the DWT wavelets and it has provided ease of use in implementations. It is also considered to be the original point in the research of the DWT and also considered to be the simplest and shortest. The diverse capabilities of the DWT support in detecting components of an image at a resolution that may not be detected by another wavelet. The DWT can be applied in several ranges of applications such as denoising, compression, watermarking, and fusion. Many researchers have shown that the usage of DWT with different filters has proved the superiority of the given results.

4.5 Future work

For a prospective work, this search can be broadly developed to achieve more accuracy by raising the level of transformations. According to the work of this chapter, the techniques used can be enhanced to watermarking, fusion, denoising, and compression.

References

[1] Hashem, M., Vellappally, S., Fouad, H., Luqman, M., and Youssef, A. E., "Predicting neurological disorders linked to oral cavity manifestations using an optimized IoMT-based neural networks." *IEEE Access*. 2020; 8: 190722–190733.

[2] Liu, J., Ma, J., Li, J., Huang, M., Sadiq, N., and Ai, Y. "Robust watermarking algorithm for medical volume data in internet of medical things." *IEEE Access*. 2020; 8: 93939–93961.

[3] Ding, Y., Wu, G., Chen, D., *et al.* "DeepEDN: A deep learning-based image encryption and decryption network for internet of medical things." *IEEE Internet of Things Journal*. 2020; 1–13.

[4] Gatouillat, A., Badr, Y., Massot, B., and Sejdić, E. "Internet of medical things: A review of recent contributions dealing with cyber-physical systems in medicine." *IEEE Internet of Things Journal*. 2018; 5(5): 3810–3822.

[5] Raj, R. J. S., Shobana, S. J., Pustokhina, I. V., Pustokhin, D. A., Gupta, D., and Shankar, K. "Optimal feature selection-based medical image classification using deep learning model in internet of medical things." *IEEE Access*. 2020; 8: 58006–58017.

[6] Khan, M. A., and Algarni, F. "A healthcare monitoring system for the diagnosis of heart disease in the IoMT cloud environment using MSSO-ANFIS." *IEEE Access*. 2020; 8: 122259–122269.

[7] Ud Din, I., Almogren, A., Guizani, M., and Zuair, M. "A decade of internet of things: Analysis in the light of healthcare applications." *IEEE Access*. 2019; 7: 89967–89979.

[8] Limaye, A., and Adegbija, T. "A workload characterization for the internet of medical things (IoMT)." *2017 IEEE Computer Society Annual Symposium on VLSI (ISVLSI)*. Bochum. 2017. pp. 302–307.

[9] Sundararajan, D., *Discrete wavelet transform a signal processing approach*. John Wiley & Sons; 2015.

[10] Mallat, S. *A wavelet tour of signal processing the sparse way*. Elsevier Inc.; 2009.

[11] Sayood, K. *Introduction to data compression*. Morgan Kaufmann; 2017.

[12] Satapathy, L. M., Das, P., Shatapathy, A., and Patel, A. K. "Bio-medical image denoising using wavelet transform." *International Journal of Recent Technology and Engineering (IJRTE)*, 2019; 8(1): 2479–2484.

[13] Ferzo, B. M., and Mustafa, F. M. "Image denoising in wavelet domain based on thresholding with applying wiener filter." *2020 International Conference on Computer Science and Software Engineering (CSASE)*, Duhok, Iraq. 2020. pp. 106–111.

[14] Lyakhov, P. A., Valueva, M. V., Nagornov, N. N., Chervyako, N. I., and Kaplun, D. I. "Low-bit hardware implementation of DWT for 3D medical images processing." *2020 IEEE Conference of Russian Young Researchers in Electrical and Electronic Engineering (EIConRus)*. St. Petersburg and Moscow, Russia. 2020. pp. 1396–1399.

[15] Guo, Q., Zhang, C., Zhang, Y., and Liu, H. "An efficient SVD-based method for image denoising." *IEEE Transactions on Circuits and Systems for Video Technology*, 2016; 26(5): 868–880.

[16] Zhang, Y., Zhang, M., Yang, Q., and Wang, T. "An improved medical image denoising algorithm based on one-dimensional heat transfer equation," *2018 IEEE 7th Data Driven Control and Learning Systems Conference (DDCLS*). Enshi. 2018. pp. 623–626.

[17] Zhang, Y., Zhang, M., Yang, Q., and Wang, T. "An improved medical image denoising algorithm based on one-dimensional heat transfer equation." *2018 IEEE 7th Data Driven Control and Learning Systems Conference (DDCLS)*. Enshi. 2018. pp. 623–626.

[18] Soni, N., and Kirar, K. "Transform based image denoising: A review." *2017 International Conference on Recent Innovations in Signal* Processing *and Embedded Systems (RISE*). Bhopal. 2017. pp. 168–171.

[19] Rambabu, T. G., and Kishore, K. V. K. "Medical image denoising using KPCA with local pixel grouping." *2015 International Conference on Computer Communication and Informatics (ICCCI*). Coimbatore. 2015. pp. 1–5.

[20] Kadam, C., and Borse, S. B. "An improved image denoising using spatial adaptive mask filter for medical images." *2017 International Conference on Computing, Communication, Control and Automation (ICCUBEA*). Pune. 2017. pp. 1–5.

[21] Li, L., Wu, J., and Jin, X. "CNN denoising for medical image based on wavelet domain." *2018 9th International Conference on Information Technology in Medicine and Education (ITME)*. Hangzhou. 2018. pp. 105–109.

[22] Lopes De Faria, M. L., Cugnasca, C. E., and Amazonas, J. R. A. "Insights into IoT data and an innovative DWT-based technique to denoise sensor signals." *IEEE Sensors Journal*. 2018; 18(1): 237–247.

[23] Chauhan, R., Dahiya, R., and Bansal, P. "Optimal choice of thresholding rule for denoising ECG using DWT." *2017 4th International Conference on Signal Processing*, Computing and Control (ISPCC), Solan. 2017. pp. 288–292.

[24] Panguluri, S. K., and Mohan, L. "Discrete wavelet transform based image fusion using unsharp masking." *Periodica Polytechnica Electrical Engineering and Computer Science*. 2020; 64: 211–220.

[25] Ravi, R. V., Sujith, M. V., Shafeen, K. M., Asharaf U. T. A., Sajidh, C. T., and Mohan, M. T. S. "Performance analysis of wavelet functions in fusion of MRI and CT images." *2020 7th International Conference on Smart Structures and Systems (ICSSS)*. Chennai, India. 2020. pp. 1–6.

[26] Ashwanth, B., and Swamy, K. V. "Medical image fusion using transform techniques." *2020 5th International Conference on Devices, Circuits and Systems (ICDCS)*, Coimbatore, India. 2020. pp. 303–306.

[27] Sreekrishna Yadav, S., and Mittal, V. "FPGA implementation of segmented feature fusion in MRI images using wavelet." *2019 International Conference on Communication and Electronics Systems (ICCES*). Coimbatore, India. 2019. pp. 1946–1951.

[28] Singh, R., Nigam, S., Singh, A. K., and Elhoseny, M.. *An overview of medical image fusion in complex wavelet domain in: Intelligent wavelet based techniques for advanced multimedia applications*. Springer, Cham; 2020.

[29] Kulkarni, J. S., and Bichkar, R. S. *Comparative analysis of image fusion using DCT, DST, DWT, walsh transform and kekre's wavelet transform*. Springer; 2018.

[30] Satone, K. N., Deshmukh, A. S., and Ulhe, P. B. "A review of image compression techniques." *2017 International Conference of Electronics, Communication and Aerospace Technology (ICECA*). Coimbatore. 2017. pp. 97–101.

[31] Kumar, S., and Kumar, D. "Comparative analysis and performance evaluation of medical image compression method for telemedicine." *2nd International Conference on Data, Engineering and Applications (IDEA)*. Bhopal, India. 2020. pp. 1–5.

[32] Kanagaraj, H., and Muneeswaran, V. "Image compression using HAAR discrete wavelet transform." *2020 5th International Conference on Devices, Circuits and Systems (ICDCS)*, Coimbatore, India. 2020. pp. 271–274.

[33] Krishnaswamy, R., and Nirmala Devi, S. "*Efficient medical image compression based on integer wavelet transform*." *2020 Sixth International Conference on Bio Signals, Images, and Instrumentation (ICBSII*). Chennai, India. 2020. pp. 1–5.

[34] Xiaoqi, L., and Baohua, Z. "Selective compression algorithm of facial images based on DWT." *2008 International Conference on Advanced Computer Theory and Engineering*. Phuket. 2008. pp. 685–689.

[35] Meng, M., and Zong, M. "A new zerotree structure for color image compression based on DWT and VQ." *2010 2nd IEEE International Conference on Information Management and Engineering*. Chengdu. 2010. pp. 339–342.

[36] Mulla, A., Baviskar, J., Wagh, S., Kudu, N., and Baviskar, A. "Probabilistic triangular shuffling approach in DWT based image compression scheme." *2015 International Conference on Communication, Information & Computing Technology (ICCICT)*. Mumbai. 2015. pp. 1–6.

[37] Nagaraj, P., Rajasekaran, M. P., Muneeswaran, V., Sudar, K. M., and Gokul, K. "VLSI implementation of image compression using TSA optimized discrete wavelet transform techniques." *2020 Third International Conference on Smart Systems and Inventive Technology (ICSSIT)*. Tirunelveli, India. 2020. pp. 667–670.

[38] Barbhuiya, A. H. M. J. I., Laskar, T. A., and Hemachandran, K. "An approach for color image compression of JPEG and PNG images using DCT and DWT." *2014 International Conference on Computational Intelligence and Communication Networks*. Bhopal. 2014. pp. 129–133.

[39] Nashat, A. A., and Hassan, N. M. H. "Image compression based upon wavelet transform and a statistical threshold." *2016 International Conference on Optoelectronics and Image Processing (ICOIP)*. Warsaw. 2016. pp. 20–24.

[40] Surabhi, N., and Unnithan, S. N. "Image compression techniques: A review." *International Journal of Engineering Development and Research*. 2017; 5: 585–589.

[41] Reddy, M. R., Sandeep, P. S., Mounika, C. H., Gopi, J., and Suganya, S. D. "Optimized watermarking technique using PCA-DWT and log filter." *2017 International Conference on Networks & Advances in Computational Technologies (NetACT)*, Thiruvanthapuram. 2017. pp. 126–130.

[42] Shekaramiz, K., and Naghsh, A. "Embedding and extracting two separate images signal in salt & pepper noises in digital images based on watermarking." *2017 3rd International Conference on Pattern Recognition and Image Analysis (IPRIA)*. Shahrekord. 2017. pp. 32–37.

[43] Mestiri, A., Kricha, A., Sakly, A., and Mtibaa, A. "Watermarking for integrity, authentication and security of medical imaging." *2017 14th International Multi-Conference on Systems, Signals & Devices (SSD)*. Marrakech. 2017. pp. 475–481.

[44] Hassan, B., Ahmed, R., Li, B., and Hassan, O. "An imperceptible medical image watermarking framework for automated diagnosis of retinal pathologies in an eHealth arrangement." *IEEE Access*. 2019; 7: 69758–69775.

[45] Su, G.-D., Chang, C.-C., and Lin, C.-C. "Effective self-recovery and tampering localization fragile watermarking for medical images." *IEEE Access*. 2020; 8: 160840–160857.

[46] Bal, S., Das, S., Dutta, S., Das, S., Biswas, D., and Basu, A. "FPGA realization of medical image watermarking." *2018 IEEE Electron Devices Kolkata Conference (EDKCON)*. Kolkata, India. 2018. pp. 523–528.

[47] Qasim, A. F., Aspin, R., and Meziane, F. "Integration of digital watermarking technique into medical imaging systems." *2019 10th International Conference on Dependable Systems, Services and Technologies (DESSERT)*. Leeds, UK. 2019. pp. 202–207.

[48] Venkateswarlu, L., Rao, N. V., and Reddy, B. E. "A robust double watermarking technique for medical images with semi-fragility." *2017 International Conference on Recent Advances in Electronics and Communication Technology (ICRAECT)*. Bangalore. 2017. pp. 126–131.

[49] Kunhu, A., Al-Ahmad, H., and Taher, F. "Medical images protection and authentication using hybrid DWT-DCT and SHA256-MD5 hash functions." *2017 24th IEEE International Conference on Electronics, Circuits and Systems (ICECS)*. Batumi. 2017. pp. 397–400.

Chapter 5

Intelligent agents system for medical information communication

Mariya Evtimova-Gardair[1] *and Evangelos Pallis*[2]

This chapter describes the importance of the agents when making an intelligent web communication system for medical information. The state-of-the-art of the agents is performed and the implementation of mobile agents is highlighted with their positive and negative features when gathering information in the Internet. In this respect, an intelligent system with a mobile agent model for communication of medical information is proposed that could be implemented in the wireless structure. The platform for the creation of the medical communication information system that is proposed is a part of intelligent wireless communication. The result from using mobile agent is generated which is also important when working with big data.

5.1 Introduction

In the recent years, there has been growing interest in studying multiagent systems for searching of information. Modern scientific researches which are related with multiagent systems are directed mainly into coordinating agent behavior and distribution and unification of decisions that are applied for information retrieval systems, especially when searching information in a big volume of data. Nowadays, multiagent systems are one developed direction that is applied to systems for searching while preparing theoretical and laboratory research. Agents may have a number of features, such as autonomy, reactivity, proactivity, and social capabilities. These characteristics determine their behavior and interaction with the surrounding environment, which is important for their application in information retrieval systems. The environment is everything external to the agent and the agent is seen in continuous connectivity and interaction with it. The multiagent system is determined by the number of agents interacting with one another. It can be considered that agents interact on behalf of their users with different purposes and

[1]Center for Teaching and Research in Computer Science (CERI), Avignon University, Avignon, France
[2]Department of Electrical and Computer Engineering, Hellenic Mediterranean University, Heraklion, Crete, Greece

motivations. In order to achieve effective interaction with each other, agents must be able to cooperate, coordinate, and negotiate with the people. Multiagent system and its feature determine their application for an information search system:

- Agents have incomplete and contradictory information about the environment and are limited in their capabilities
- Distributed multiagent system management
- Decentralized data
- Asynchronous calculations

Recent research on multiagent systems is mainly related with coordination of the agent behavior and distributing and unifying decisions that are critical when working with big data. Multiagent systems are used in many modern information retrieval systems. The problem of the need to continuously improve the quality of search systems implies the standard for all new information systems to evaluate the quality of returns and compare the result with the quality of returns of other created search engines.

5.2 Analysis of the agent technologies

Searching systems that are on the market have a number of disadvantages and do not always meet the need of the medical information of the user who have a health problem. A solution for an effective significant search system from medical information and Internet is a searching system that uses artificial intelligence.

Multiagent systems and autonomous agents provide a new method for analyzing, designing, and implementing complex applications because they are a part of distributed artificial intelligence (DAI). Today, the majority of applications require distributed tasks between autonomous agents to achieve their goals in an optimal way [1]. This enables intelligent agents to give users accurate and relevant (up-to-date) results. This is the result of the ability of intelligent agents to help the user find and filter the information on the web. This necessitates the use of intelligent agents in search systems, which allows information systems to work with a great deal of information and to enable search systems to learn from their environment in order to reproduce accurate and relevant results [2].

5.2.1 Meaning of the agent when searching information

Nowadays, thousands and millions of people are engaged in searching for information each day when they use a web search engine or search in their Internet mail. Currently, the global search information space consists of millions of HTML pages on the Internet. A big volume of variable, semi- or unstructured heterogeneous data are additionally available in related databases, file systems, multimedia database systems, and software applications. This includes, for example, bibliographic inputs, photos, speaking text, and video data [3–5].

The impacts of data, system, and data heterogeneity of information overloading to the user are numerous. This is particularly true due to potentially significant differences in the data model, data structures, data representation, using ontologies and dictionaries, query languages, and operations to search, retrieve, and analyze information in an appropriate context.

Information agent technologies are emerging as a major part of intelligent software agent technologies. The leading idea for information agent technologies is the development and efficiency, efficient use of standalone computing software objects called intelligent information agents who have access to multiple, heterogeneous, and geographically distributed information sources, both on the Internet and in a local intranet. The primary task of such agents is to proactively search, maintain, and mediate relevant information with users and other agents. This includes skills such as searching, analyzing, manipulating, and distributing heterogeneous information, as well as visualizing and directing the user to the possible individual information space [6].

5.2.2 Intelligent agents for information: definitions and basic features

Information agents can be classified by type of information software agents. Software agent technologies come from distributed systems with artificial intelligence. The term agent is widespread in the literature. It can be seen as a tool for analyzing systems, not a complete feature that separates the world from agents and nonagents. Intelligent agents are usually assumed to exhibit autonomous behavior that is determined by their:

- proactivity, which means taking an initiative to meet the goals of a design and displaying purposeful behavior;
- reactive and advisory actions mean environmental perception and timely change of management to match design goals; and
- social interaction in groups with other agents and human users if necessary.

This depends on the specific field of application and the potential for solution of a problem, with what type of information agent should be used in practice. Agents are deployed in various environments such as industrial control, Internet search engine, personal assistant, network management, games, distributed software, and more [1,7–11]. Multiagent technologies have standards for software architectures and applications such as OMG MAF and FIPA specifications. Intelligent Internet agents are commonly referred to as an information agent. Information agent can be defined as autonomous, existence of computing software (intelligent agent) that has access to one or more heterogeneous and geographically distributed information resources that proactively use, mediates, and maintains relevant information on behalf of users or preferably exactly during working hours. Information agents are supposed to satisfy one or more of the following requirements:

- Information acquisition and management: There is an opportunity to provide clear access to one or many different information resources for information retrieval, analysis and filtering of data, monitoring of sources, and updating of relative information as part of its user or other agents.

The acquisition of information encompasses a wide range of scenarios including the advancement of information search in databases and also the purchase of relative information from the supplier from the electronic shopping places.

- Information synthesis and presentation: Agents are able to merge heterogeneous data and provide unified, multidimensional views of relative information to the user.
- The agent can dynamically adapt to changes in user preferences as well as in the network environment and provide intelligent and interactive help to the typical users who support their information business on the Internet. In this context, using the intelligent user interface as a believable, realistic hero can significantly increase not only the user's awareness of his personal information agent but also the way information is interactively previewed. Many systems with information agents have been created or are currently being developed in academic and commercial research laboratories, but they still have to wait to create a realistic world of broad Internet users. The ambitious and required goals to meet all the requirements listed above are supposed to be completed over the next 10 years.

5.2.3 Classification of information agents

Information agents can be characterized in several different classes according to one or more of the following:

1. Noncooperative or cooperative information agents, depending on the ability of agents, cooperate with each other for the performance of their tasks. Several protocols and methods are available to achieve cooperative autonomous information agents and various scenarios, such as hierarchical delegation of tasks, contracts, and decentralized contracting.
2. Adaptive information agents have the ability to abdicate only the changes in the network and the information environment. An example of such a type of agent is the training of personal assistant on the Internet.
3. Rational information agents are utilitarian in the economic sense. They play and can even cooperate to increase their own benefits. One main application domain for such agents is automated trading and e-commerce on the Internet.
4. Mobile information agents have the ability to travel automatically over the Internet. Such agents can allow, for example, dynamic load balancing in large-scale networks, a reduction in data transfer between information server applications, and small business logic migration with middle-range enterprise search intranets [8,12–16].

Advantages of mobile agents [16]:

1. Reduce network load
2. Overcoming network latency
3. Encapsulation of protocols
4. Asynchronous and autonomous implementation

5. Dynamic adaptation
6. Natural heterogeneity and strength
7. Resistance to failure

According to the definition and classification of information agents, they can be divided into communication agents, knowledge, cooperation, and low-level tasks. Figure 5.1 lists the basic skills of the information agent. The communication agent's communication skills define a communication with information systems and databases, human users, or other agents. For the last case, the use of language for the communication agent should be considered at the top of, for example, platforms or specific applications. The information agent acquires and maintains the knowledge of himself and his surroundings in presentation and processes of ontological knowledge and metadata, profiles and input languages, translation of the data format, and application of machine learning techniques.

A high level of cooperation of the information agent with other may lead, for example, to mediation, coordination, negotiation, and cooperation (social) filtering. The intelligent agent, working with its human users, applies techniques that came from human–computer interaction and effective computing.

5.2.4 Basic features of the intelligent agents for information

According to the main properties of the intelligent agents for information, basic supporting technology are defined in the following cases whether they cooperate with other agents or not.

– Access to heterogeneous distributed information systems and resources on the Internet. This includes standardized platforms as well as efficient client techniques of server-side web-based applications.
– Searching for and filtering relative information from any type of digital international environment, such as content-based, rich media, and complex language for searching of information,
– Management of metadata and the process of the ontological processing of knowledge that facilitate the reconciliation of the semantic heterogeneity of retrieved data and information derived from multiple heterogeneous sources.
– Visualization of the information.

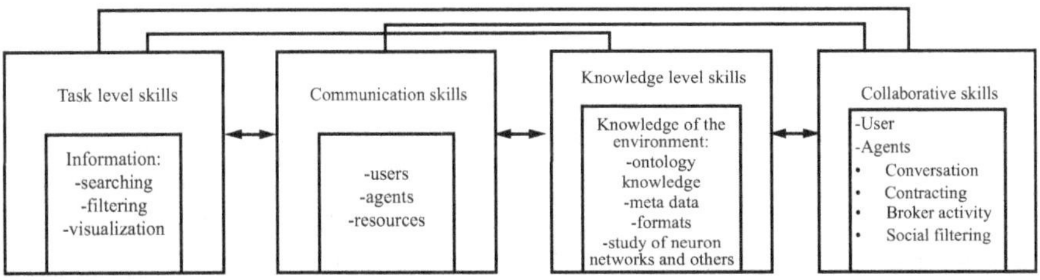

Figure 5.1 Basic skills of the intelligent agent for information

5.3 Usage of multiagent system for extraction of information

5.3.1 The possibilities for the creation of mobile agents

The growing number of applications that serve millions of users and use terabyte data require a paradigm for faster processing. Nowadays, there is a growing interest in large data analysis. Big data analysis is becoming a very important aspect of productivity growth, reliability, and quality of service (QoS). For processing big data, using a powerful machine is not an effective solution.

In the beginning, the semantic web was designed to share knowledge from distributed, dynamic, and heterogeneous sources, in which content is expressed in machine-readable formats through languages such as RDF and OWL [17]. This is the way information is shared on the web. By using agent technology, agents play an integral role using machine-readable formats for gathering and presenting knowledge, as well as reasoning to manage and reach new facts. Together with their ability to process semantic web content, agents contribute features such as distribution, autonomy, and distributed knowledge bases. There are many tools that are designed to manage standard sources of knowledge, but they are usually centralized and in a static environment where overall control is centralized. But such achievements cannot successfully integrate into such an environment as the semantic Internet, which is an open, dynamic, and often chaotic environment [18].

Distributed decentralized systems are characterized by components with the same roles and the ability to exchange knowledge and services directly with each other [19]. P2P systems are equal networks with equal roles and capabilities, and such a system is proposed that uses P2P technology to share and search for a large amount of data [20,21].

5.3.2 Mobile agents

When a distributed web crawler is based on the migration of a web agent or migrating agent [2,22–24], the process of selecting or filtering of the web documents can be performed on the web servers, and not from the side of web searching agents that can reduce network load derived from web robots [23,24]. The mobile agent [25] is a type of software agent with features such as autonomy, social abilities, learning, and mobility. It is a composition of computer software and data that have the ability to migrate from one computer to another autonomously and continue their processes on a remote computer. In the migrating web robots, the mobile code generated from the side of the searching system is executed on the web servers, the environment that is controlled from the other side.

Mobility allows the agent to move or jump between agent platforms. Migration agents include computational software processes that can walk through the most big part of the network, as well as the Internet, interacting with foreign hosts, gathering information on behalf of their owner, and returning once they perform the duties defined by their owner. The agent is an autonomous entity that acts on behalf of others autonomously. Nwana [25] identifies seven types of agents, that is, collaborative agents, interface agents, reactive agents, hybrid agents, and intelligent agents. Basic features of mobile agents are [2,25] migration, data collection, route determination, and communication.

The agent platform provides a computing environment in which the agent operates. The platform from which the agent originates is defined as the starting platform and is usually the most secure environment for the agent. One or two hosts may include an agent platform, and the agent platform can support multiple computing environments or meeting locations where the agents can interact. They can collaborate or communicate with other agents by locating some internal objects and methods known to other agents without sharing all of the information.

Mobile agents ask their environment to get the information they need to reach their goals. This information needs to be filtered locally by the agent in advance or stored by the agent or forwarded to a receiving destination. When the agent once ends up with the network node, it must decide where to move afterward.

The ability of agents to communicate is fundamental for mobile systems. Advantages of mobile agents are [2,24,25] efficiency, traffic, latency, performance of asynchronous tasks, strength, tolerance to failures, support form for heterogeneous environments, support for e-commerce, paradigm for easy development, and equal (P2P) [20,21,26,27].

When big data are stored on remote hosts, these data must be processed locally instead of being transferred to the network. The basic concept is that it is better to move the calculations to the data than the data to the calculations.

When migrating to the resource location, the mobile agent can interact faster with the resources than over the network.

While the agent acts on behalf of the client on a remote site, the client may perform other tasks. Instead of being online for a long term, the mobile user can create an agent request while it is off, then the agent can be launched during a short session and the agent can accept the results later.

5.3.3 Comparison of the standard model of searching system and searching systems with mobile agents

Web searching system in Figure 5.2 represents a set of programs that can read every page that can be found on the Internet, creates an index of the information that found, compares the information with that of the user's query, and finally returns the results again to the user. It is a searchable database that collects information from web pages, indexes the information, and then saves the result in a big database where it can be searched quickly. Web search engines are a link between the web user and web documents. Without the help of search engines, endless information on web pages remains inaccessible to the user. Web search system basically consists of three parts: web bot, index, and query mechanism.

The web robot represents a module that searches web pages from the web. These are small programs that read on the web on behalf of the web search engine and follow the links to reach the different pages. As it is started from selected URLs, robots retrieve URLs that appear in retrieved pages and keep pages in a

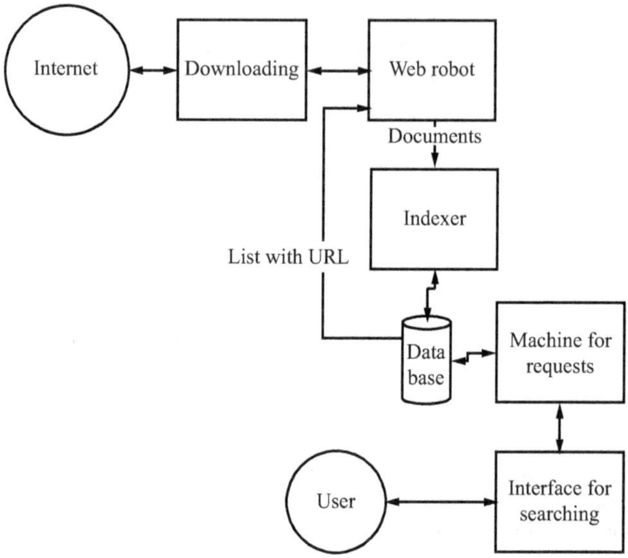

Figure 5.2 Principal architecture of the web search

warehouse database. The index extracts all unusual words from each page and registers the URL to which each new word is found. The index retrieves all unusual words from each page and registers the URL where each word is found. The result is stored in a large list containing URLs pointing to pages in a database where a word is encountered. The machine for querying is responsible for receiving and completing user search queries. It relies on the indexes and the database. Because of the size of the Internet, and the fact that users usually put one or two keywords, the result of the information found is usually quite big. Methods with distributed web robots with migrating agents [2,22,23,28] allow packet wrapping and distribution to specific hosts where interaction can take on the same place. Migrating agents are also useful in reducing the flow of raw data on the network. When a big amount of data is stored on remote hosts, these data must be handled locally where data are located and not transmitted over the network. The basic concept is the movement of calculations to data rather than data to calculations. When migrating to the resource location, the mobile agent can interact with the resource much faster than over the network, and also reduces network traffic.

A distributed web-based method with migrating agents is presented in Figure 5.3, which uses web-based website management, which assigns migrating web robots to web servers a list of URLs on the respective web servers. Migrating robots, when they reach the server, crawl the pages, choose the best of their collection pages, and return to the web search engine with the collection. This reduces the unnecessary amount of information and therefore unnecessary pages of the web search engine. The size of the collection can be further reduced by filtering the required specialized web pages and even compressing them.

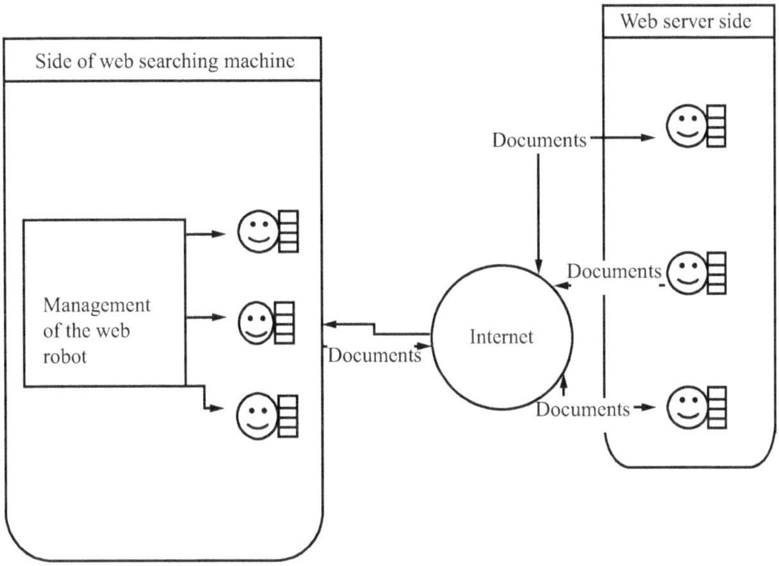

Figure 5.3 Web searching system based on agent migration

5.3.4 Features and benefits of mobile agents

A mobile agent is a software abstraction that can migrate during execution through a heterogeneous or homogeneous network. It has the ability to stop its performance according to some factors and to resume it in another machine.

Mobile agent features: There are several features that can determine the structure of the mobile agent.

Condition: A basic feature of the mobile agent. It can stop the execution of a machine and resume the execution of another machine.

The condition depends on two factors:

1. Execution state, which is a state during work that includes the counter in the program and the stack.
2. The status of the object that stores the current values of its variable.

A mobile agent is a software abstraction that can migrate during execution through a heterogeneous or homogeneous network. It has the ability to stop its performance according to some factors and to resume it in another machine. Mobile agent features: There are several features that can determine the structure of the mobile agent:

Condition: A basic feature of the mobile agent. It can stop the execution of a machine and resume the execution of another machine.

Condition depends on two factors:

1. Performance state, which is a state during work that includes the program counter and the stack.

2. The status of the object that stores the current values of its variable.

Execution: This is a program code that defines the behavior of the tasks. If JADE is used as a platform for mobile agents, the classes represent the execution code. There are two ways to make the classes available to the mobile agent [26]:

1. Getting all the required classes along its route and using it any-time anywhere.
2. Taking a part of the required classes and when the mobile agent needs a class that is not available, he can retrieve it from the remote location. This operation is called code-on-demand technique and is a common technique for distributed network systems.

Interface: The mobile agent cooperates with other agents to handle the assigned job. Interface is needed to make communication possible between agents.

Unique identifier: It identifies the agent during its life. It is used as a key that needs to be especially relevant to a specific agent when it is traveling on the net.

Route: This is a group of addresses created once when the mobile agent starts, which defines the agent's route on the network.

Principles: This is the information about the person, organization, or corporation to which the mobile agent belongs. The principles are needed to authenticate the mobile agent who is traveling to several destinations on the network.

Advantages of mobile agents [29]: There are many advantages when using mobile agents for solving problems in distributed application.

Reducing network traffic: Collaboration in a distributed system is often achieved through the use of communication protocols. These protocols transfer a large amount of data stored in remote hosts through the network to a central processing location resulting from high network traffic. In this case, the mobile agent uses alternative communication protocols.

Offline tasks: The network connection may fail at any time.

Agents can solve this problem by performing offline tasks and sending results to a server application when it is online again.

Support for heterogeneous environments: Mobile agents can work on the top of any operating system that has the same mobile framing.

Failure: Mobile agents react dynamically and autonomously to changes in their environment. If a host is stopped or a platform does not work, all agents running on this machine will be alerted and given time to send themselves and continue their operation to another host on the network.

Encapsulation protocol: The encapsulation protocol allows the components of the distributed system to communicate and coordinate their actions. Mobile agents provide a solution to the problem of improving the protocol code at all locations in the distributed system.

The result of using mobile agents in a personalized search engine is to improve the speed of the performance of the system.

5.4 Big data impact when searching

The data have been rising rapidly over the last decades, and "big data" is becoming a popular topic. Big data not only refers to massive data but also to a series of techniques that turn the data stream into valuable information [30]. Big data is related to a large volume, complex, and growing data set with multiple, autonomous sources [31]. "Big data" is characterized by volume, variety, and speed. Big data is not only large in volume but also varied in type. They can be structured data, unstructured data, or semistructured data. Speed describes data flow that requires timeliness in data analysis. The ultimate goal of data analysis is to obtain valuable information from the data stream.

5.4.1 Challenges when analyzing big data

1. Content and search when making analysis

   Traditional data analysis methods developed for structured databases are path analysis, time analysis, charting and network analysis, and what if analysis [30]. Big data are developed rapidly in every aspect of data processing, that is, data collection, data mining, data storage, data modeling, data processing, and data interpretation. With the development of theoretical analyses and technologies, big data will be used in more and more areas (i.e., health, smart grids, the economy, and social networks). In addition to the challenges outlined above, there are still other issues [30]. The security of big data is a difficult problem that needs to be defined by the law and protected by technical measures [32].
2. Diversity and heterogeneity of data

   There are traditional tools, such as SQL, to handle structured databases, but when databases are semistructured or unstructured, such as audio, video, text, and web pages, traditional tools are not effective. Therefore, initially, the data should be structured so that data can be processed more efficiently. To customize medicine, clinical information (such as medical diagnosis, medical images, and patient history) and biological data (i.e., gene, protein sequences, functions, biological process, and pathways) have to be managed and integrated with a variety of formats and are generated from different heterogeneous sources. Over the last decade, it is challenging to work with big data, but today it is important to focus on developing tools and techniques to get a better sense of data and make use of knowledge to find knowledge. Despite the overall availability of data arrays, interoperability is still lacking. In order to personalize data in the basic is the integration of data and the use of different data sources.
3. Volume and size of data

   Big data expands its volume faster than the computational resources. Processors follow the Moore's law until the size of the data leads to an explosion. But there is a limit to Moore's law. Because the size of the chips is becoming smaller and smaller, the quantum effects are so significant that they cannot be ignored. As a result, a solution needs to be found to overcome the problem with the fast-growing data. The huge volume of data itself does not solve the problems with the quality of the returned results from the searching system. Data must be summarized or retrieved in a meaningful way in order to transform the data into information, knowledge, and finally into wisdom. The big volume of data needs to be effectively explored so that to be used to make a decision. For the processing of big data, tools such as Hadoop is used that can increase the velocity of processing data and queries.

4. Velocity and timeliness of the data

It is accepted that as much bigger the volume of the data is, as much longer it takes to analyze the data. Over time, the data value is destroyed and in some cases, the timeliness of the data processing is important (i.e., bank credit verification). In other words, velocity is a constant that constantly changes and develops when performing healthcare data. These rapid changes in data give a significant challenge in creating the relevant models in the on-demand domain so that to be useful for searching, browsing, and analyzing content in real-time. This requires solving the following questions:

1. Ability to filter, prioritize, and classify data (which refers to a domain or a case)
2. Ability to receive data quickly
3. Ability to select, create, and refine relevant knowledge [33]

Free space, uncertainty, and incomplete data are critical for applications with big data. For data that are rare, the number of data points is too small to produce reliable conclusions. This is usually a compilation of data size problems where data in high-dimensional space (such as more than 1000 dimensions) do not clearly show directions or distributions. For most machine learning algorithms and algorithms for data mining, high-dimensional free data significantly degrades the reliability of the models derived from the data. Such studies are created to use the reduction of the dimensions or the selection of properties to reduce the size of data or to carefully additional examples to reduce the lack of data such as generic uncontrolled training methods when searching in data. Uncertain data are a special type or real data, where each data field is no longer determined but is subject to sporadically appearance of error. This is mainly related to domain-specific applications with inaccurate data and collections. The use of complex data is a major challenge in big data applications because every two sides in a complex network are potentially mutually interested in making a social connection. Data from application that engages in dynamic environments have many sources, massive, heterogeneous, and dynamic characteristics. This determines the important features of big data, such as petabyte (PB) calculations and even the exa-byte (EB), a level of complex calculation process [31,34].

The concept of big data quickly expands in many sciences and engineering fields. Free space, uncertainty, and incomplete data are critical elements for big data usage. Consequently, the use of parallel computing infrastructure and, respectively, programming language support and software models for efficient analysis and searching of distributed data are essential objectives for big data processes to switch from "quantity" to "quality" [14,35].

5.5 Model of the multiagent searching system

5.5.1 Requirements of the searched system

The proposed searched system is designed to provide a quality search for medical information that corresponds to the individual needs and needs of people of all ages who would like to receive information on a health problem depending on their medical complaints. Because English is the most widely used language and could be found the most and best quality text sources in any field, the system will be developed and tested in English for the time being. The approaches, methods, and algorithms used in it could, in principle, be implemented in any natural language, but this will require serious efforts because of the significant differences between human languages and, moreover, the effectiveness of the system depends on the text and sources, meaning that it would be most effective in English at the moment.

Main system requirements are as follows:

- Ability to display user information on given search words
- Adaptability—easily adapted to the needs of the particular user
- Expandability—it is easy to add new components to the system
- Work effectively with vague and inaccurate information from the user
- Ability to work with big data
- Precision of Internet search according to the specific case
- Increase of the recall results from the search according to the specific case
- User-friendly interface
- Platform independence requirements of the searched system

5.5.2 The choice of multiagent platform for system realization

An agent could be defined as an entity that performs one or several tasks to be able to distinguish their goals. Some of the special characteristics of the mobile agent are move between several machines changing its state, possibility to find by themselves the way on the network, and migration from one execution environment to another. Personalization of the searching results could be performed with semantic rules. The choice of a multiagent software platform is important for the creation of multiagent software. Some of the agent platforms are no longer maintained and others continue to be used.

There is a variety of software for multiagent systems that are either commercial or open-source [29]. Also, there is, in Wikipedia, comparative analysis of some basic characteristics of multiagent systems [36].

One of the basic characteristics is reviewed in the literature sources for comparison of multiagent software [29,37–39]. Concerning the development of a multiagent system appropriate for the field is reviewed in the following criteria:

- Programming language for the creation of the specific MAS
- Standards requirements supported by the MAS
- Tools for communication
- Mobility of the agents—strong (possibility of the system to migrate code or condition of execution from the code); weak (migration only of the code)—clean and effective method for migration, threads must be recreated or restarted from waiting daemon.
- Transport of the messages
- Security
- Accessibility
- Model
- Elements
- Level of activity
- GUI tools
- Specifics

5.5.2.1 Aglets multiagent platform

Aglets (http://aglets.sourceforge.net/) was created by IBM in 1997 and is open-source since 2001, which is a very popular mobile agent platform. But there is no new release until 2004, so the future is not clear. One of the advantages is that it follows MASIF specification. Aglet is created from the one thread model of agents and a communication infrastructure based on the transmission of message [38].

- Java
- Integrated standard MASIF works with CORBA
- Sockets: exchange of messages, maintain proxy, but not dynamic proxy; maintain synchronous and asynchronous communication
- Weak mobility
- Asynchronous transmission of messages; during synchronous transmission of messages deadlock is possible
- Basic security mechanism: security during inter-platform communication
- Open source from IBM
- Events model
- Elements: contexts, Aglets, and Tahiti
- Level of activity—very weak
- GUI tools available
- ATP (maintain HTTP sets, problems with firewall; one-way without sending); initialization

5.5.2.2 Voyager multiagent platform

Voyager (http://www.recursionsw.com/) was created in the beginning from ObjectSpace in 1997 and now from Recursion Software, the computing platform is distributed, that is directed mainly in simplification of the management of remote communications of the traditional CORBA and RMI protocols [38].

- Java
- CORBA and RMI protocols
- Maintain proxy; doesn't maintain messages
- Strong mobility
- Synchronous and asynchronous communication
- Security mechanism (management of the security and so on)
- Commercial product: paid
- Procedural model
- Elements: servers; agents;
- Level of activity: high;
- No available GUI tools
- Features: multicast; publish/ recording; and dynamic aggregation

5.5.2.3 Grasshopper multiagent platform

Grasshopper is developed from IKV++ in 1999 and then it became a part of the commercial Enago Mobile and now its development is left [10].

- Java
- Maintain the following standards: MASIF, FIPA, and CORBA working (VisiBroker and Orbix)
- Perform weak mobility with the ability to maintain strong mobility
- Communication: ACL, synchrony, asynchrony, dynamic, multicast communication, and different transport protocols (sockets, RMI, and IIOP)
- Basic security: policy for security: external (X.509, SSL-confidentiality, integrity of the data, and general certificate) and internal security (based on the mechanism provided from JDK)
- The product is not available
- Procedural model
- Elements-: places, regions, and agents
- No level of activity
- There are graphical tools
- Features: MASIF, FIPA, and multicast

5.5.2.4 JADE platform

JADE is developed from Telecom Italia Lab in July 1998 and is open source from February 2000. This is a very popular FIPA helpful agent platform. The agent consists of different concurrent behavior that can be added dynamically. One of the benefits is that there is a wide range of tools and can be integrated with other software such as Jess (rules generator). It must be mentioned that it maintains the development for the representation of knowledge of the agents [38]. Furthermore, it is applied to the full FIPA communication model. There are software instruments for the correction of mistakes during development. Configuration could be changed during implementation by moving agents from one computer to another [1].

Mobility is not a key element from JADE and maintains mobility between containers.

JADE maintains mobility between containers in the same JADE platform (like the idea for containers in Grasshopper) [38].

- Java
- Integrated standards: FIPA, works with CORBA (Orbacus)
- Communication: ACL, maintains for interplatform messages with the plug-ins MTPs (RMI, IIOP, HTTP, and WAP), ACL XML codec for messages
- Mobility: weak mobility
- Policy for security: Object manager in Jade provides validation of the connection, customer validation, and RPC encryption of the messages. JADE socket proxy agent holds like bi-directional way between the JADE platform and the TCP/IP connection
- Open source platform
- Model: behavior
- Elements: containers, basic containers, platforms, agents, DF, AMS, and MTS
- Level of activity: high
- Available GUI tools
- Features: FIPA, JESS, JADEX, maintains ontologies

Quality estimation compared to other platforms shows the advantages of the JADE Aj(Amax) = 33, Aj/Amax. 100, and % = 84.6%. Comparative analysis of the existing mobile agent platforms that are most famous define that SPRING have the best parameters [40], but one of the basic disadvantages is that it does not support FIPA standard. Concerning to the report [40], it is defined as more appropriate for the choice of mobile platform Glasshoper, JADE, and AGLETS, as it follows the line of order. But for Glasshoper, it must be added FIPA, so that it is able to use it and AGLETS does not maintain FIPA standard. JADE platform is very well documented and has strong support from the industry and also a broad user community [39]. Practical usage of JADE: a lot of universities and companies including INRIA, Nice-Sophia-Antipolis, ACACIA research team, ATOS Sophia Antipolis agency in the European COMMA project, KOD project IST-12503, CSELT, KPN and Starlab in DICEMAN project, and business and technology research laboratories.

JADE

JADE, the multiagent system, provides the application that will develop the possibility to work in nonpredictably changed web environment. JADE provides efficient and standardized way for mutual cooperation and coordinates the work of the agents through the exchange of standardized ACL messages. JADE clearly differentiates the process for building from the process for using the profile. JADE gives relative autonomy to the system for building the profile and in the same time takes and controls all nondeterministic and time and resource consuming operations and in this way provides effective work of the search system and the possibility to react and return the found results in acceptable consumer time. Basic disadvantage in

JADE is that it use ontologies only for providing a common dictionary in the process of communication between the agents and cannot automatically develop ontologies. So that to be able to use JADE for dynamic creation of web ontologies, it is important to put extension of the multiagent platform that adds the possibility to work with protégé ontologies [41]. Each container in JADE includes directory to facilitate DF agent, remote monitoring interface (RMI), and agent management system (AMS). In this way, agents can find other agents dynamically using DF agent and can communicate with each other using peer to peer. JADE agents communicate using asynchronous transmission of messages that is mostly widespread for distributed and free connected communications. JADE defends security of the agents who provide a serious certified mechanism to confirm the rights of each agent. Messages that are exchanged between agents use agent communication language (ACL) defined by FIPA. Even so, JADE maintains the implementation of multiple parallel tasks with one thread. This characteristic maintains the resource limit of the environment. This maintains agent mobility that let the agent transfer her code and also the condition of the remote hosts. Mobility in JADE is well known as "not-so-weak" as steak and program counters are not able to be saved in Java. JADE has integrated security features. JADE object manager provides connection, validation of the customer, and RPC encryption of the message. JADE socket proxy agent maintains a bidirectional entrance between the JADE platform and the common TCP/IP connection [38].

JADE maintains a protocol contract net to facilitate a difficult multiagent application. JADE can manipulate internal security with other agents using standard FIPA.

5.6 Conceptual schema of the proposed searching system

The conceptual schema defined in Figure 5.4 can find relevant medical information that corresponds to the customer health problem [42–44]. That conceptual schema is presented in two basic parts. The first part includes the definition of the query that is requested from the user with personalization. And the second part is the mechanism of searching information from Internet sources when using a mobile coordinating agent. Agents represent software elements that have their goals, and can operate autonomously in a particular environment and can have communication with another agent or group of agents. Using agent-based programming has several advantages. The agent can communicate with the environment that leads to update the user preference dynamically. In addition, the system requires a lot of interactions with the user and the system to define the query and return appropriate results. Furthermore, the agents are recommended to be used because of their effective communication and level of abstraction. In the proposed conceptual schema, the user first creates a profile where he can add common information, medical information, and health condition and after that, the user enters text questions into the interface. After that, the user request is modified with the content of the user profile so that the results become personalized concerning the user request.

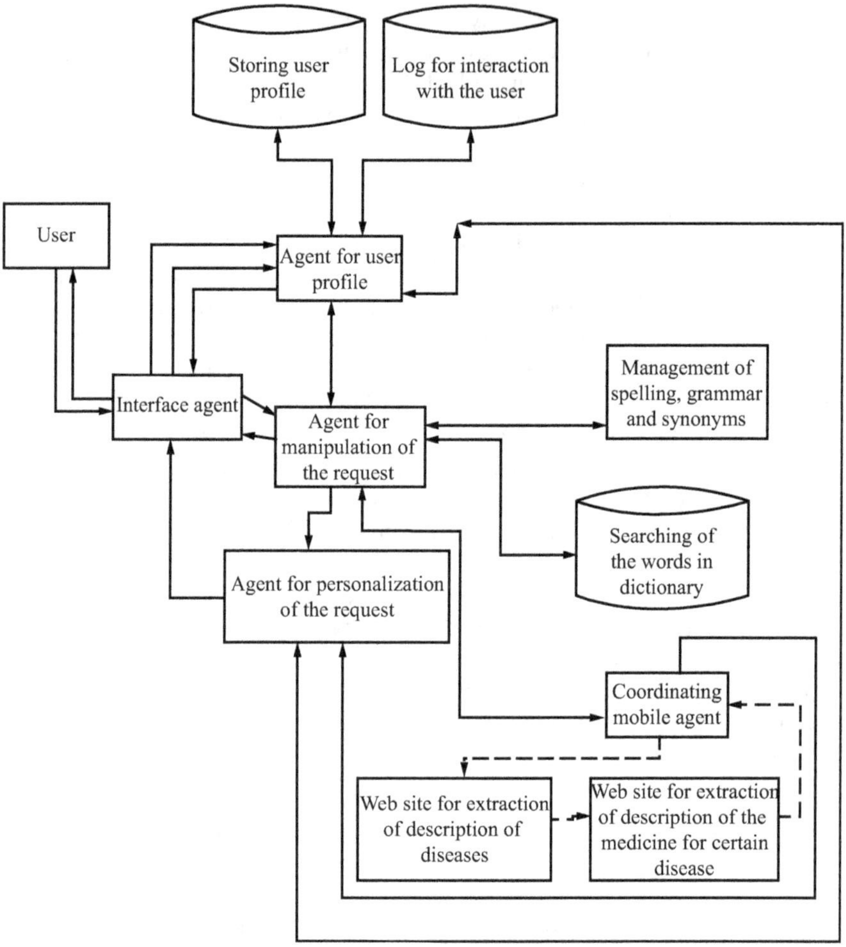

Figure 5.4 Conceptual schema of the proposed searching system with mobile agent

The conceptual scheme consists of five agents. The interface agent that interacts with the user, agent for user profile that captures and maintains user preferences, agent for manipulation of the query, agent for personalization of the returned results, and a mobile agent that searches information on the websites. In Figure 5.4, the conceptual scheme is shown that is proposed in this chapter.

1. Interface agent

   The interface agent makes the interaction with the user to retrieve the user input and display the retrieved results. The entered data can have the preferences explicitly entered from the user. These preferences can be transmitted to the user profile agent to update the user profile. The entered data could be the user requests that are transmitted to the agent that defines the request to perform the relevant adjustment of the query. Agent for manipulation of the query will communicate back to the user, whether it is necessary to review the user query and get any missing information or spell correction. Moreover, the interface agent shows the user results and formulates the personalized user results. User interactions may be either explicit or implicit. Explicit interactions can be captured by asking the user about his feedback on the results, while implicit interactions can be captured by monitoring user behavior of results.

2. User profile agent

   Agent of the user profile needs to manage the profile. The user is asked to fill out a form that reflects user preferences. Although many users do not like it, they need to take time to fill in such forms. So that it helps to process the results for the personalization of the results from the agent.

3. Agent for manipulation of the results

   Agent manipulates the results which is required to enrich and process the user request. After receiving the user request, the agent determines the language because each language has its own syntax and processing. The word request is then divided and then the spelling of the query is checked using spell check and synonym manager services. Spellcheck and synonyms manager provides two basic services. One of them verifies the spelling and the other provides synonyms for each term. After that, the query is classified using the word search service in the dictionary. The dictionary for searching words is a predefined term storage that helps to identify stop words and the relationship between the terms. This is a multilingual vocabulary where there is a list of terms in each language that is used to search for and match the terms of the query. After that, the unnecessary words are filtered using a word search dictionary that has a list of redundant words such as- between, do, on, etc. Other terms are then identified as possible links between these terms, using a Word Search Dictionary that matches predefined terms. The agent also gets the synonyms of the spell checker terms and the synonym manager and links them to the terms of the user request. After that, the information is submitted to the mobile agent to find the relevant terms in the request on the relevant web site with information.

   The information gathered from the mobile agent sends the data to the agent to personalize the results. If the user request does not match, then the agent reconsiders the request using the user profile. The profile is retrieved by the agent for two reasons:

   1. To reconsider the request
   2. To add more user information to the request

   Finally, the application annotation is produced and sent to the coordinator agent.

4. Agent for personalization of the results

   The agent responsible for personalization of the results is important for the returned results to the user. Firstly, it receives an annotated request from the agent for manipulation of the query and then applies the appropriate algorithm to extract the appropriate result of the requested query. In this agent, results are further processed by allocating for conflicts, aggregating similar results, ranking, and sorting after receiving the user preferences taken by the profile agent. Finally, the results are personalized and sent to the interface agent to display to the end user.

5. Coordinating mobile agent

   The function of the mobile agent is to get the requested query from the user and o find the relevant information on the web for this request from the user and then to send this information to the agent for the personalization of the request. The goal of the mobile agent is to get the updated information about a user request and to search for relevant information from websites. After the agent for modification of the request receives a message, it searches to send the request to the coordinating mobile agent. The coordinated mobile agent retrieves the desired categories from the content of the request and searches for all the locations in those categories. Once they are discovered, the movement begins from one place to another. Each place has a corresponding web site, so when the agent arrives at the site, it starts collecting the data from the relevant page and gathers them in a database and then moves to the next place. This action is repeated until all places are visited and then the mobile coordinating agent informs the agent for modification of the query that the data has been collected.

5.6.1 Searching with a personalized search system of information on the web when using coordinating mobile agent

Functionality of the agent includes learning, planning, and searching for the current information in the Internet. Collecting information is a difficult process depending on the type of information that needs to be collected, but researchers try to improve current methods or even find new ones. The coordinating mobile agent accepts the user request from the agent, modifies the query, finds the required category, and transfers the request to the appropriate web agent and that should be used if the agent was not mobile. This diagram is shown in Figure 5.5. In Figure 5.6, a schema is shown with coordinating mobile agent that moves through all locations in these categories. When the agent arrives in a definite place, the mobile agent extracts all the requested information from the relevant web page. The scheme with the mobile coordinator agent is shown in Figure 5.6.

Using a coordinating mobile agent saves the creation of web agents when making the system model simpler.

5.6.2 Presentation of the static and mobile agent when searching for information on the Internet

The sniffer agent in JADE allows real-time visualization about the agent interaction to solve a problem. This sequence is described in Figure 5.7 and it presents the search of information on the Internet when using static agent.

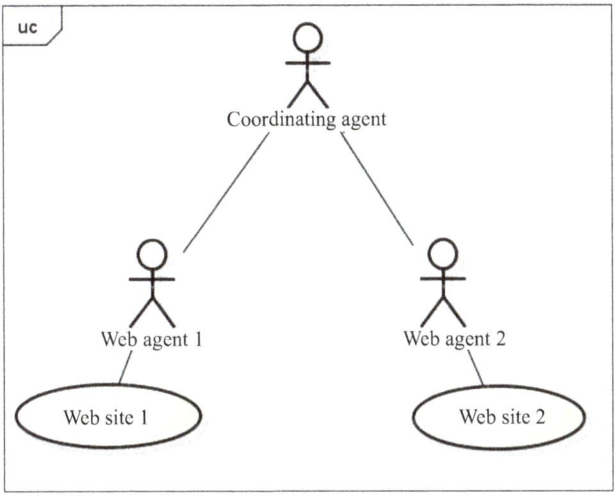

Figure 5.5 Collecting information with static agent

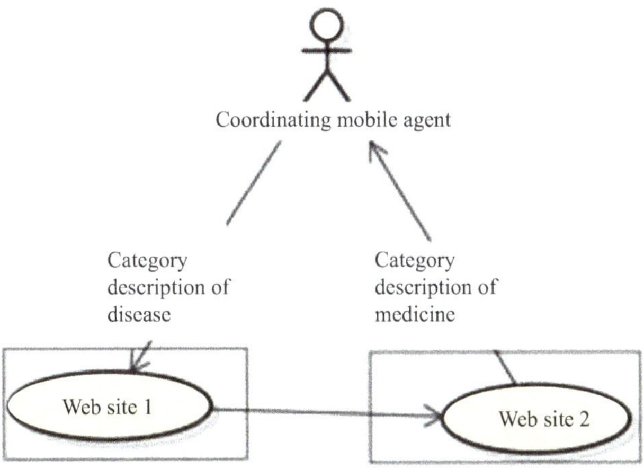

Figure 5.6 Collecting information with mobile agent

When using a mobile agent, its functionality replaces that of the coordinating agent and web agents, making the scheme simpler. As it could be observed in Figure 5.7, when the agent for modification of the request accepts a request from the user. It searches for an agent that offers the coordination service through the request message and receives a response by informing the communication activity.

After that, it sends a request to a coordinating agent who will ask for the agents located in the desired categories. And then it sends a request to a coordinating agent who will ask for the agents located in the desired categories. Once found, the

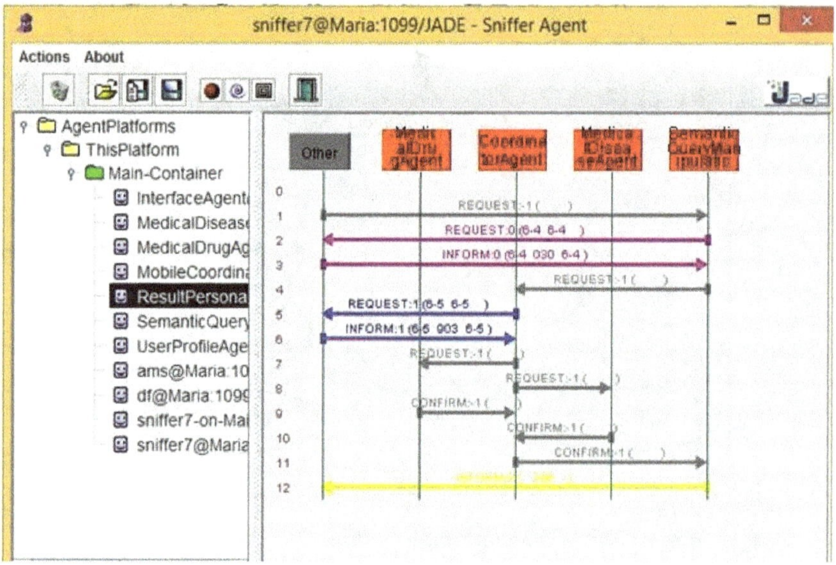

Figure 5.7 The sniffer agent tracks the sequence of the communication between the agents when using static agent

coordinator will send requests to those funds awaiting their confirmation. When all of them have confirmed that they will send a confirmation message to the agent for modification of the request and there is the possibility to learn when the request is completed.

5.7 Implementation of the searching system in Internet

The application is implemented using Java technologies such as JADE and HtmlUnit. The JADE function for agent execution is used. During startup, each agent registers at AMS and then registers his service with DF yellow pages. The services are distinguished by their description so that each web agent is registered with the name of his category, and the agents register with their services. In this way, agents can easily find interaction and communication to achieve their goals.

When registering agent services, "type" is the type of the agent service and "name" is the name of the service. The mobile coordinating agent can register as a "coordinator." The mobile coordinator agent replaces the role of the web agent (using a static coordinating agent) and retrieves information from a specific web page. This process is done using the getResults() method of the class of the corresponding web page.

This method uses HtmlUnit class libraries to fill the search form of the page, click the send button, and then analyze the resulting HTML page to collect the data.

The JADE platform is used to written the following code in the JAVA language. JADE demonstrates the algorithm that searches for the www.medicine.com disease description site.

The collection of data is presented using HtmlUnit with the following code:

```
public void getFinalres(){
WebClient1 webClient1= new WebClient1();
webClient1.setJavaScriptEnabled(false);
HtmlPage1 pages=null;
try{
pages= webClient1.getPage(webSite1);
}catch(Exception el){
el.printStackTrace();
}
HtmlForm1 forms=(HtmlForm1)pages.getElemById("search_fields");
HTMLElement1 destTB=forms.getElementById("DiseaseSearch");
HtmlElement1 checkbox= forms.getElemById("ViewList");
HtmlElement1 submitBut-ton=forms.getElemById("submitButton");
destinationTB.setAttribute("value", disease);
HtmlPage1 finalPage=null;
try{
finalPage=submitButton1.click();
}catch(IOException el)
{
el.printStackTrace();
}
}
```

In the same way, information about the medicine for the relevant disease is also collected from http://www.drugs.com/.

The coordinating mobile agent combines the role of the coordinating static agent and web agent. The coordinating mobile agent first finds all available places in the platform through GetAvailableLocations behavior by submitting a request to the agent management service.

After receiving the request, he will filter the list of places and choose only those who need to use the method for filtering the locations, and then start moving to the websites using the move() method. The method that is responsible after the movement tells the agent what to do after the agent arrives.

Depending on the place where it arrives, it will begin collecting data from the web site. This is repeated until all places are visited and he informs the agent handling the request that he has to complete the task.

5.8 Implementation of the intelligent system in Internet

In order to justify the use of mobile agent when searching for information, a short comparative analysis with results is presented. This is shown in Table 5.1. These

Table 5.1 Comparative results

	Searching with static agent	**Searching with mobile agent**
Time for response	80544 ms	71123 ms

results are obtained by performing the searching of data in the Internet with a mobile or static agent. For system testing, a computer with the following system configuration is used: Intel Core i7-10510U @2.20GHz, 16GB RAM. A detailed description of the results obtained is presented in the following paragraphs.

The total time it takes to get a response from static agent architecture will be the maximum time it takes for a web agent to collect all the data from the web page plus the time for agent communication. Compared to this, the total time required for mobile agent architecture will be the sum of the times needed to collect the data from each web page. The time required for communication is reduced in mobile agent architecture because of the reduced number of agents involved.

At first glance, it seems obvious that the system works better with static agents, but actually when making analysis it observed different results. The larger number of agents running the system at the same time on the same platform, even those waiting, presents a large number of computer processes that need to be handled, which makes it harder for agents to work and filter web pages when looking for the desired information. This aspect makes the period of time used to collect data grows, resulting in a slower system. Based on these results, mobile agent architecture is more preferred because of the need for resource management. Another advantage that can be added is the security of the communication channel communicated by the agents. The test results can be changed when the application of a different system configuration is started at another speed of the Internet connection. So the mobile agent architecture approach has advantages in the current context: the way data are collected in combination with the system, but when information gathering is improved on the side of the information, there is a good chance that a static agent approach is better.

5.9 Conclusion

This chapter describes an intelligent system with a mobile agent for communication of medical information. The system after testing gave qualified returned results to the user compared with the other intelligent systems for information of medicine for a reasonable time when using a mobile agent. The time for searching is very important when working with big data. Nowadays, more and more people, professionals and nonprofessionals, are searching for medical information in the Internet and the proposed intelligent mobile agent system will help them to select the appropriate medical information they need.

References

[1] Mavromoustakis, C.X., Pallis, E., and Mastorakis, G. *Resource management in mobile computing environments*. Vol. 3. Modeling and Optimization in Science and Technologies Book Series. Cham: Springer; 2014. p. 642.

[2] Taiwo, K. "Dimensions and issues of mobile agent technology." *Villanova Journal of Science, Technology and Management, Nigeria*. 2019; 72–86.

[3] Al-Shayea, T. K., Mavromoustakis, C. X., Batalla, J. M., and Mastorakis, G. "A hybridized methodology of different wavelet transformations targeting medical images in IoT infrastructure." *Measurement*. 2019. p. 148.

[4] Al-Shayea, T. K., Batalla, J. M., Mavromoustakis, C. X., and Mastorakis, G. "Embedded dynamic modification for efficient watermarking using different medical inputs in IoT." *24th International Workshop on Computer Aided Modeling and Design*. IEEE; 2019.

[5] Al-Shayea, T. K., Mavromoustakis, C. X., Batalla, J. M., Mastorakis, G., and Pallis, E. "Medical image watermarking in four levels decomposition of DWT using multiple wavelets in IoT emergence." *Convergence of artificial intelligence and the Internet of things*. India: Springer; 2020. pp. 15–31.

[6] Qasim, A., Ameen, H. M. B., Aziz, Z., and Khalid, A. "Efficient performative actions for e-commerce agents." *Applied Computer Systems*, 2020; 25: 19–32.

[7] Han, Y., Zhiqi, S., Cyril, L., Chunyan, M., and Victor, L. *A survey of multi-agent trust management systems. IEEE Access*. 2013; 1: 35–50.

[8] Godfrey W., Jha S., and Nair S. "On a mobile agent framework for an internet of things." *International Conference on Communication Systems and Network Technologies*. India: IEEE; 2013. pp. 345–350.

[9] Toumbas, L., Mavromoustakis, C. X., Mastorakis, G., and Dobre, C. "Recent advances in remote assisted medical operations: Ambient assisted living and enhanced living environments." *Principles, Technologies and Control*. Elsevier Inc.; 2016. pp. 123–145.

[10] Dasgupta S., Aroor A., Shen F., and Lee Y. "Smartspace: Multiagent based distributed platform for semantic service discovery." *IEEE Transactions on Systems, Man, and Cybernetics:Systems*. 2014; 44(7): 805–821.

[11] Gao Q., and Cho Y. "A multi-agent personalized ontology profile based query refinement approach for information retrieval." *13th International Conference on Control, Automation and Systems*. South Korea: IEEE; 2013. pp. 543–547.

[12] Raja, A., and Raj, E. "MAD-ARM: Mobile agent based distributed association rule mining." *International Conference on Computer Communication and Informatics*. India: IEEE; 2013. pp. 1–5.

[13] Higashino, M., Hayakawa, T., Takahashi, K., Kawamura, T., and Sugahara, K. "Management of streaming multimedia content using mobile agent technology on pure P2P-based distributed e-learning system." *27th International Conference on Advanced Information Networking and Applications*. Spain: IEEE; 2013. pp. 1041–1047.

[14] Vidyalakshmi B., Raymond, W., and Chi-Hung, C. "Decentralized trust driven access control for mobile content sharing." *International Congress on Big Data*. USA: IEEE; 2013. pp. 239–246.

[15] Adel, B., Benatchba, K., and Balla, A. "DA5DCSWS: A distributed architecture for semantic web services discovery and composition." *8th International Conference for Internet Technology and Secured Transactions*. UK: IEEE; 2013. pp.182–187.

[16] Essa, Y., Attiya G., and Ayman, E.-S. "Mobile agent based new framework for improving big data analysis." *International Conference on Cloud Computing and Big Data*. China: IEEE; 2013. pp. 381–386.

[17] Sakarkar, G., Deshpande, S., and Thakare, V. "Comparative study of user's profile in web based systems." *International Journal of Computer Engineering and Applications*. 2015; 171–183.

[18] Morente-Molinera, J., Pérez, I., Ureña, M., and Herrera-Viedma, E. "Building and managing fuzzy ontologies with heterogeneous linguistic information." *Journal of Knowledge-Based Systems*. 2015; 88: 154–164.

[19] Mchamundeeswari, T. M. "Agent based personalized semantic web information retrieval system." *International Journal of Advanced Computer Science and Applications*. 2014; 5(8): 103–110.

[20] Zhang, X., Xin-Ya, G., Qian, Z., and Jia, J. "Research on the rough extension of ontology description language of SWRL." *Journal of Electrical and Computer Engineering*. 2016; (1): 1–8.

[21] Zhang, L. "OWL-S based web service discovery in distributed system." *IEEE Workshop on Electronics, Computer and Applications*. Canada: IEEE; 2014. pp. 882–885.

[22] Russell, S., and Norvig, P. *Artificial intelligence: A modern approach*. Upper Saddle River, NJ: Prentice Hall Press; 2009.

[23] Gayathri, J., and Chandar, S. "Personalized information retrieval using agent based approach." *International Journal On Engineering Technology and Sciences*. 2015; 2(1): 7–12.

[24] Aundhakar, S., and Pokale, N. "A survey on user personalization using ontology." *International Journal of Science and Research*. 2015; 4(1): 696–699.

[25] Mohd, A., and Ansari, M. "A survey: Soft computing in intelligent information retrieval systems." *12th International Conference on Computational Science and Its Applications*. Brazil: IEEE; 2012. pp. 26–34.

[26] Maeva, A., Laurent, P., Fabrice, H., and Francois, B. "A generic API for load balancing in structured P2P systems. *26th International Symposium on Computer Architecture and High Performance Computing Workshops*. France: IEEE; 2014. pp. 138–143.

[27] Ilya, R., and Ivan, J. "Semantic P2P search engine." *Proceedings of the Federated Conference on Computer Science and Information Systems*. Poland: IEEE; 2011. pp. 991–995.

[28] Raju, Y., and Babu, D. "A novel approaches in web mining techniques in case of web personalization." *International Journal of Research in Computer Applications and Robotics*. 2015; 3(2): 6–12.

[29] Brian, B., Robert, G., Katsuhiro, M., David, K., George, G. C., and Daniela, R. "Mobile agents in distributed information retrieval." *Intelligent information agents*. Heidelberg: Springer-Verlag; 1999. Chapter 15. pp. 355–395.

[30] Hui, L., and Xin, L. "Challenges and trends of big data analytics." *Ninth International Conference on P2P, Parallel, Grid, Cloud and Internet Computing*. China: IEEE; 2014. pp. 566–567.

[31] Xi, C., Huajun, C., Ningyu, Z., Jiaoyan, C., and Zhaohui, W. "OWL reasoning over big biomedical data." *International Conference on Big Data*. USA: IEEE; 2013. pp. 29–36.

[32] Jerome, J. "Big data: Catalyst for a privacy conversation." *Indiana Law Review*. 2014; 48(1): 213–242.

[33] Maryam, P., Vahid, T., Ashutosh, J., and Jyotishman, P. "Empowering personalized medicine with big data and semantic web technology: Promises, challenges, and use cases." *Proceeding of IEEE International Conference Big Data*. IEEE. 2014.

[34] Daniel, L., and Trevor, M. "Managing vagueness with fuzzy in hierarchical big data." *Journal of Procedia Computer Science*. 2015; 53: 19–28.

[35] Xindong, W., Xingquan, Z., Gong-Qing, W., and Wei, D. "Data mining with big data." *IEEE Transactions on Knowledge and Data Engineering*. 2014; 26(1): 97–107.

[36] Comparison of agent-based modeling software. Available from http://en.wikipedia.org/wiki/Comparison_of_agent-based_modeling_software [Accessed 12 September 2018].

[37] Naik, Y.M., Tarihal, S., Swami, R., Purandare, A., and Adike, K. "Distributed information retrieval using mobile agent". *American International Journal of Research in Science, Technology, Engineering & Mathematics*. 2013. pp. 355–395.

[38] Trillo, R., Ilarri, S., and Mena, E. "Comparison and performance evaluation of mobile agent platforms." *Autonomic and Autonomous Systems*. Greece: IEEE; 2007. p. 41.

[39] Gupta, R., and Kansal, G. "A survey on comparative study of mobile agent platforms." *International Journal of Engineering Science & Technology*. 2011; 3(3): 1943.

[40] Nguyen, G., Dang, T. T, Hluchy, L., Laclavik, M., Balogh, Z., and Budinska, I. "Agent platform evaluation and comparison." *Institute of Informatics*. Slovak. 2002. Available from www.pellucid.ui.sav.sk.

[41] Ivanova, T. "Agent development platforms -tools, languages, methodologies and application areas." *Third International Bulgarian-Turkish Conference "Computer Science'06"*. Istanbul; 2006. pp. 381–387.

[42] Ahmed, A.-N., and Tarek, H. "Al-mulhem mohammed, 'user's profile ontology-based semantic framework for personalized food and nutrition recommendation."*5th International Conference on Ambient Systems, Networks and Technologies*. Vol. 32, Elsevier; 2014. pp. 101–108.

[43] Loukia, K., Isambo, K., and Sally, M. C. "Uncertainty reasoning for the "big data" semantic Web." *15th International Conference on Information Reuse and Integration*. USA: IEEE; 2014. pp. 147–154.

[44] Fernando, B., and Umberto, S. "Optimising fuzzy description logic reasoners with general concept inclusion absorption." *Journal of Fuzzy Sets and Systems*. 2016; 292: 98–129.

Chapter 6

Intelligent Internet of things in wireless networks

Mona Bakri Hassan[1], Elmustafa Sayed Ali[1,2] and Rashid A. Saeed[1,3]

Nowadays, intelligence Internet of things (IoT) wireless networks are the most promising technologies for intelligent future applications, since it takes the advantages of artificial intelligence for sensing and data analysis to develop a new generation of smarter IoT applications around everywhere. Next-generation wireless networks are expected to support extremely high data rates and radically new applications, which require a new wireless radio technology paradigm. The challenge is that of assisting the radio in intelligent adaptive learning and decision-making so that the diverse requirements of next-generation wireless networks can be satisfied. Machine learning is one of the most promising artificial intelligence tools, conceived to support smart radio terminals. The main challenge for future IoT networks is how to cope with such complex systems, where a huge number of devices compete for limited wireless resources and where heterogeneity is ever-increasing. There is an urgent need for more intelligent networks that lead to more interoperable solutions and that can make autonomous decisions on optimal operation modes and configurations. In IoT wireless networks, the AI can insure to collect only adequate data in order to reduce the processing time of big data amount sent by sensors. To manage big data and store it in IoT networks, AI processes play a major role in data integration, selection, classification, and mining, as well as assessment of information patterns. It offers unique solutions in decision-making, measurement, and contribution to the construction of IoT networks with a new, fast, and intelligent character. In this chapter, we will provide intelligence IoT wireless networks in addition to the AI contribution in programming and configuring of IoT network devices. This chapter will also introduce the algorithms and strategy of intelligence related to cognitive IoT networks and quality of service, in addition to the benefits of cognitive and SDN to IoT operations in heterogeneous networks.

[1]Department of Electronics Engineering, Sudan University of Science and Technology (SUST), Khartoum, Sudan
[2]Department of Electrical and Electronics Engineering, Red Sea University (RSU), Portsudan, Sudan
[3]Department of Computer Engineering, Taif University (TU), Al-Taif, Saudi Arabia

6.1 Introduction

Low latency and efficient spectrum have gained significant demand recently for ultra-high data rates intelligent applications. Decision-making techniques for intelligent learning are very important due to the requirements of intelligent communications because of the diverse needs of wireless networks. The use of machine learning will provide artificial intelligence tools to support radio systems for 5G networks besides the use of cognitive radios, UM-MIMO in small cells, energy harvesting, and heterogeneous networks which are required in an intelligent wireless network for future applications [1]. In the Internet of things (IoT) wireless networks, the coexistence of different network technologies are major considerations especially in IoT-based 5G networks. This coexistence will flood a massive data generated by a large number of IoT devices which require to make efficient decisions for handling these big data. Artificial intelligence (AI) can provide a means of technologies able to analyze the data to extract patterns and make sense of the data to prescribe action to the end devices [2].

Intelligent technologies in future IoT-based cellular networks can help to let networks able to be self-organized and learn from the environment in order to make decisions in real-time by using different kinds of AI technologies such as game theory, machine learning, and optimization algorithms. Spectrum management is also requiring an intelligent mechanism to manage the exchange of a huge generation of data on the communication channel and manage bandwidth consumption specially in the IoT end devices using fog computing [2,3]. As an example, in NB-IoT cellular-based radio access technology, the configurations of selecting uplink resource in NB-IoT requires a real-time dynamic traffic adjustment through selecting the configurations at the Evolved Node B (eNB) in the multiple CE groups scenarios. The use of the reinforcement learning (RL) technique can provide a solution for such requirements by enabling an automatic updating to the uplink resource configuration based on the communication procedures in NB-IoT [3]. Automotive machines and robotics can work in IoT networks through an artificial intelligence approach to create a model of self-learning behavior without constant human supervision. Artificial intelligence makes machines and robotics acting independently alongside Industrial IoT (IIoT) applications, promising to develop new IoT models that enable the transition from traditional sensor networks to the future networks which are consist of smart sensors with automated mechanisms [4]. It is also considered as a step in the development of communication networks to create mechanisms that enable access to any place smartly through the Internet with independent decisions in the management in addition to sending decisions to decision-making centers, by giving intelligence to smart sensors making them the ability to act in an intelligent way enabling IIoT to better respond to critical situations in real-time.

6.2 IoT networks

Nowadays, there is an extensive access to the Internet and millions of people are engaged in online social networks in a regular way. In addition, smart devices and sensors are becoming a part of people's lives for performing intelligent tasks. Intelligent IoT (IIoT) networks should able to add a new paradigm to add intelligence to the things that are connected to the Internet. Furthermore, intelligent objects can act as a service and be discovered through the network due to smart adaption between users. Moreover, the interaction between humans and robots will become a more real mixture by intelligent mechanisms that reside in the IoT networks [5]. The IoT networks need to be developed by the context-awareness concept to make the smart sensors and devices to determine their context. Context itself is a means to make sensors intelligent and develop automated supervised and sensing systems in the IoT. By this means, sensors in IoT systems can able to understand the surroundings and can able to be adapt and improve their functionality [6]. This will be done intelligently performing the interconnection between massive sensory devices to do the fusion between the physical world and the data world. It is considered that IoT architectures accommodate the large and fast data processing requirements for extracting deep insights from data using cognitive computing capabilities in IoT architectures. According to the IoT network architecture, the application layer must recognize IoT to create a smart environment such as smart buildings, smart home, smart health, and smart industry. In addition, this layer of application ensures data integrity, authenticity, and confidentiality. Intelligent mechanism should be considered in the network layer to interact with different communication technologies [7,8].

In IoT networks, the configuration of a massive number of sensors and devices manually maybe impossible and needs high operating costs which means there is a need to move forward to the intelligent configuration. Sensors and devices in IoT network can able configuring themselves due to the concept of self-configuration by using programs applying to the network which will help in maintaining the stability of the IoT network processes [9]. Another concept related to the intelligent IoT networks is self-adaptation and self-management. The IoT systems must able to adapt itself to different resources taking into consideration the system errors and charging in maintaining the process. Sensors and other objects and devices in the IoT network should be self-managed in order to calibrate its functionality by tuning some parameters to correspond to the environment where it was deployed [10]. The configurations must ensure correct functionality and management for the duration of the deployment of the devices in IoT networks. In addition, the use of decision-making functions in IoT network communications can improve its stability and throughput.

6.2.1 LPWAN IoT networks

In cellular IoT networks, several techniques are used to provide communication between devices in the IoT. Networks such as LTE and NB-IoT are used to implement the real idea of the IoT and make the interactions between different devices possible through cellular concepts. The technologies such as massive MIMO, D2D, Wireless SDN, Radio Access techniques, Green communication, big data, and mobile cloud computing are all can be used to achieve high data rate, high quality of service (QoS), high capacity, and low latency for IoT networks [11]. Resource management techniques in different communication modes of a cellular IoT network such as D2D, Het Net, NOMA, and dynamic spectrum access-based modes can able to optimize the communication session setup and management in addition to channel allocation and interference management in the IoT networks [12]. Many IoT applications require to save energy for both short- and long-range communications. Network technology such as low power wide area networks (LPWAN) (see Figure 6.1) can provide low power operation for long-distance communications by limiting the data rate and energy consumption [13].

LoRa is one of the most popular LPWA technologies that operate on unlicensed spectrum. It operates on a coverage range of 20 km, supports millions of nodes, and provides a data rate of 50 kbps. LoRa WAN technology can provide massive IoT applications and solutions to some of the largest IoT challenges on a large-scale including energy and resources management. The increasing number of LoRa devices in IoT networks leads to spectral interference which requires an intelligent coordination mechanism to manage spectrum communications, especially for unauthorized spectrum in LoRa, in addition to other application aspects such as transmission distance and number of nodes [13,14]. In LoRaWAN, protocol specifies a number of mechanisms that ensure reliable, smart, and secure communication. The use of adaptive data rate (ADR) mechanism in LoRa WAN IoT networks will dynamically manage end IoT device's link parameters in order to increase the packet delivery ratio. Other IoT technology is SigFox which can offer end to end connectivity for M2M communications with a small amount of data transmission and low bandwidth. In Sigfox, base stations can be configured with cognitive software-defined radio technique to servers for the connection operating on IP-based network architectures [15]. The Sigfox networks can operate the end

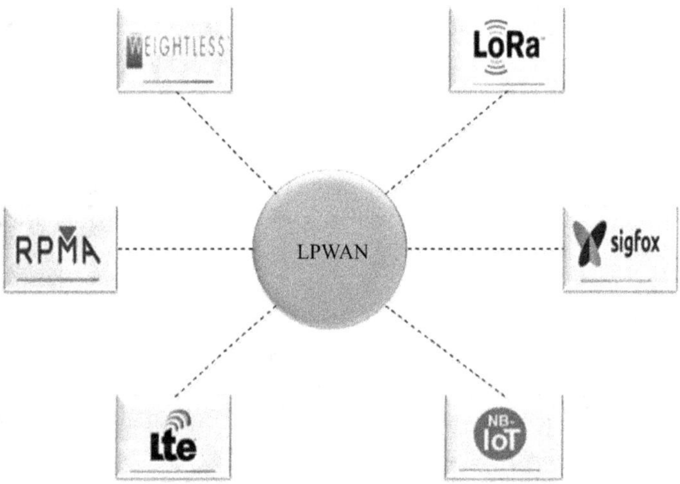

Figure 6.1 LPWAN technologies

IoT devices to the base stations enabling them to provide highly sensitive and ultra-low power consumption for long ranges which contributes to utilize the IoT in efficient and high throughputs [15,16].

The LPWAN IoT network can provide a variety of features, such as extensive coverage of IoT everywhere, fast IoT upgrading, low power consumption, battery life assurance, high coupling, high reliability, and high security for the corporate network [16]. In addition to standardized management of the business platform, narrow-band IoT (NB-IoT) perfectly matches the requirements of the LPWAN market and manages traditional businesses such as intelligent metering and tracking, and also opens additional industrial opportunities, for example, Smart City or E-Health. NB-IoT makes possible to deliver more things, but managing the commercial value of the resulting big data is an important task [17]. Weightless is a new wireless IoT technology that works in both unlicensed and licensed spectrums for different bands and low power consumption. This technology uses TV white spaces and cognitive radio technology to enable IoT devices to use these bands by opportunistic users without causing interference to the user's primary devices in TV white space network.

6.2.2 Cognitive IoT networks

In IoT wireless networks, it is important to select appropriate resources related to the network devices. Cognitive IoT network can provide a means of constantly searching for resources opportunistically that are better suited for IoT devices to boost the overall performance of the network. Cognitive IoT networks can let the IoT devices to share the allocated resources by dividing them into the primary and secondary devices and give the resources of the primary device to the secondary one when the primary is absent or not activated. When primary device is activated, the secondary one must vacate the channel resources. Moreover, it can integrate and improve performance and achieve intelligence [18]. CIoT networks are used to analyze perceived information based on prior knowledge, make intelligent decisions, and implement adaptive and control actions [18,19]. In order to apply cognitive IoT networks, there is a need for intelligent resource allocation way to consider primary devices' activity and QoS requirements as well as for secondary devices in addition to wireless propagation and network parameters like channel sensing, detection, and acquisition. The use of machine learning algorithm optimization gives the ability to improve performance based on existing semantic models to provide the system with self-learning capabilities [20]. For any cognitive IoT networks, the architecture must consist of a cognitive computing layer to produces an algorithm based on chosen feature selection according to data pre-processing, data analysis, cognitive traits extraction, and machine learning which enable to provide more solutions for IoT network applications (see Figure 6.2). Merging both cognitive traits and machine learning models can let to deliver cognitive computing.

Some IoT applications require to make devices able to move such as in the military, industrial, and human applications. Mobile IoT (MIoT) is the IoT mobility

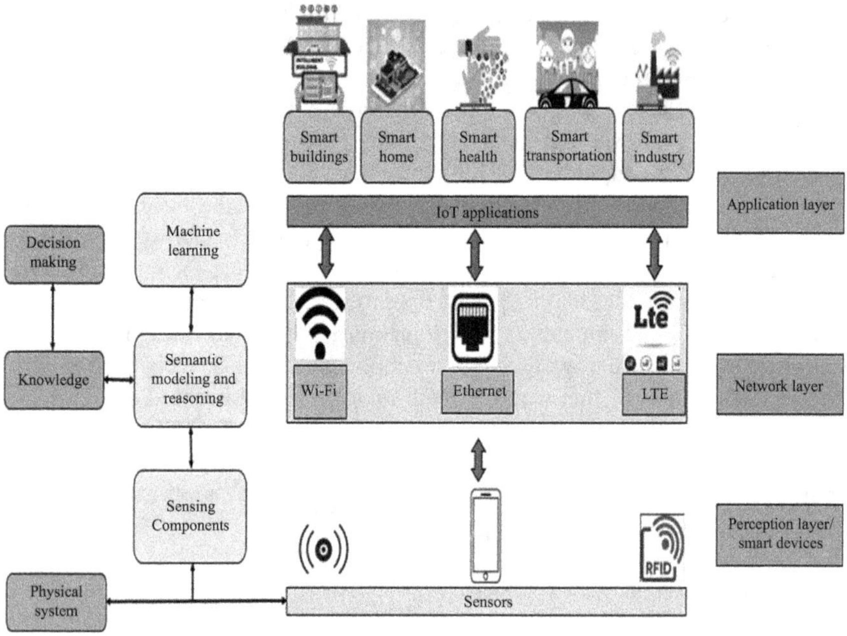

Figure 6.2 Cognitive IoT (CIoT) network architecture

that can provide a service in mobile devices and physically be transferred from one location to another. Due to the mobility of devices in IoT networks, additional control information is shared between network management entities and resource management in MIoT represents a greater challenge to be handled for IoT networks which require an intelligent management scheme. In addition to these challenges, the application context is also an important parameter to consider for resource management [21].

6.2.3 Dynamic IoT networks

Cloud computing plays an important role in the formation of IoT networks, but it can become more complex in the case of interconnection with physical systems, especially in the management of big data generated from Internet devices which is a great challenge for how to collect, process, and sense and exchange within IoT networks. One of the major challenges facing the collection and processing of information from these IoT devices is the detection and configuration of sensors and associated data flows [22]. The use of automated detection mechanisms and intelligent mapping capabilities are essential for network management. Dynamic IoT network can provide a model for detecting and configuring sensors to give it the ability to work with sensors that are operated by different communication protocols. This method assigns sensors and deployment platforms that allow software systems to retrieve data from sensor devices when needed to enable communication between the sensors, resulting in improved system performance and increased adaptation to the configuration environment during deployment [23].

6.2.4 Semantics intelligent IoT networks

Semantics provide a representation of knowledge and the use of models to create common designs, data formats, and labels to provide a homogeneous environment within the sensor network. It is able to identify scalable solution processes that operate in difficult contexts of big data, where sources can be very large and heterogeneous. Semantics for IoT networks allow the collection of important metadata from network components and use them to provide redundancy and reliability so that this information is used more to develop a more reliable network system and to create or implement concepts that improve overall network performance or intelligence [24]. In IoT networks, intelligent configuration can be done for an improved upgrade of configuration typically applied to stable and specific technological environments. It can also provide easily interchangeable information and independent processes in a way adapted to the classic configuration to meet new requirements. In addition, it also predicts the requirements for the formation of the architectural model or ontology that are used to design the IoT where the configuration can depend on the number of IoT sensors, devices heterogeneity, communication scheduling, data acquisition, and network adaptation [25].

An intelligent IoT network configuration should be able to rapidly configure a significant number of sensors autonomously and able to configure devices from different brands that usually communicate differently as most devices use different techniques for measurement and overall functioning. IoT communication is determined by the user's requirement and relative to the frequency in which the sensors need to generate data or exchange information [25]. For IoT data exchange, it is important to take correct methods of measurement result for better efficiency in addition to the capability of handling the adaptation of the network of sensors' data to the environment.

6.3 Reprogrammable and reconfigurable of IoT devices

With the demands growing of IoT networks, there is a need that IoT devices should be smaller, smarter, and increasingly connected so that they are deployed in enormous numbers according to the required application, and trusting and securing access to billions of heterogeneous devices which is a major challenge. Such networks, in addition to the large consumption of energy, require the design of a new generation of intelligent IoT devices taking into account the constraints of the budget of safety, performance, and energy [26]. In many IoT applications, a large number of sensors are deployed that may require basic programming to interact with heterogeneous resources effectively to address a large number of application requirements. Software-defined networks (SDNs) and network functions virtualization (NFV) can use effective models to provide on-demand services and manage network functions and lifecycles. Most IoT architecture maintains cloud servers as a reference, so heterogeneity of servers and IoT devices is a major obstacle to creating an integrated programmable environment. Therefore, there must be a mechanism for reprogramming and reconfiguring of IoT devices that can handle the huge number of devices and the huge data traffic. Programmable IoT devices can be used to enable devices to be scalable and reconfigurable and have the ability to be integrated into the cloud edge to support the programmability of programmable cloud sensors [27]. It can provide a service management framework for IoT devices with real-time data generated plus an unreal time management in a cloud structure.

In IoT networks, the design requirements of the IoT software platform must be taken into account to offer scalability and interoperability between heterogeneous IoT devices and their business models. In addition, providing IoT hardware technology with a mechanism of communicating with others in real-time mode needs a type of an event-driven process decentralized software architecture [28]. The service-oriented architecture (SoA) can provide scalability and interoperability based on the provision of heterogeneous technologies in a single platform and can consist of four layers as shown in Figure 6.3. The sensor layer integrated sensors of the state of objects and the network layer that connect things together and collect data from the hardware infrastructure. The service layer, which creates and manages the services requested by users or applications in addition to an interface layer that enables methods of interaction with applications or end-users [29].

The software SoA for IoT middleware must provide between objects equipped with sensors and applications for the advantage of stripping objects, service administrations, and service configuration through a secure network. In IoT networks, this control strategy can be used as an application on IoT devices that enable connection to the occupancy information database depending on the application

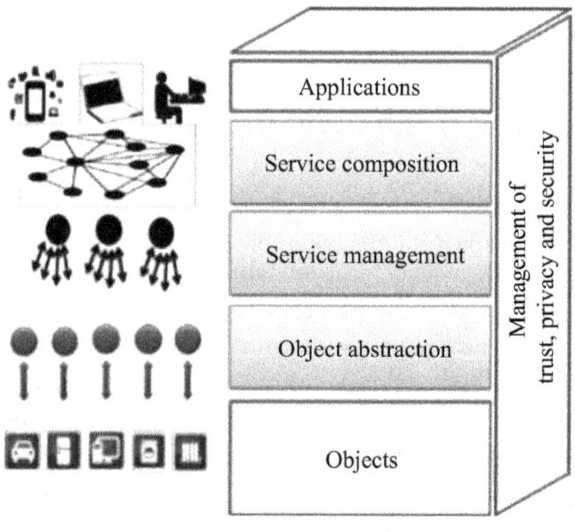

Figure 6.3 Service-oriented architecture for middleware IoT devices. Adapted from [29]

[29]. This control strategy works only with the IoT system, which enables users of different levels of authorization to control the devices as well as implement an automatic control strategy to save energy using information collected by smart devices for operation in the application environment.

6.4 Open source platforms in IoT networks

Various IoT platforms are available today to be used to develop the IoT solution. Google, IBM, Microsoft Azure, and digital service clouds are the most famous provider of IoT platforms. Open IoT platform such as digital service cloud can support IoT innovators to own their customers in a way that customers own their products so that it supports product startup, global technology brands, and product innovators. Moving to an intelligent open-source IoT platform, Zetta platform is an example of a developed platform for creating IoT servers that run across geo-distributed servers and the cloud [30]. Its combines API, reactive programming, and web sockets which are suitable for assembling many devices into data-intensive and real-time applications (see Figure 6.4). The architecture of such a platform is optimized for data-intensive and real-time applications [31], which can able to support monitoring of IoT device and system behavior in code and use visualization tools for actionable insights and provide data flow to machine analysis platforms in an intelligent way that enables to handle IoT projects which consist of multiple devices across multiple locations running multiple applications developed by various companies.

Cognitive transformation of IoT applications allows to use an optimized and intelligent solution using an immersive technology and provides intelligent activities for a wide range of IoT devices, software platforms, and services. Technologies such as virtual and augmented reality are able to make robotic things interact with one another and with IoT platform systems. Cognitive IoT will provide an embedded intelligence into systems and processes enabling the IoT devices to interact with people and other things in the network [32]. Information exchanged in the IoT networks is managed by IoT platforms that use cognitive systems that can be experienced, analyzed, and intelligently extracting the business ecosystem to generate new better services. In cognitive IoT, there is a need for integration of IoT platforms to provide edge device control for operation, communication, and device monitoring and management, in addition to the ability of firmware updates, data visualization, and security. IoT platforms are considered as an intelligent layer which is acting to let things connected to the IoT networks and abstract the applications from the things to enable the development of the network services. Moreover, platforms facilitate communications, data exchanges, and device managements in addition to application functionality [32]. The functionality of IoT platform covers the digital value chain from sensors and actuators and other system hardware to connectivity cloud and applications. Open-source IoT platforms are emerging in the customer IoT space and able to integrate the set of end devices and protocols making it closer for proprietary platform providers. New technologies

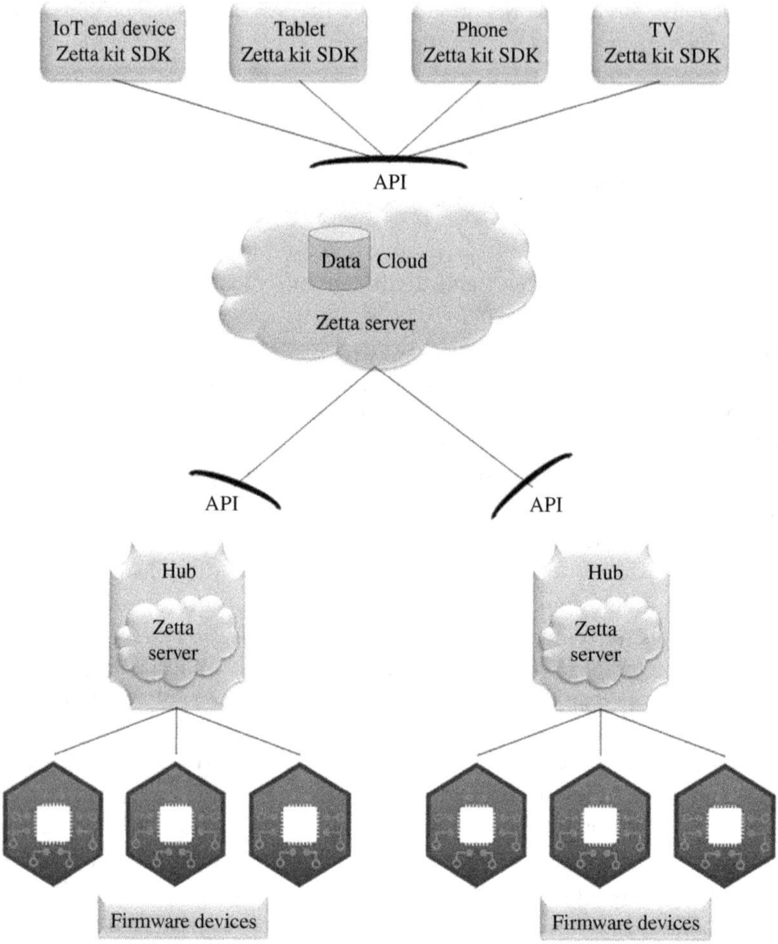

Figure 6.4 Zetta open platform deployment in IoT network

known as container-based virtualization have gained attention in recent years, and the opportunities for their use have been increasing to move from traditional IoT to intelligent IoT concepts. Open-source software (OSS) for container management such as Docker and Kubernetes are used to execute intelligent techniques to support parallel distributed processing for large-scale data models in cognitive IoT networks [33].

6.4.1 Cognitive IoT-based LPWAN

Recently, the continuous development of IoT communication technologies and the progressive maturation of AI have led to a strong cognitive computing capability. Cognitive computing in IoT network platforms plays an important role in enabling users to access intelligent services that are efficient and convenient in many intelligent applications of heterogeneous networks. LPWAN technology is used for providing efficient communication along with wide-area IoT networks with low energy consumption. The use of cognitive computing with LPWAN devices will extract the capabilities of intelligent transmission at the network level and big data analysis in the cloud. Cognition LPWAN realizes mixing a variety of LPWAN technologies and provides users with more efficient and convenient intelligent services [34]. In cognitive LPWAN, the use of cognitive engine will enable the deployment of high-performance intelligence algorithms and store a lot of user data and IoT business flows which will turn to provide high precision calculation and data analysis and provide cloud support for LPWAN communication technology selection. In cognitive IoT networks, multimedia data flow is processed in real-time in a network environment with data analysis, automated processing power in a smart way, and realization of business and resource data through a variety of cognitive calculation methods [35,36]. This includes data mining, machine learning, deep learning, and artificial intelligence. Computing, communication, and network resources of heterogeneous IoT are perceived by the cognitive resource engine in addition to edge and remote cloud to provide real-time feedback of comprehensive resource data to the cognitive data engine [37]. The use of cognitive technology with LTE-M can possibly let IoT platforms deployed in the femtocell which is a small cellular base station designed for smart home and environments IoT applications. In addition, the femtocells are more interested within the telecommunication industry due to the unique benefits they offer, both for the operators as well as the end-users. In such networks, self-optimization and auto-configuration is the main challenge for femtocell operations in addition to configuration and management. Furthermore, the optimization of the coverage of femtocells, integration, and interoperability with the core network are considered to be driven intelligently to provide high data rate services. Thus, integration and interoperability with the operator's existing network and services are important concerns for the operators [35].

For cellular IoT networks, as in NB-IoT, capabilities to perform sophisticated spectrum sensing methods over long durations are required. Therefore, it is important to reduce the spectrum sensing overhead while maximizing the NB-IoT network-level throughput [38]. Two intelligent technique can be used for such a problem: dynamic spectrum access and network management. Dynamic spectrum access technique can be used to increase spectrum efficiency in cognitive IoT LPWAN networks by real-time radio resource control through local spectrum sensing and investigation, to enable self-establishment of local wireless communications between nodes and IoT knowledge networks. This process is conceived by the radio cognition of real-time spectrum auctions between different constituencies, from which the spectrum is allocated to different devices and platforms in the IoT so that it allows to optimize the use of resources and enhance the spectrum [39].

The infrastructure of LPWAN IoT networks requires complex radios that are often very energy consuming and too expensive to sense applications that require only a few bits of data along with a wide area network. Due to these purposes, IoT platforms must meet the requirement of network management to keep enhancing the resource and power consumption and enable to simplify the design for deploying IoT devices across long-distance communications. Open Chirp Platform as an example can allow multiple devices and groups the ability to provision and manage battery-operated transducers across large area applications such as in smart cities. This open-source platform architecture (see Figure 6.5) can provide structured data with supporting meta-information and services such as a web interface and storage. The platform software architecture can expose an application layer allowing users to register devices, describe transducer properties, transfer data, and retrieve historical values in an intelligent way.

Open Chirp is built using LoRa WAN and it enables to support meta information and services such as a web interface and storage. The separation between the platform API and the internal implementation of various services can contribute to create a modular architecture that also allows us to experiment with various components of the infrastructure. The use of cognitive radio with such platform can enable to operate the controlled information and optimize services in response to dynamic changes in end-user requirements. Network management, in addition, makes use of unused and temporarily available spectrum within the licensed

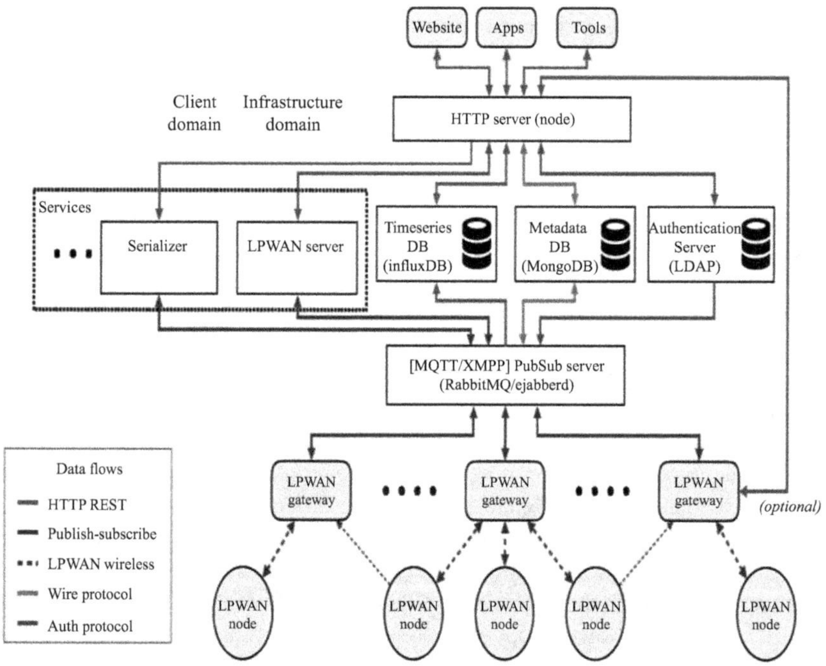

Figure 6.5 Open chirp LPWAN system architecture. Adapted from [40]

spectrum by deploying cognitive radio network (CRN) networks to take advantage of these transport opportunities through cognitive communication techniques [40].

6.4.2 SDN over IoT network

In wireless sensor networks, SDN controller will able to serve as a central collection point that receives the network information and configures the decisions for network resource behaviors. In the side of cloud computing, SDN is a principle of fog computing that enables to control data to compute the services in the edge of the networks, in addition to provide network storage and services between end devices and cloud computing data centers. The benefits gained from using the fog computing is to enabling dense geographical distribution for end-users and local resource pooling to achieve better QoS [41]. Moreover, it can use multiple cloud platforms to enable interaction between fog and cloud. For IoT networks, fog networking can support connectivity between different end-users with dense IoT devices. The concept of software-defined sensor networks (SDSNs) is based on taking the features of wireless IoT network and SDN technologies to make the intelligent IoT network nodes capable to be programmed in demand according to the uploaded application. The use of SDSNs will allow programming and easy managing of the APIs in IoT systems based on the IoT application implemented. Furthermore, the use of SDN in cellular IoT networks can offer to provide several subscribers, user mobility, and real-time adaptation for huge innovation in mobile applications.

In IoT networks, SDN architecture model will centralize control of the network by a high-level software application that allows quick network management. In addition, it can help to control the network traffic behavior without requiring an individual configuration of hardware devices. SDSN architecture model (see Figure 6.6) for IoT network will give the management control to the sensor control server which needs to reprogramming the sensor nodes with the code that a user requires to be run within the network. SDSNs consist of three main software-defined layers: the data layer that consists of sensor nodes and architecture model to control the media access [43]. Control or networking layer which is responsible for data transmission across the network and the control of SDN. Finally, the applications layer manages the programs and performs sensing tasks in addition to operating system for sensor node controlling.

In low power IoT networks, SDN can provide a flexible and scalable data collection, actuation, and data dissemination for multiple networks connected across a backbone network, and protocols such as 6LoWPAN. It is able to allow IoT networks to reconfigure when adding new sensors, actuators, and network capabilities with relative ease without updating firmware [44]. The applying of SDN architecture in low power IoT network will allow to virtualize network functions such as routing, security, and data aggregation, enabling IoT sensor networks to take advantage of greater computing resources and give the IoT network the ability to dynamically serve multiple applications with varying QoS requirements [45].

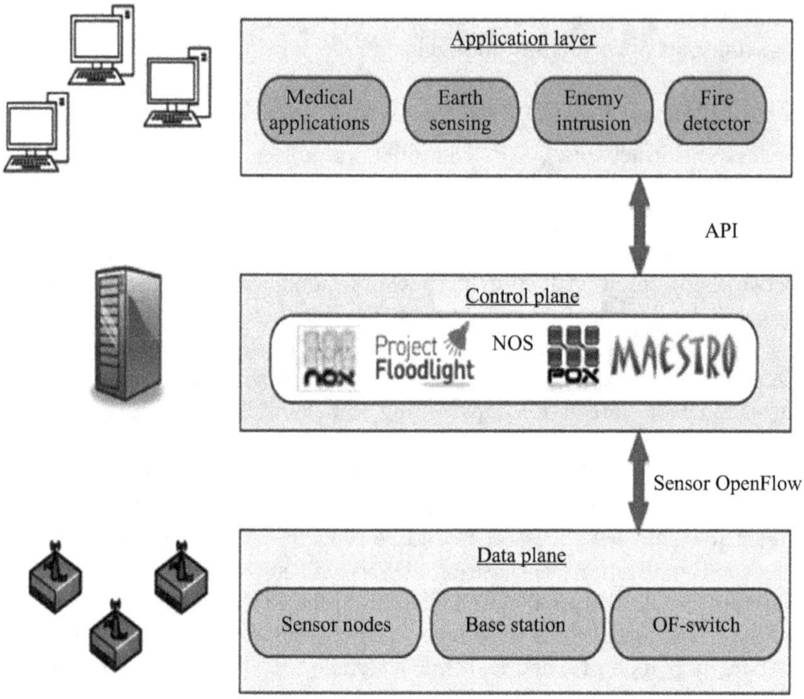

Figure 6.6 SDSN architecture layers. Adapted from [42]

6.5 Analysis IoT network context

The evaluation of relationship between artificial intelligent (AI) and IoT is related to the AI key components for the realization of future development of IoT networks in different aspects including data protection and privacy. IoT network context analysis will liberate the full potential of the IoT due to identifying the standardization processes, sensing and communications, remote control, and security in IoT networks. AI will add an important key benefit for developing promote safe use of IoT network and for data processing through learning techniques in addition to ensuring optimization and autonomy [46]. Technologies such as cloud computing, big data, and 5G can be integrated with the AI to develop intelligent IoT networks and able to provide the more advanced feature. Besides, the considerations of intelligent network connectivity, cloud computing, and AI, trust in accessing the IoT networks is very important and plays an essential role between all network components [47]. IoT network authentication, confidentiality, availability, integrity, and attack detection are all concerns technological aspects related to AI. Trust is an important feature in the decision-making process of various IoT applications and services. For example, in industrial applications, the combination of artificial intelligence, data, and networks has begun to emulate human intelligence.

Context analysis of IoT networks instantly updates the information associated with the target devices in the network, helping to enhance human behavior by collecting and visualizing user context data distributed from IoT distributed devices and providing behavior support across distributed actuators [48]. The control in IoT sensor networks can be conducted based on context-perception due to the information of current network status to support decision making according to the user-centric context and system-centric context. In a user-centric context, IoT networks act as a conduit for capturing user context based on a system that can recognize the circumstances and needs of the user accordingly and provide intelligent behavior enhancement while in a system-centric context, it depends on enabling the IoT system to understand system conditions and self-condition [49].

In IoT networks, device context analysis information and data are determined by portable sensors that match the context of the IoT application. These sensors can transmit data to neighboring devices according to the appropriateness of information with them. The data from nearby sensors can be used collectively to reduce noise. The implementation of AI in IoT networks is required to comprehensively evaluate the infrastructures related to collecting data from the physical layer to the highest layers (see Figure 6.7) and its association with location, self-configuration, data management, and operation. Even so, AI techniques cannot be placed at IoT in each network level [50].

To provide information management for mobile objects (MO) and devices in IoT networks, the device management (DM) model is remotely managing connected devices by an efficient way to configuration many aspects related to maintenance and management, user preferences, provisioning, software management in addition to fault detection, query, and reporting. According to the importance of lifting the intelligence abilities of IoT network configurations, the placing of AI in the context of IoT networks can be added to network servers because of their computing power [50]. The AI placing in IoT networks is based on three communications schemes and they are preliminary communication which describes the types of data collected from the environment by the IoT devices and sent from a variety of IoT systems. The context communication introduces the type of data processed by systems and already has an appropriate context to make the IoT networks to take an appropriate decision, while internal communication describes the channel of communication in IoT networks. In the context of mobile devices, it must ensure that information is cached in objects while not accessing the web [51] and all data, in the context of a given cycle, already present in the main IoT systems are prepared for external AI systems. For internal communication, the importance of contextual communication is related to gather enough data from multiple devices and simultaneously to send them to the upper layers of the system.

6.6 Intelligent IoT network algorithms and strategy

The evolution in IIoT networking promises to give IoT objects the ability to understand their surroundings and make decisions independently. Moreover, it

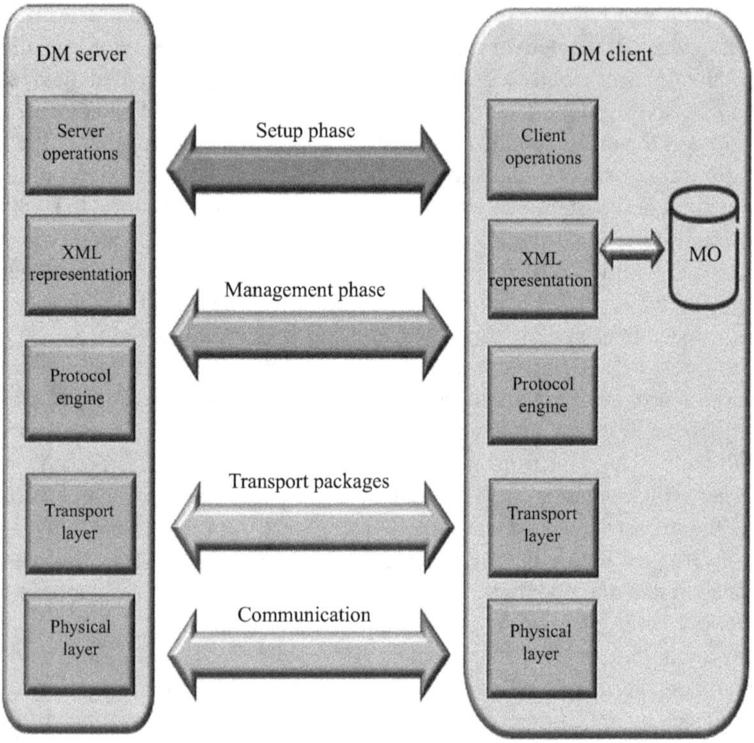

Figure 6.7 IoT networks communication management architecture. Adapted from [50]

provides sensors the ability to act according to the stimulus observed which enables the IIoT to respond better to time-critical situations because the decisions are made without a centralized style. Due to the growing demand for intelligence behaviors into IoT networks, there is a tendency to augment devices with sensing, computing, and communication functionalities, connecting them together to form what is known as intelligent IoT network and make use of their collective capabilities. Intelligent algorithms can be used in the IoT wireless networks such as machine learning (ML) and intelligent sensing. In machine learning, various algorithms and statistical models can be used to analyze data for learning the IoT networks process. Many strategies in the IoT can offer a combination of computational intelligent sensing capabilities and the emergence of sensor systems capable of inferring intelligence from raw sensed data, through the use of sensing techniques and data mining [52].

The incorporating of sensing data can extract a new technology that is able to fusion sensors in intelligent manner and makes it able to collect data in an intensive way in IoT platforms which reduces privacy exposure when collecting data. Real-time processing technologies develop and deploy updated software that can separate programmable and reconfigurable software components such as data analytics algorithms or virtualized data simulators and their ability to quickly adapt to dynamic updates and requirements in IoT network platforms. Dynamic processing can enable data itself to undergo a contextual conversion to extract transparent or hidden information behind the collected data. It also provides emerging technologies for the context of data, enhanced data analytics, real-time awareness, and situation prediction. The self-configuration strategy with real-time situation awareness concept can detect the degree of IoT congestion and extract additional environmental context information for IoT sensors, making it easier to cross-reference between platforms that support different services as an additional context and predict location based on historical and real-time data [53]. This strategy will make IoT gateways at the edge layer filter lightweight data and fragment private personal information before sending data to the platform, enabling IoT gateways to perform key data context tasks in the platform. Consciousness of the real-time situation can also be transferred to the edge and from it high-level contextual information is reported to the cloud for historical data analyzes. In IoT networks, the integration of data obtained from different sensors can be done via sensor fusion technology, which provides solutions for more comprehensive knowledge about things in the IoT environment. In addition, intelligent IoT network control signals can provide sensor information describing human mobility and network usage information, and collect information on the fusion of sensors such as number and type to handle them more efficiently and to manage sensor resources and use them widely for multiple purposes for different applications such as in smart city [54].

6.7 Heterogeneous IoT-based 5G networks

Heterogeneity in IoT networks is considered as a main challenge and that is required to be handled by intelligence interoperability. The IoT network components and elements comprising devices, communication, services, and applications must seamlessly cooperate and communicate with each other to realize the full potential of an intelligent IoT network. Due to intelligent wireless network innovation, 5G radio access will drive economic and societal growth entirely. 5G can enable access to existing wireless technologies such as LTE, HSPA, GSM, and Wi-Fi. It will generate networks that are capable of providing zero-distance connectivity between people and connected machines. It will be a cornerstone for realizing breakthroughs in the transformation of ICT network infrastructure and provide ultra-broadband and intelligent network features that achieve near-instantaneous connectivity between people and machines. For IoT networks, the implementation of 5G wireless technologies will provide support for massive capacity and massive connectivity as well as for an increasingly diverse set of services, applications, and users, in addition to efficient use of all available spectrums for wildly different IoT applications [55]. Moreover, it will enable an efficient mobile IoT network adding means of IoT network access for person-to-person and person-to-machine connectivity with providing advances in the quality of service, reliability, and security [56]. The use of 5G with IoT will depend on ultra-wide bandwidth with submillisecond latencies to support wide range of mobile IoT services performance requirements. According to this benefit, the use of intelligent IoT-based 5G M2M communication will allow different machines and objects on IoT networks to exchange data, check the availability of resources, discover how to compose complex services, handle device registration, and offer a standardized output to any IoT applications.

For interoperability of IoT-based 5G networks, SDN was introduced for mobile IoT networks to enable to shift the intelligent part from physical nodes to the cloud which will help in turn to improve the performance of IoT-based 5G networks by coordinating the use of available small cells and interference aware spectrum allocation. The use of SDN can provide cognitive decisions on network configurations by using data from lower layers but it requires to have knowledge about the overall IoT network performance to be intelligent. The improvement in SDN performance can be done by taking into consideration the parameters related to traffic statistics for each cell, service requirements for each user, efficiency of spectrum utilization, and signal quality indicators of each IoT [57]. These parameters information must be obtained in real-time and being stored for long period for analysis. The software-defined 5G IoT network can be consist of four layers (see Figure 6.8): they are radio access network (RAN) layer which is responsible to

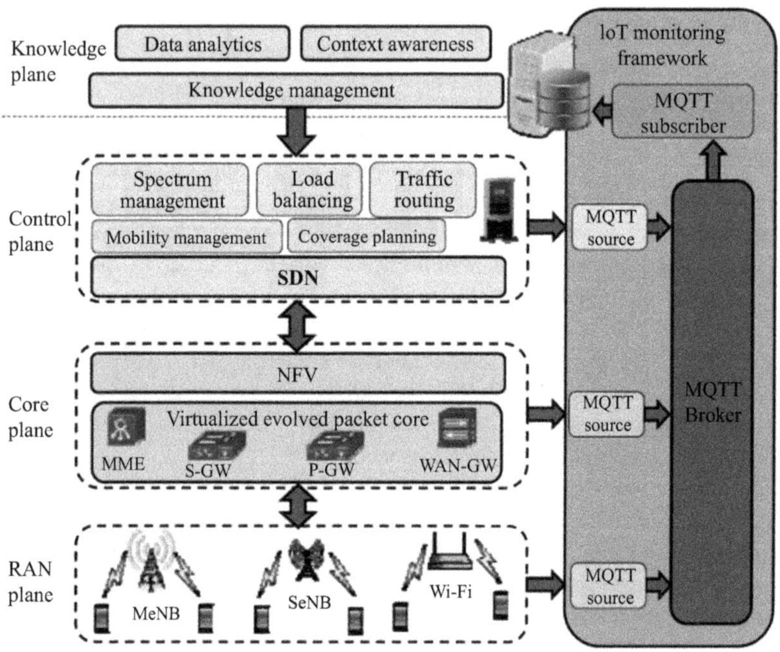

Figure 6.8 Architecture of SDN over 5G network with IoT monitoring framework. Adapted from [58]

enable wireless access to the end-users. Core layer will provide the quality assurance functions in addition to billing and secured access and depends on network function virtualization (NFV) technology in the cloud [59]. The operations of spectrum management, load balancing, traffic routing, mobility management, and coverage planning for 5G IoT network are all done through the control layer and knowledge layer. The knowledge management is responsible for data analysis and context awareness is placed in the last layer known as the knowledge layer.

The IoT network data acquisition can be monitored by IoT monitoring framework which is used to collect data from network things and object and interact with knowledge layer. The IoT framework can collect any type of data in a cloud database which will open an opportunity for operators to customize the monitoring system according to their demands on their IoT applications. The use of network function virtualization (NFV) technology will utilize virtualization technologies to separate physical network equipment from the functions it works on and it can implement various virtual network functions and deploy them to one or more physical servers to facilitate 5G to support new services of IoT domain which demands multitenancy, low cost, efficient resource utilization, and low power consumption. The 5G IoT network traffic management can successfully be applied to data centers and commercial networks by the use of SDN technology [60]. It distinguishes between the data level and controls the execution of all control functions in a central network controller by transferring the control function to software-based entities, eliminating the use of a vendor's devices and promoting the use of commodity keys in the data level on proprietary devices. The convergence of mobile IoT and intelligent wireless systems in the 5G technology can lead to tremendous growth in resource-intensive and computational applications, covering extensive monitoring, smart e-health, intelligent transport, and vehicle Internet (IoV). Complementing various cloud resources and bringing the account closer to smart devices and things is a very promising enabling technology to reap the potential of the IoT in the fifth generation.

6.8 IoT network adaptive quality of service

Adaptive QoS will contribute to improving the performance of systems on IoT networks. In addition, the quality of service information in real-time of the network contributes to making IoT applications making informed adaptation procedures. The quality of IoT services depends on network behavior, decisions, and modifications based on various design criteria, such as the required transmission rate, data exchange mechanism, and so on. Therefore, it must be ensured that the IoT constantly improves reliability [61]. Because of the heterogeneity of IoT network components, IoT gateways play an important role in ensuring the quality of the integrated service. Delays in any layer of the IoT structure, including sensors can lead to some quality of service problems in different IoT applications. In addition, there are other factors that can cause poor quality of service such as security and reliability [62]. Therefore, there are approaches that can be used to address quality requirements on each layer and methods that are able to evaluate themselves according to an appropriate quality model, and such methodologies are Q-learning algorithm and deep generative network.

6.8.1 Q-learning algorithm

In heterogeneous wireless networks, Q-learning can provide to make an intelligent decision depending on the network performance status and enable to perform an action according to targeted objectives. The targeted objectives can be considered as Q-function and are updated in an iterative approach after collecting the performance status of the network. This process should be done by exploring the network environment in addition to the status of the operation, which will maximize overall efficiency in order to identify better options for future decisions. The concept of Q-learning in wireless network environments can be considered as an example like a tradeoff between high and low channel collision probabilities whereas Q-values resemble a more sophisticated and prophetic verdict of how acceptable or not acceptable about the particular division state. The optimum decision actions can be selected based on Q-value findings and the states of a maximum Q-value are preferred [63]. In IoT networks, the QL algorithm can be used without having complete information about the state of the network devices' environment, so that the IoT devices will make business decisions by extracting new network parameters according to the previously collected parameters as seen in Figure 6.9.

In this scenario, IoT device would like to plan its future decisions based on the given state (the IoT network and environment action), along with its intelligent sensing and network utilization. Other considerations are related to the use of QL-based channel access scheme which can be used to guide densely deployed IoT devices and allocate radio resources more efficiently. QL algorithms can serve as a means of monitoring and learning the environment for optimal policy choice for dense wireless networks in IoT-based systems [64]. In dense IoT networks, the channel collision is the most critical issue that causes network performance degradation. So addressing MAC collision problems using the QL algorithm can be used to model the optimal competition window in the regression mechanism based on channel monitoring. Some studies have been conducted on how to use the

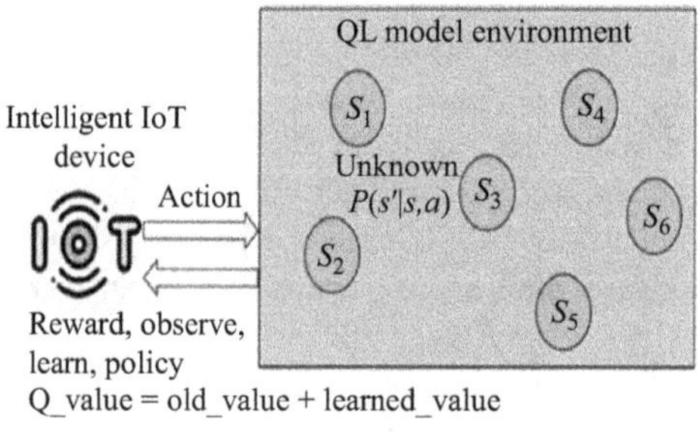

Figure 6.9 Q-learning environment for IoT intelligent devices. Adapted from [63]

genetic algorithm to balance the load in foggy IoT networks and manage resource allocation intelligently.

6.8.2 Deep generative network

Dynamic and interactive environment communications can be implemented by using intelligent IoT devices that can able to sense the environment and make control decisions to react. In autonomous IoT networks (see Figure 6.10), the network computation performance such as communication delay, power consumption, and network reliability will have important impact on the control performance of the AIoT systems. Separation between the control of IoT sensors, devices, and communications resources control will enhance and optimize the learning mechanism [65]. The deep learning agents in AIoT systems can reside in the IoT devices, edge and fog servers, or in cloud servers and it depends on the time-sensitivity of the IoT application. In mobile or flying things IoT applications, agent must be residing locally on the typical vehicle to make fast decisions, instead of transmitting the sensory data to the cloud and return the predictions back to the same vehicle.

Deep generative learning in IoT networks will contribute to achieve the capability of exploiting unlabeled data to learn useful patterns by using restricted Boltzmann machine with generative adversarial network (GAN). The use of generative adversarial network will reduce the computational and time complexity of the trained model to be used for various tasks and can achieve multiple objectives without model retraining in addition to make storage, analysis, and computation of data [65,66]. GAN will be used as a compromised game between the captured distributed data in the network and the estimates of the probability that a sample comes from the training data. The importance of GAN in IoT networks is it can be used to develop IoT devices with individual identity authentication as an effective

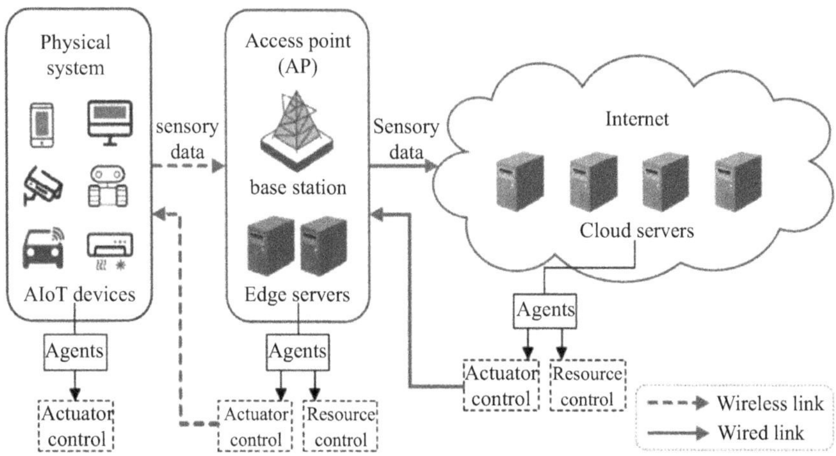

Figure 6.10 Autonomous IoT system. Adapted from [65]

way to enhance the security of IoT. This can be achieved by detecting and recognizing the wireless signals from different IoT network devices for training and classification by extracting the features representing the individual identification to let intelligent IoT network able to detect the identification of the new assigning of IoT devices to find out the comparison between the legal and illegal IoT devices [66]. Moreover, GAN can provide a strategy for defining a game between two competing networks in heterogeneous IoT networks.

6.9 Summary

Next-generation wireless networks are expected to support extremely high data rates and radical new applications, which require a new model of wireless radio technology. The challenge is to help the radio adaptive intelligent learning and decision-making so that the diverse requirements of wireless next-generation networks can be met. Machine learning is one of the most promising artificial intelligence tools, designed to support intelligent radio stations. In this chapter, we give a brief idea about the use of machine learning in the amazing applications of IoT 5G networks including cognitive radios and heterogeneous cells, smart grid, energy harvesting, and communication from one device to another. Through the chapter, the ways to optimize the number of IoT devices offered by adjusting uplink resource configurations in real-time according to dynamic traffic are illustrated. Moreover, the deep learning is proposed as a promising solution, as the DL agent for example in IoT cellular-based network implements updates in the uplink resource configuration by interacting with the environment. This chapter also investigates some artificial intelligence methods such as deep generative networks to solve the challenges in the IoT.

References

[1] Liu, Y., Kashef, M., Lee, K. B., Benmohamed, L., and Candell, R. "Wireless network design for emerging IIoT applications: Reference framework and use cases." *Proceedings of the IEEE.* 2019; 107(6): 1166–1192. doi: 10.1109/JPROC.2019.2905423.

[2] Javaid, N., Sher, A., Nasir, H., and Guizani, N. "Intelligence in IoT-based 5G networks: Opportunities and challenges." *IEEE Communications Magazine.* 2018; 56(10): 94–100. doi: 10.1109/MCOM.2018.1800036.

[3] García, L., Miguel Jiménez, J., Taha, M., and Lloret, J. "Wireless technologies for IoT in smart cities." *Network Protocols and Algorithms.* 2018; 10(1). doi: https://doi.org/10.5296/npa.v10i1.12798.

[4] Arsénio, A., Serra, H., Francisco, R., Nabais, F., Andrade, J., and Serrano, E. "Internet of intelligent things: Bringing artificial intelligence into things and communication networks." In F. Xhafa, and N. Bessis (eds.), *Inter-cooperative Collective Intelligence: Techniques and Applications. Studies in*

Computational Intelligence, vol. 495. Springer, Berlin, Heidelberg, 2014. Available from https://doi.org/10.1007/978-3-642-35016-0_1.

[5] Lampropoulos, G., Siakas, K., and Anastasiadis, T. "Internet of things in the context of industry 4.0: An overview." *International Journal of Entrepreneurial Knowledge.* Centre for International Scientific Research of VSO and VSPP. 2019; 7(1): 4–19.

[6] Carlesa, G., Stefanob, C., Anthonyc, F., Georged, R., and Davye, P. "Internet of things for enabling smart environments: A technology-centric perspective." *Journal of Ambient Intelligence and Smart Environments*. 2019; 11(1): 23–43.

[7] Park, Jh., Salim, M. M., Jo, J. H., *et al.* "CIoT-Net: A scalable cognitive IoT based smart city network architecture." *Human Centric Computing and Information Science*. 2019; 9: 29. Available from https://doi.org/10.1186/s13673-019-0190-9.

[8] Al-Shayea, T. K., Mavromoustakis, C. X., Batalla, J. M., and Mastorakis, G. "A hybridized methodology of different wavelet transformations targeting medical images in IoT infrastructure." *Measurement*. 2019; 148: 106813. Available from https://doi.org/10.1016/j.measurement.2019.07.041.

[9] Athreya, A. P., DeBruhl, B., and Tague, P. "Designing for self-configuration and self-adaptation in the internet of things." *9th IEEE International Conference on Collaborative Computing: Networking, Applications and Work sharing*. Austin, TX, 2013, pp. 585–592. doi: 10.4108/icst.collaboratecom.2013.254091.

[10] Lee, S. K., Bae, M., and Kim, H. "Future of IoT networks: A survey." *Applied Sciences*. 2017; 7(10): 1072. Available from https://doi.org/10.3390/app7101072.

[11] Alsulami, M. M., and Akkari, N. "The role of 5G wireless networks in the internet-of-things (IoT)." *1st International Conference on Computer Applications & Information Security (ICCAIS)*, Riyadh, 2018, pp. 1–8. doi: 10.1109/CAIS.2018.8471687.

[12] Yu, C., Doppler, K., Ribeiro, C. B., and Tirkkonen, O. "Resource sharing optimization for device-to-device communication underlaying cellular networks." *IEEE Transactions on Wireless Communications*. 2011; 10(8): 2752–2763. doi: 10.1109/TWC.2011.060811.102120.

[13] Queralta, J. P., Gia, T. N., Zou, Z., Tenhunen, H., and Westerlund, T. "Comparative study of LPWAN technologies on unlicensed bands for M2M communication in the IoT: Beyond LoRa and LoRaWAN." *Procedia Computer Science*. 2019; 155: 343–350. Available from https://doi.org/10.1016/j.procs.2019.08.049.

[14] Haxhibeqiri, J., De Poorter, E., Moerman, I., and Hoebeke, J. "A survey of LoRaWAN for IoT: From technology to application." *Sensors*. 2018; 18(11): 3995. Available from https://doi.org/10.3390/s18113995.

[15] Morin, É., Maman, M., Guizzetti, R., and Duda, A. "Comparison of the device lifetime in wireless networks for the internet of things." *IEEE Access*. 2017; 5: 7097–7114. doi: 10.1109/ACCESS.2017.2688279.

[16] Mekki, K., Bajic, E., Chaxel, F., and Meyer, F. "Overview of cellular LPWAN technologies for IoT deployment: Sigfox, LoRaWAN, and NB-IoT." *2018 IEEE International Conference on Pervasive Computing and Communications Workshops (PerCom Workshops)*, Athens, 2018, pp. 197–202. doi: 10.1109/PERCOMW.2018.8480255.

[17] Mwakwata, C. B., Malik, H., Mahtab Alam, M., Le Moullec, Y., Parand, S., and Mumtaz, S. "Narrowband internet of things (NB-IoT): From physical (PHY) and media access control (MAC) layers perspectives." *Sensors*. 2019; 19(11): 2613. Available from https://doi.org/10.3390/s19112613.

[18] Moon, B. "Dynamic spectrum access for internet of things service in cognitive radio-enabled LPWANs." *Sensors (Basel)*. 2017; 17(12): 2818. doi: 10.3390/s17122818. PMID: 29206215; PMCID: PMC5750700.

[19] Katzis, K., and Ahmadi, H. "Challenges implementing Internet of things (IoT) using cognitive radio capabilities in 5G mobile networks." In J. M. B. Hamed, C. X. Mavromoustakis, and G. Mastorakis (eds.), *Internet of things (IoT) in 5G mobile technologies*. 2016.

[20] Joshi, G. P., Nam, S. Y., and Kim, S. W. "Cognitive radio wireless sensor networks: Applications, challenges and research trends." *Sensors (Basel)*. 2013.

[21] Hussain, F., Hassan, S. A., Hussain, R., and Hossain, E. "Machine learning for resource management in cellular and IoT networks: Potentials, current solutions, and open challenges." *IEEE Communications Surveys and Tutorials*. 2020; 22(2): 1251–1275. doi: 10.1109/COMST.2020.2964534

[22] Nikoloudakis, Y., Panagiotakis, S., Markakis, E., Pallis, E., Mastorakis, G., and Ma, C. X. "A fog-based emergency system for smart enhanced living environments." *IEEE Cloud Computing*. 2016; 3(6): 54–62. doi: 10.1109/MCC.2016.118.

[23] de Silva, J. C., Rodrigues, J. J. P. C., Al-Muhtadi, J., Rabêlo, R. A. L., and Furtado, V. "Management platforms and protocols for internet of things: A survey." *Sensors*. 2019; 19(3): 676. Available from https://doi.org/10.3390/s19030676.

[24] Ali, S., Qaisar, S. B., Saeed, H., Khan, M. F., Naeem, M., and Anpalagan, A. "Network challenges for cyber physical systems with tiny wireless devices: A case study on reliable pipeline condition monitoring." *Sensors*. 2015; 15(4): 7172–7205. Available from https://doi.org/10.3390/s150407172.

[25] Dang, L. M., Piran, M. J., Han, D., Min, K., and Moon, H. "A survey on internet of things and cloud computing for healthcare." *Electronics*. 2019; 8: 768. Available from https://doi.org/10.3390/electronics8070768.

[26] Mavromoustakis, C. X., Kormentzas, G., Mastorakis, G., Bourdena, A., and Pallis, E. "Context-oriented opportunistic cloud offload processing for energy conservation in wireless devices." *IEEE Globecom Workshops*. 2014.

[27] Nguyen, T. M. C., Hoang, D. B., and Dang, T. D. "Toward a programmable software-defined IoT architecture for sensor service provision on demand." *27th International Telecommunication Networks and Applications Conference (ITNAC)*; Melbourne, VIC, 2017, pp. 1–6. doi: 10.1109/ATNAC.2017.8215419.

[28] Markakis, E. K., Politis, I., Lykourgiotis, A., Rebahi, Y., and Mastorakis, G. "Efficient next generation emergency communications over multi-access edge computing." *IEEE Communications Magazine*. 2017; 55(11): 92–97. doi: 10.1109/MCOM.2017.1700345.

[29] Khajenasiri, I., Estebsari, A., Verhelst, M., and Gielen, G. "A review on internet of things solutions for intelligent energy control in buildings for smart city applications." *Energy Procedia*. 2017; 111: 770–779. Available from https://doi.org/10.1016/j.egypro.2017.03.239.

[30] Agarwal, P., and Alam, M. "Investigating IoT middleware platforms for smart application development." *arXiv*. 2018; 1–14.

[31] Nakhuva, B., and Champaneria, T. "Study of various internet of things platforms." *International Journal of Computer Science & Engineering Survey (IJCSES)*. 2015; 6(6): 61–74. doi: 10.5121/ijcses.2015.6605.

[32] Vermesan, O., Bröring, A., Tragos, E., *et al*. "Internet of robotic things – Converging sensing/actuating, hyper connectivity, artificial intelligence and IoT platforms." *Cognitive Hyperconnected Digital Transformation – Internet of Things Intelligence Evolution* (pp. 97–155). River Publishers, 2017.

[33] Kato, J.-Y., Mitake, H., Suda, A., Yamada, H., Okitsu, K., and Kanai, S. "Open source software efforts to transform IoT/AI services." *NTT Technical Review*. 2018; 16(2): 1–6.

[34] Nurelmadina, N., Kamrul Hasan, M., Memon, I., *et al*. "A systematic review on cognitive radio in low power wide area network for industrial IoT applications." *Sustainability*. 2021; 13: 338. Available from https://doi.org/10.3390/su13010338.

[35] Chen, M., Miao, Y., Jian. X., Wang, X., and Humar, I. "Cognitive-LPWAN: Towards intelligent wireless services in hybrid low power wide area networks." *IEEE Transactions on Green Communications and Networking*. 2019; 3(2): 409–417. doi: 10.1109/TGCN.2018.2873783.

[36] Atsali, G., Panagiotakis, S., Markakis, E., Mastorakis, G., and Mavromoustakis, C. X. "A mixed reality 3D system for the integration of X3DoM graphics with real-time IoT data." *Multimedia Tools and Applications*. 2018; 77: 4731–4752. Available from https://doi.org/10.1007/s11042-017-4988-z.

[37] Mavromoustakis, C. X., Batalla, J. M., Mastorakis, G., Markakis, E., and Pallis, E. "Socially oriented edge computing for energy awareness in IoT architectures." *IEEE Communications Magazine*. 2018; 56(7): 139–145. doi: 10.1109/MCOM.2018.1700600.

[38] Abdelmonem, M. A., Nafie, M., Ismail, M. H., *et al*. "Optimized spectrum sensing algorithms for cognitive LTE femtocells." *Journal on Wireless Communications and Networking*. 2012; 6. Available from https://doi.org/10.1186/1687-1499-2012-6.

[39] Alonso, R. M., Plets, D., Pupo, E. F., *et al*. "IoT-based management platform for real-time spectrum and energy optimization of broadcasting networks."

Wireless Communications and Mobile Computing. 2018; 2018: 14. Available from https://doi.org/10.1155/2018/7287641.

[40] Dongare, A., Hesling, C., Bhatia, K., *et al*. "OpenChirp: A low-power wide-area networking architecture." *IEEE International Conference on Pervasive Computing and Communications Workshops (PerCom Workshops)*; Kona, HI. 2017, pp. 569–574. doi: 10.1109/PERCOMW.2017.7917625.

[41] Wang, A., Zha, Z., Guo, Y., and Chen, S. "Software-defined networking enhanced edge computing: A network-centric survey." *Proceedings of the IEEE*. 2019; 107(8): 1500–1519. doi: 10.1109/JPROC.2019.2924377.

[42] Fernández, J. A. P., Villalba, L. J. G., and Kim, T.-H. "Software defined networks in wireless sensor architectures." *Entropy*. 2018; 20: 225. Available from https://doi.org/10.3390/e20040225.

[43] Bera, S., Misra, S., Roy, S. K., and Obaidat, M. S. "Soft-WSN: Software-defined WSN management system for IoT applications." *IEEE Systems Journal*. 2018; 12(3): 2074–2081. doi: 10.1109/JSYST.2016.2615761.

[44] Batalla, J. M., Gajewski, M., Latoszek, W., Krawiec, P., Mavromoustakis, C. X., and Mastorakis, G. "ID-based service-oriented communications for unified access to IoT." *Computers & Electrical Engineering*. 2016; 52: 98–113. Available from https://doi.org/10.1016/j.compeleceng.2016.02.020.

[45] Tayyaba, S. K., Shah, M. A., Khan, O. A., and Ahmed, A. W. "Software defined network (SDN) based internet of things (IoT): A road ahead." *Proceedings of the International Conference on Future Networks and Distributed Systems (ICFNDS'17)*. Association for Computing Machinery, New York, USA, Article 15, 1–8. Available from https://doi.org/10.1145/3102304.3102319.

[46] Guo, K., Lu, Y., Gao, H., and Cao, R. "Artificial intelligence-based semantic internet of things in a user-centric smart city." *Sensors (Basel, Switzerland)*. 2018; 18(5): 1341. doi:10.3390/s18051341.

[47] Granat, J., Batalla, J. M., Mavromoustakis, C. X., and Mastorakis, G. "Big data analytics for event detection in the IoT-multicriteria approach." *IEEE Internet of Things Journal*. 2019.

[48] Kryftis, Y., Mavromoustakis, C. X., Mastorakis, G., Pallis, E., and Batalla, J. M. "Resource usage prediction algorithms for optimal selection of multimedia content delivery methods." *IEEE International Conference on Communications (ICC)*. London, 2015, pp. 5903–5909. doi: 10.1109/ICC.2015.7249263.

[49] Xiao, B., Rahmani, R., Li, Y., Gillblad, D., and Kanter, T. "Intelligent data-intensive IoT: A survey." *2nd IEEE International Conference on Computer and Communications (ICCC)*. Chengdu, 2016, pp. 2362–2368. doi: 10.1109/CompComm.2016.7925122.

[50] de Silva, J. C., Rodrigues, J. J. P. C., Al-Muhtadi, J., Rabêlo, R. A. L., and Furtado, V. "Management platforms and protocols for internet of things: A survey." *Sensors*. 2019; 19(3): 676. Available from https://doi.org/10.3390/s19030676.

[51] Poniszewska-Maranda, A., and Kaczmarek, D. "Selected methods of artificial intelligence for Internet of things conception." *Proceedings of the Federated Conference on Computer Science and Information Systems. ACSIS.* 2015; 5: 1343–1348. doi: 10.15439/2015F161.

[52] Bica, I., Chifor, B.-C., Arseni, Ş.-C., and Matei, I. "Multi-layer IoT security framework for ambient intelligence environments." *S*ensors (Basel). 2019; 19(18): 4038. doi:10.3390/s19184038.

[53] Hadi, M. S., Lawey, A. Q., El-Gorashi, T. E. H., and Elmirghani, J. M. H. "Big data analytics for wireless and wired network design: A survey." *Computer Networks.* 2018; 132: 180–199. Available from https://doi.org/10.1016/j.comnet.2018.01.016.

[54] Silva, L. A., Leithardt, V. R. Q., Rolim, C. O., González, G. V., Geyer C. F. R., and Silva, J. S. "PRISER: Managing notification in multiples devices with data privacy support." *Sensors.* 2019; 19(14): 3098. https://doi.org/10.3390/s19143098.

[55] Mavromoustakis, C. X., Bourdena, A., Mastorakis, G., Pallis, E., and Kormentzas, G. "An energy-aware scheme for efficient spectrum utilization in a 5G mobile cognitive radio network architecture." *Telecommunication Systems.* 2015.

[56] Hadjioannou, V., Mavromoustakis, C. X., Mastorakis, G., and Batalla, J. M. "Security in smart grids and smart spaces for smooth IoT deployment in 5G." In C. Mavromoustakis, G. Mastorakis, and J. Batalla (eds.), *Internet of things (IoT) in 5G mobile technologies*, vol. 8. Springer, Cham, 2016. Available from https://doi.org/10.1007/978-3-319-30913-2_16.

[57] Alam, I. , Sharif, K., Li, F., *et al.* "IoT virtualization: A survey of software definition & function virtualization techniques for internet of things." *Networking and Internet Architecture, arXiv:*1902.10910. 2019; 1–29.

[58] Maksymyuk, T., Dumych, S., Brych, M., Satria, D., and Jo, M. "An IoT based monitoring framework for software defined 5G mobile networks." *Proceedings of the 11th International Conference on Ubiquitous Information Management and Communication (IMCOM'17). Association for Computing Machinery*; New York, USA, Article 105, 1–4. 2017. Available from https://doi.org/10.1145/3022227.3022331.

[59] Ullah, I., Ahmad, S., Mehmood, F., and Kim, D. "Cloud based IoT network virtualization for supporting dynamic connectivity among connected devices." *Electronics.* 2019; 8: 742. Available from https://doi.org/10.3390/electronics8070742.

[60] Ding, A. Y., and Janssen, M. "Opportunities for applications using 5G networks: Requirements, challenges, and outlook." *Proceedings of the Seventh International Conference on Telecommunications and Remote Sensing (ICTRS'18).* Association for Computing Machinery; New York, USA, 27–34, 2018.

[61] Ezdiani, S., Acharyya, I. S., Sivakumar, S., and Al-Anbuky, A. "An IoT environment for WSN adaptive QoS." *IEEE International Conference on*

Data Science and Data Intensive Systems; Sydney, NSW, 2015, pp. 586–593. doi: 10.1109/DSDIS.2015.28.

[62] White, G., Nallur, V., and Clark, S. "Quality of service approaches in IoT: A systematic mapping." *Journal of Systems and Software*. 2017; 132: 186–203. Available from https://doi.org/10.1016/j.jss.2017.05.125.

[63] Ali, R., Qadri, Y., Bin Zikria, Y., *et al*. "Q-learning-enabled channel access in next-generation dense wireless networks for IoT-based eHealth systems." *Journal on Wireless Communications and Networking*. 2019; 178. Available from https://doi.org/10.1186/s13638-019-1498-x.

[64] Marchese, M., Moheddine, A., and Patrone, F. "IoT and UAV integration in 5G hybrid terrestrial-satellite networks." *Sensors*. 2019; 19: 3704. Available from https://doi.org/10.3390/s19173704.

[65] Lei, L., Tan, Y., Zheng, K., Liu, S., Zhang, K., and Shen, X. "Deep reinforcement learning for autonomous internet of things: Model, applications and challenges." *Machine Learning, arXiv:1907.09059*. 2020; page 1:39.

[66] Zhao, C., Shi, M., Cai, Z., and Chen, C. "Research on the open-categorical classification of the internet-of-things based on generative adversarial networks." *Applied Science*. 2018; 8: 2351. Available from https://doi.org/10.3390/app8122351.

[67] Bedogni, L., Felice M. D., and Bononi, L. "Dynamic spectrum access for machine to machine communications: Opportunities, standards, and open issues." In W. Zhang (ed.), *Handbook of Cognitive Radio*. Springer, Singapore, 2019. Available from https://doi.org/10.1007/978-981-10-1394-2_57.

Chapter 7

Impact of jamming signal on system performance in downlink of IoT network relying on nonorthogonal multiple access

Thi-Anh Hoang[1], Chi-Bao Le[1], Dinh-Thuan Do[1], Imran Khan[2], Constandinos X. Mavromoustakis[3], George Mastorakis[4], Evangelos Pallis[5] and Evangelos K. Markakis[5]

7.1 Introduction

With the popularity of smartphones and the Internet of things, there is an explosive demand for new services and data traffic for wireless communications. The capacity of the fourth-generation (4G) mobile communication system is insufficient to satisfy such a demand in the near future. The development of the fifth-generation (5G) mobile communication system has been placed on the agenda with higher requirements in data rates, latency, and connectivity. Recently, high spectral efficiency is provided in the nonorthogonal multiple access (NOMA) technique which is considered as one of the promising multiple access (MA) schemes for the 5G in wireless communication system [1,2]. This NOMA scheme operating in the power domain is significantly different from other conventional orthogonal multiple access (MA) technique [3]. To implement NOMA architecture, superposition coding (SupC) is employed at the base station (BS) side to multiplex the signals from multiple users in the power domain. To eliminate the inter-user interference at the users side, successive interference cancellation (SIC) is required. As a distinctive characterization of NOMA, user fairness is a further feature [4]. In this MA

[1]Department of Communications Engineering, Faculty of Electronics Technology, Industrial University of Ho Chi Minh City (IUH), Ho Chi Minh City, Vietnam
[2]Department of Electrical Engineering, University of Engineering and Technology, Peshawar City, Pakistan
[3]Department of Computer Science, University of Nicosia and University of Nicosia Research Foundation (UNRF), Nicosia, Cyprus
[4]Department of Management Science and Technology, Hellenic Mediterranean University, Agios Nikolaos, Crete, Greece
[5]Department of Electrical and Computer Engineering, Hellenic Mediterranean University, Heraklion, Greece

scheme, users with worse channel conditions are allocated with a higher proportion of power while less power is assigned to users with higher channel gains. As a result, to implement the NOMA communication system, it is necessary to guarantee a trade-off between the performance and user fairness. Different from the conventional orthogonal MA protocols, several studies have shown that NOMA exhibits significant advantages such as faster traffic and higher throughput. In order to further enhance the reliability of NOMA, this technique has been extended to cooperative transmission scenarios [5–15]. In cooperative NOMA (C-NOMA) networks, to combat channel fading and improve the performances, users and relays need to cooperate together. The users can utilize the diversity techniques, such as the maximal-ratio combining (MRC) to process the two independent signals received from the direct and relaying links. Therefore, improvements significantly in the cooperative NOMA system can be achieved thanks to the qualities of services (QoS) for the users.

The escalating high data rate requirements and the unprecedented increase of mobile devices call for the 5G wireless communication systems to address challenging issues, such as spectrum scarcity and massive connectivity [9,10]. Nonorthogonal multiple access (NOMA) is a promising technique to address these challenges [11]. NOMA can provide high spectral efficiency (SE) and simultaneously serve multiple users. Unlike orthogonal multiple access (OMA), successive interference cancellation (SIC) techniques are required to reduce the mutual interference among different users due to the exploitation of nonorthogonal resources [12]. NOMA has received increasing attention since it can achieve a significant performance gain in terms of SE and energy efficiency (EE) compared with OMA [13–15].

In cooperative jamming, to interfere with an eavesdropper, a jamming signal is transmitted by a cooperating node. Most of the previous works on a single cooperating node transmit a jamming signal to form cooperative jamming techniques [16,17]. To minimize the secrecy outage probability (SOP), a relay and a cooperating node are selected among multiple intermediate nodes in [16]. In [17], a destination act as a cooperating node to transmit a jamming signal. Recently, the secrecy performance can be improved since cooperative jamming from multiple cooperating nodes is proposed [18–20]. In [18,19], they introduced a model with multiple cooperating nodes which is selected among relays to forward the data signal. In addition, in [20], the network including a source and destination serves as a cooperating node, but this work considers a single decode-and-forward (DF) relay which needs to decode its received signal first.

The broadcast nature of relaying the Internet of things network (IoT) allows sensor nodes (SNs) within the coverage range of a transmitter to capture its signals. However, such an IoT is vulnerable to eavesdropping and signal interception; therefore, security in the IoT is of significant interest, and this issue has been addressed over many years. However, no work has studied the outage performance under the impact of a friendly jammer. Thus, this chapter considers a model of IoT downlink and outage probability is analyzed. The considered model includes nonorthogonal multiple access (NOMA). We also propose an exact outage performance to enhance physical layer secrecy and we demonstrate the related impacts on performance. The analysis of the simulation results supports our hypothesis, which is in line with Monte Carlo simulations.

This chapter is organized as follows. Section 7.2 introduces the fundamentals of downlink NOMA and the two-user NOMA scheme is considered under impacts of the jamming signal. In Section 7.3, we investigate the outage performance with fixed power allocation for the scenario of two-user NOMA. The outage performance of these users in downlink of the NOMA scheme is evaluated in Section 7.4 via simulations, and the conclusion is drawn in Section 7.5.

Notations: The mathematical notations are used to provide convenience analysis. $Pr(.)$ is the probability of event, and $f_M(x)$ and $F_M(x)$ are the probability density function (PDF) and cumulative distribution function (CDF) of the random variable M, respectively.

7.2 Consideration on IoT system under the impact of a jamming signal

Figure 7.1 illustrates IoT system including a base station (BS), two sensor nodes (D_1, D_2), and some eavesdropper make the jamming signal. To implement NOMA transmission in downlink, BS forwards signal $x = \sqrt{a_1}x_1 + \sqrt{a_2}x_2$ to D_1 and D_2. The channels from BS to D_1 and D_2 are denoted as h_{SU1} and h_{SU2}, and the channels from the jamming signal to D_1 and D_2 are denoted $f_{n,1}$ and $f_{n,2}$. The channel gains follow the independent complex Gaussian distribution with mean equal to zero and

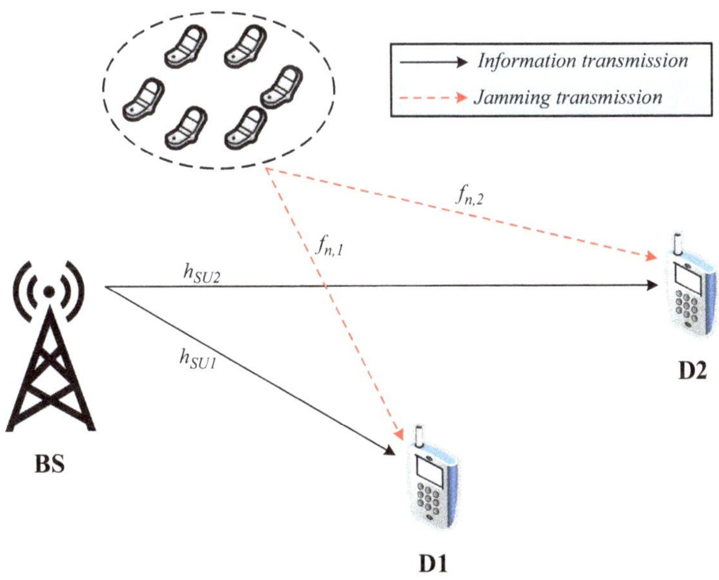

Figure 7.1 System model of IoT system under the impact of jammer

variances equal to λ_{SU_p} and Ω_p $(p = 1, 2)$. In particular, these channels are computed considering channel distribution as $h_{\mathrm{SU_p}}\mathcal{CN}(0, \lambda_{SU_p})$ and $f_{\mathrm{n,p}} \sim \mathcal{CN}(0, \Omega_p)$

The transmit power of the BS and eavesdropper sources are denoted as P_S and P_n, respectively. The noise terms (w_1, w_2) are corrupted by each receiver modeled as the additive white Gaussian noise (AWGN) with mean equal to zero and variance equal to N_0. For simplicity, the signal to noise (SNR) can be denoted as $\rho = \frac{P_S}{N_0} = \frac{P_n}{N_0}$.

It needs to be considered the signal received at D_1 considering transmitting signal x from BS and transmitting from the jamming signal (z) is

$$y_1 = h_{\mathrm{SU1}}\sqrt{P_S}x + f_{n,1}\sqrt{\theta P_n}z + w_1 \tag{7.1}$$

where $\theta \in (0, 1] : D_p$ is affected by jamming signal. $\theta = 0$: D_p can operate normally.

The signal received at D_2 when transmitting from BS to D_2 and when transmitting from the jamming signal is

$$y_2 = h_{\mathrm{SU2}}\sqrt{P_S}x + f_{n,2}\sqrt{\theta P_n}z + w_2 \tag{7.2}$$

The received signal to interference plus noise (SINR) at D_1 to detect $D_2's$ message x_2 is formulated by

$$\gamma_{2\to1} = \frac{|h_{\mathrm{SU1}}|^2 a_2\rho}{|h_{\mathrm{SU1}}|^2 a_1\rho + |f_{n,1}|^2\theta\rho + 1} \tag{7.3}$$

After performing imperfect SIC, the SINR to decode x_1 is given by

$$\gamma_1 = \frac{|h_{\mathrm{SU1}}|^2 a_1\rho}{|g|^2\varepsilon\rho + |f_{n,1}|^2\theta\rho + 1} \tag{7.4}$$

where $\varepsilon = 1$ and $\varepsilon = 0$ stand for imperfect SIC and perfect SIC, respectively, we set $g \sim \mathcal{CN}(0, \lambda_g)$.

The SINR to decode $x2$ at D_2 is given by

$$\gamma_2 = \frac{|h_{\mathrm{SU2}}|^2 a_2\rho}{|h_{\mathrm{SU2}}|^2 a_1\rho + |f_{n,2}|^2\theta\rho + 1} \tag{7.5}$$

The CDF and PDF for D_p when eavesdropper transmits jamming to main transmission from BS are

$$F_{|h_{\mathrm{SD}i}|^2}(x) = 1 - e^{-\frac{x}{\lambda_i}}, \quad i \in \{1, 2\} \tag{7.6}$$

and

$$f_{|h_{\mathrm{SD}i}|^2}(x) = \frac{1}{\lambda_i}e^{-\frac{x}{\lambda_i}}, \quad i \in \{1, 2\} \tag{7.7}$$

In this scenario, we only concern about the impact of a strong channel related to eavesdroppers who make crucial degradation on system performance. There is an eavesdropper in NOMA-based wireless network and it is passive and intends to wiretap the main transmission signal in all the data-bearing sub-channels. In addition, the wireless channels between these nodes are assumed to be perfectly measured. The strong channel is given by the corresponding index as

$$n^* = arg \underbrace{max}_{1 \le n \le N} \left|f_{n,i}\right|^2, \quad i \in \{1,2\} \tag{7.8}$$

The CDF and PDF for D_p when transmitting from the jamming signal are:

$$f_{\left|f_{n^*,i}\right|^2}(x) = \frac{n}{\Omega_i} \sum_{n=1}^{N} \binom{N}{n} (-1)^{n-1} e^{-\frac{nx}{\Omega_i}}, \quad i \in \{1,2\} \tag{7.9}$$

and

$$F_{\left|f_{n^*,i}\right|^2}(x) = 1 - \sum_{n=1}^{N} \binom{N}{n} (-1)^{n-1} e^{-\frac{nx}{\Omega_i}}, \quad i \in \{1,2\} \tag{7.10}$$

7.3 Outage probability and throughput analysis

7.3.1 The outage probability of the D_1

In this section, the closed-form expression of the secure outage performance is derived, followed by the computation of the SINR of each user, and then it studies the analytical derivation in order to consider the secure performance of NOMA system in case of imperfect SIC (ipSIC) for the user D_1 as

$$\begin{aligned} \mathrm{OP}_1^{ip\mathrm{SIC}} &= \Pr\left(\frac{1}{2}\log_2(1 + \min(\gamma_1, \gamma_{2\to1})) < R_1\right) \\ &= 1 - \Pr\left(|h_{\mathrm{SU1}}|^2 > \frac{\xi_1\left(\left|f_{n^*,1}\right|^2\theta\rho + 1\right)}{(a_2 - \xi_1 a_2)\rho}\right) \\ &\quad \times \Pr\left(|h_{\mathrm{SU1}}|^2 > \frac{\xi_1(|g|^2\varepsilon\rho + |f_{n^*,1}|^2\theta\rho + 1)}{a_1\rho}\right) \end{aligned} \tag{7.11}$$

where $\xi_1 = 2^{2R_1} - 1$, and R_1 and R_2 are secure target rates for the user D_1 and D_2.

To solve such an expression, we divide it into several cases.

If $\frac{\xi_1}{a_1\rho} \geq \frac{\xi_1}{(a_2-\xi_1 a_2)\rho}$ and $\varepsilon = 1$, $\mathrm{OP}_1^{ip\mathrm{SIC}}$ can be written as

$$\begin{aligned}
\mathrm{OP}_1^{ip\mathrm{SIC}} &= 1 - \Pr\left(|h_{\mathrm{SU1}}|^2 > \frac{\xi_1\left(|g|^2\varepsilon\rho + |f_{n^*,1}|^2\theta\rho + 1\right)}{a_1\rho}\right) \\
&= 1 - \int_0^\infty f_{|g|^2}(x)\int_0^\infty f_{|f_{n^*,1}|^2}(y)\bar{F}_{|h_{SU1}|^2}(h(x,y))dxdy \\
&= 1 - \sum_{n=1}^{N}\binom{N}{n}\frac{n(-1)^{n-1}(\lambda_{SU1}a_1\rho)^2}{(\lambda_{\mathrm{SU1}}a_1 n + \Omega_1\xi_1\theta)} \\
&= \times\frac{e^{-\frac{\xi_1}{\lambda_{\mathrm{SU1}}a_1\rho}}}{(\lambda_{\mathrm{SU1}}a_1\rho^2\varepsilon + \lambda_g\xi_1)}
\end{aligned} \tag{7.12}$$

where $\bar{F}_{|h_{\mathrm{SU1}}|^2}(h(x,y)) = 1 - F_{|h_{\mathrm{SU1}}|^2}\left(\frac{\xi_1}{a_1\rho}(x\varepsilon\rho + y\theta\rho + 1)\right)$ if $\frac{\xi_1}{a_1\rho} \geq \frac{\xi_1}{(a_2-\xi_1 a_2)\rho}$ and $\varepsilon = 0$, and $\mathrm{OP}^{1p\mathrm{SIC}}$ can be formulated by

$$\begin{aligned}
\mathrm{OP}_1^{p\mathrm{SIC}} &= 1 - Pr\left(|h_{SU1}|^2 > \frac{\xi_1}{a_1\rho}\left(|f_{n^*,1}|^2\theta\rho + 1\right)\right) \\
&= 1 - \int_0^\infty f_{|f_{n^*,1}|^2}(x)\bar{F}_{|h_{SU1}|^2}(h(x))dx \\
&= 1 - \sum_{n=1}^{N}\binom{N}{n}\frac{n(-1)^{n-1}\lambda_{SU1}a_1 e^{-\frac{\xi_1}{\lambda_{SU1}a_1\rho}}}{(n\lambda_{SU1}a_1 + \Omega_1\xi_1\theta)}
\end{aligned} \tag{7.13}$$

where $\bar{F}_{|h_{SU1}|^2}(h(x)) = 1 - F_{|h_{SU1}|^2}\left(\frac{\xi_1}{a_1\rho}(x\theta\rho + 1)\right)$

If $\frac{\xi_1}{a_1\rho} < \frac{\xi_1}{(a_2-\xi_1 a_2)\rho}$ and $\varepsilon = 1$, and $\mathrm{OP}_1^{ip\mathrm{SIC}}$ can be computed as

$$\begin{aligned}
\mathrm{OP}_1^{ip\mathrm{SIC}} = \quad & 1 - \underbrace{Pr\left(\begin{array}{l}|h_{SU1}|^2 > \dfrac{\xi_1\left(|f_{n^*,1}|^2\theta\rho + 1\right)}{(a_2 - \xi_1 a_2)\rho}, \\ |g|^2 < \left(|f_{n^*,1}|^2\theta\rho + 1\right)\Lambda\end{array}\right)}_{A_1} \\
& + \underbrace{Pr\left(\begin{array}{l}|h_{SU1}|^2 > \dfrac{\xi_1\left(|g|^2\varepsilon\rho + |f_{n^*,1}|^2\theta\rho + 1\right)}{a_1\rho}, \\ |g|^2 > \left(|f_{n^*,1}|^2\theta\rho + 1\right)\Lambda\end{array}\right)}_{A_2}
\end{aligned} \tag{7.14}$$

where $\Lambda = \frac{a_1}{\varepsilon\xi_1}\left[\frac{\xi_1}{(a_2-\xi_1 a_2)\rho} - \frac{\xi_1}{a_1\rho}\right]$.

Then, A_1 can be calculated as

$$A_1 = \sum_{n=1}^{N} \binom{N}{n} \frac{n(-1)^{n-1} e^{-\frac{\xi_1}{\lambda_{SU1}(a_2-\xi_1 a_2)\rho}}}{\Omega_1 \lambda_g} \times \int_0^{\infty} e^{-x\left(\frac{n}{\Omega_1}+\frac{\xi_1\theta\rho}{\lambda_{SU1}(a_2-\xi_1 a_2)\rho}\right)} \left(1 - e^{-\frac{(x\theta\rho+1)\Lambda}{\lambda_g}}\right) dx$$
$$= \sum_{n=1}^{N} \binom{N}{n} \frac{n(-1)^{n-1} e^{-\frac{\xi_1}{\lambda_{SU1}(a_2-\xi_1 a_2)\rho}}}{\Omega_1 \lambda_g} \times \left[\frac{\Omega_1 \ell}{n\ell + \Omega_1 \xi_1 \theta} - \frac{\lambda_g \Omega_1 \ell e^{-\frac{\Lambda}{\lambda_g}}}{n\ell\lambda_g + \lambda_g \Omega_1 \xi_1 \theta + \Omega_1 \ell \Lambda \theta \rho}\right] \tag{7.15}$$

where $\ell = \lambda_{SU1}(a_2 - \xi_1 a_2)$

Similarly, A_2 can be given by

$$A_2 = \sum_{n=1}^{N} \binom{N}{n} \frac{n(-1)^{n-1} e^{-\frac{\xi_1}{a_1\rho\lambda_{SU1}}}}{\Omega_1 \lambda_g} \times \int_0^{\infty} e^{-x\left(\frac{n}{\Omega_1}+\frac{\xi_1\theta}{a_1\lambda_{SU1}}\right)} \int_{(x\theta\rho+1)\Lambda}^{\infty} e^{-y\left(\frac{1}{\lambda_g}+\frac{\xi_1\varepsilon\rho}{a_1\rho\lambda_{SU1}}\right)} dxdy$$
$$= \sum_{n=1}^{N} \binom{N}{n} \frac{n(-1)^{n-1}\lambda_g a_1\rho\lambda_{SU1}}{\Omega_1(a_1\rho\lambda_{SU1} + \xi_1\varepsilon\rho)} \times \frac{e^{-\frac{\xi_1}{a_1\rho\lambda_{SU1}}-\Lambda\left(\frac{a_1\rho\lambda_{SU1}+\lambda_g\xi_1\varepsilon\rho}{\lambda_g a_1\rho\lambda_{SU1}}\right)}}{\left[\frac{n}{\Omega_1} + \frac{\xi_1\theta}{a_1\lambda_{SU1}} + \theta\rho\Lambda\left(\frac{a_1\rho\lambda_{SU1}+\lambda_g\xi_1\varepsilon\rho}{a_1\rho\lambda_{SU1}}\right)\right]} \tag{7.16}$$

Substituting (7.16) and (7.15) into (7.14), $\mathrm{OP}_1^{ip\mathrm{SIC}}$ is given by

$$\mathrm{OP}_1^{ip\mathrm{SIC}} = 1 - \left\{ \sum_{n=1}^{N} \binom{N}{n} \frac{n(-1)^{n-1} e^{-\frac{\xi_1}{\lambda_{SU1}(a_2-\xi_1 a_2)\rho}}}{\Omega_1 \lambda_g} \times \left[\frac{\Omega_1 \ell}{n\ell + \Omega_1 \xi_1 \theta} - \frac{\lambda_g \Omega_1 \ell e^{-\frac{\Lambda}{\lambda_g}}}{n\ell\lambda_g + \lambda_g \Omega_1 \xi_1 \theta + \Omega_1 \ell \Lambda \theta \rho}\right] + \sum_{n=1}^{N} \binom{N}{n} \frac{n(-1)^{n-1}\lambda_g a_1\rho\lambda_{SU1}}{\Omega_1(a_1\rho\lambda_{SU1} + \xi_1\varepsilon\rho)} \times \frac{e^{-\frac{\xi_1}{a_1\rho\lambda_{SU1}}-\Lambda\left(\frac{a_1\rho\lambda_{SU1}+\lambda_g\xi_1\varepsilon\rho}{\lambda_g a_1\rho\lambda_{SU1}}\right)}}{\left[\frac{n}{\Omega_1} + \frac{\xi_1\theta}{a_1\lambda_{SU1}} + \theta\rho\Lambda\left(\frac{a_1\rho\lambda_{SU1}+\lambda_g\xi_1\varepsilon\rho}{a_1\rho\lambda_{SU1}}\right)\right]} \right\} \tag{7.17}$$

If $\frac{\xi_1}{a_1\rho} < \frac{\xi_1}{(a_2-\xi_1 a_2)\rho}$ and $\varepsilon = 0$, we consider perfect SIC situation, and then OP_1^{pSIC} can be computed as

$$
\begin{aligned}
\mathrm{OP}_1^{p\mathrm{SIC}} &= 1 - Pr\left(|h_{\mathrm{SU1}}|^2 > \frac{\xi_1}{(a_2-\xi_1 a_2)\rho}\left(\left|f_{n^*,1}\right|^2\theta\rho + 1\right)\right) \\
&\quad \times Pr\left(|h_{\mathrm{SU1}}|^2 > \frac{\xi_1}{a_1\rho}\left(\left|f_{n^*,1}\right|^2\theta\rho + 1\right)\right) \\
&= 1 - Pr\left(|h_{\mathrm{SU1}}|^2 > \Delta\left(\left|f_{n^*,1}\right|^2\theta\rho + 1\right)\right) \\
&= 1 - \int_0^\infty f_{\left|f_{n^*,1}\right|^2}(x)\left[1 - F_{|h_{\mathrm{SU1}}|^2}(\Delta(x\theta\rho + 1))\right]dx \\
&= 1 - \sum_{n=1}^{N}\binom{N}{n}\frac{n(-1)^{n-1}e^{-\frac{\Delta}{\lambda_{\mathrm{SU1}}}}}{\Omega_1}\int_0^\infty e^{-\eta x}dx \\
&= 1 - \sum_{n=1}^{N}\binom{N}{n} n\eta^{-1}(-1)^{n-1}e^{-\frac{\Delta}{\lambda_{\mathrm{SU1}}}}
\end{aligned}
\tag{7.18}
$$

where $\eta = \left(\frac{n}{\Omega_1} + \frac{\Delta\theta\rho}{\lambda_{SU1}}\right)$ and $\Delta = \max\left(\frac{\xi_1}{(a_2-\xi_1 a_2)\rho}, \frac{\xi_1}{a_1\rho}\right)$.

Remark 1:

From the analytical expressions of outage performance for user D_1, we can see that the outage performance depends on the number of eavesdroppers. Such outage is predicted to improve performance as increasing transmit SNR at the BS, ρ. However, at high SNR, it may show static performance. Furthermore, power allocation factors a_1 and a_2 lead to varying outage performance. According to the definition of the NOMA for downlink, the poor user D_1 should be quickly connected with a low data rate, while the strong one D_2 should be served opportunistically. Therefore, it needs to be considered the performance of D_2 in Section 7.3.2, which in turn results in the conclusion that the performance of NOMA systems is mostly determined by transmitting SNR.

7.3.2 The outage probability of the second user D2

In this section, we elaborate to evaluate the performance comprehensively of the second user D_2 who does not care about the situation of SIC. It further provides performance of such IoT network for general case when many users are employed. In particular, the key performance metrics including outage probability and throughput are characterized. Then, the analytical results and simulations enable us to gain more insights into the system.

$$
\begin{aligned}
\mathrm{OP}_2 &= Pr\left(\frac{1}{2}log_2(1+\gamma_2) < R_2\right) \\
&= Pr\left(\frac{|h_{SU2}|^2 a_2\rho}{|h_{SU2}|^2 a_1\rho + |f_{n^*,2}|^2\theta\rho + 1} < \xi_2\right) \\
&= 1 - Pr\left(|h_{SU2}|^2 > \frac{|f_{n^*,2}|^2\xi_2\theta\rho + \xi_2}{a_2\rho - \xi_2 a_1\rho}\right) \\
&= 1 - \int_0^\infty f_{|f_{n^*,2}|^2}(x)\bar{F}_{|h_{SU2}|^2}(g(x))dx \\
&= 1 - \sum_{n=1}^{N}\binom{N}{n}\frac{n(-1)^{n-1}(a_2\rho - \xi_2 a_1\rho)\lambda_2}{(a_2\rho - \xi_2 a_1\rho)\lambda_2 n + \xi_2\theta\rho\Omega_2} \\
&\quad \times e^{-\frac{\xi_2}{(a_2\rho - \xi_2 a_1\rho)\lambda_2}}
\end{aligned}
\tag{7.19}
$$

where $\bar{F}_{|h_{SU2}|^2}(g(x)) = F_{|h_{SU2}|^2}\left(\frac{x\xi_2\theta\rho+\xi_2}{a_2\rho-\xi_2 a_1\rho}\right)$ and $\xi_2 = 2^{2R_2} - 1$.

Remark 2:

In this section, we elaborate to evaluate the outage performance comprehensively of the IoT network for two users. In particular, the key performance metrics including the outage probability and throughput in delay-limited transmission mode are characterized. Then, these performance evaluations enable us to gain more insights into the system by varying some parameters to further evaluate system performance.

7.3.3 Throughput performance

Throughput of the considered system on each user (target rates R_1, R_2) can be given by

$$
\tau_* = (1 - OP_*)R_* \tag{7.20}
$$

where $* \in \{1, 2\}$.

7.4 Numerical results

In this section, by performing numerical simulation, we can evaluate the correctness of the derived analytical expressions related to outage performance of two NOMA users under different settings, that is, power allocation factors are $a_2 = 0.9$ and $a_2 = 0.1$, secure target rates $R_1 = 0.1$ and $R_2 = 0.1$, and channel gains are $\lambda_{SU1} = \lambda_{SD2} = 1$, $\Omega_1 = \Omega_2 = 0.1$, and $\lambda_g = 0.01$.

Figure 7.2 shows the outage probability of a pair of NOMA users versus the transmit SNR ρ. These outage probabilities are drawn to provide how SIC situation of ipSIC/pSIC makes an impact on system performance evaluation, where $N = 1, 3, 5$. It

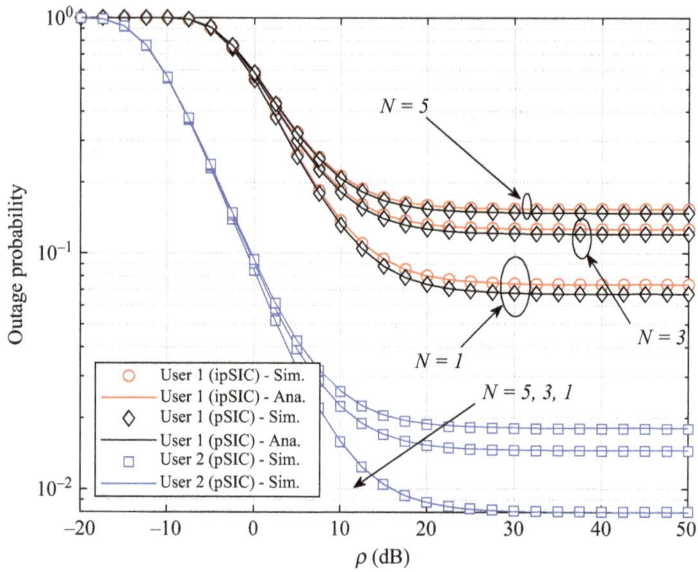

Figure 7.2 Outage performance of two NOMA users under the impact of the number of jamming source

can be seen clearly that pSIC case at user D_1 at $N = 1$ exhibits better performance compared with ipSIC case. This result confirms that less jamming source contributes to reduce the probability of outage event. In this scenario, the different values of power allocation factors are main factors to give the existence of a performance gap between two NOMA users. In particular, the outage performance of user D_2 is very low. At high SNR, such value is greater than 35 dB, and the outage event meets constant level, that is, the outage remains stable as ρ is greater than 30 dB. More importantly, the exact analytical curves for the outage probability of these NOMA users match Monte Carlo results very well, and it exhibits the exactness of our obtained expressions.

As a further evaluation of the imperfect SIC scenario, Figure 7.3 plots the impact of increasing level of ipSIC to the outage probability versus SNR with different $\lambda_g = 0.5, 0.1, 0.05$. It is intuitively as performance gap exists as a comparison between user D_1 in pSIC case and in ipSIC case. It is observed that the outage behavior of each user varies as the changing level of ipSIC λ_g.

Figure 7.4 examines the impact of power allocation factors for two NOMA users on outage performance. It can be seen clearly that the outage performance of the second user D_2 will be improved as a_2 increases while the outage behavior of user D_1 will be worse. Especially, the outage performance of the first user D_1 for two cases (ipSIC and pSIC) meets outage as a_2 approaches 1 at two values of $N = 1, 10$. It shows that outage performance can be controlled by adjusting power allocation factors for two NOMA users.

In Figure 7.5, the different outage probabilities of NOMA and OMA are shown in the entire region of the considered transmit SNR ρ. Due to the power allocation factor, the performance gap between OMA and NOMA can be observed. Furthermore, such

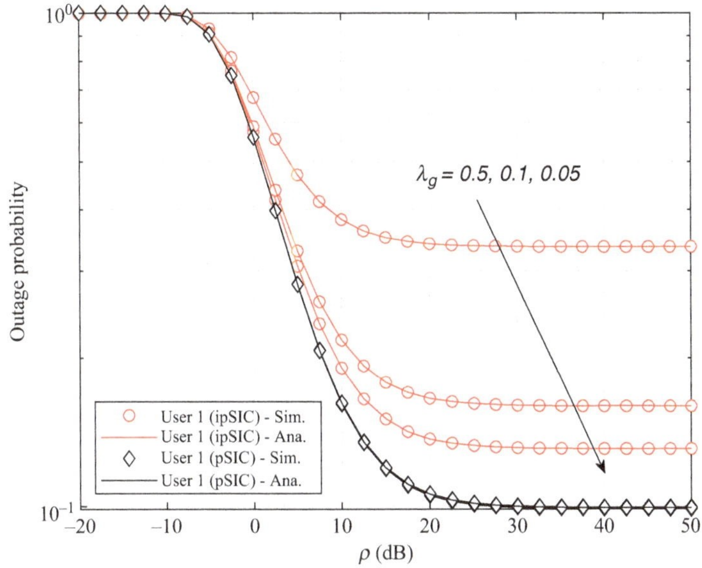

Figure 7.3 Impact of the level of imperfect SIC on outage performance in user D_1 with $N = 2$

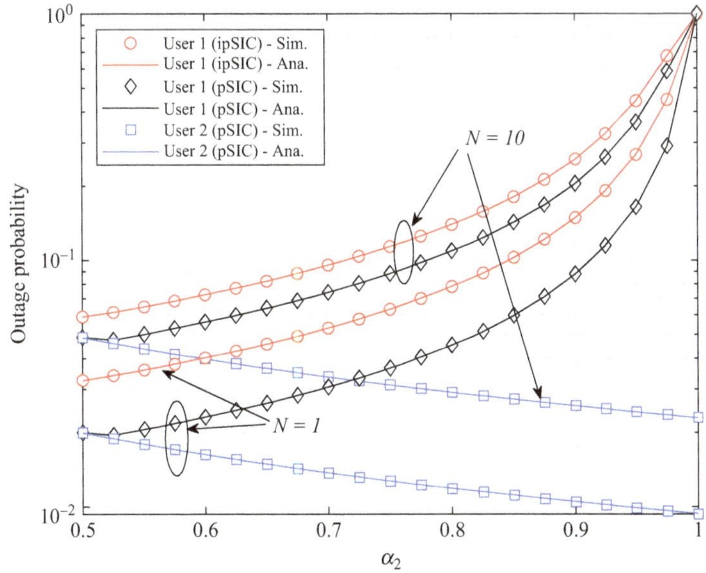

Figure 7.4 Impact of power allocation factors for two NOMA users on outage performance with $\rho = 15$(dB)

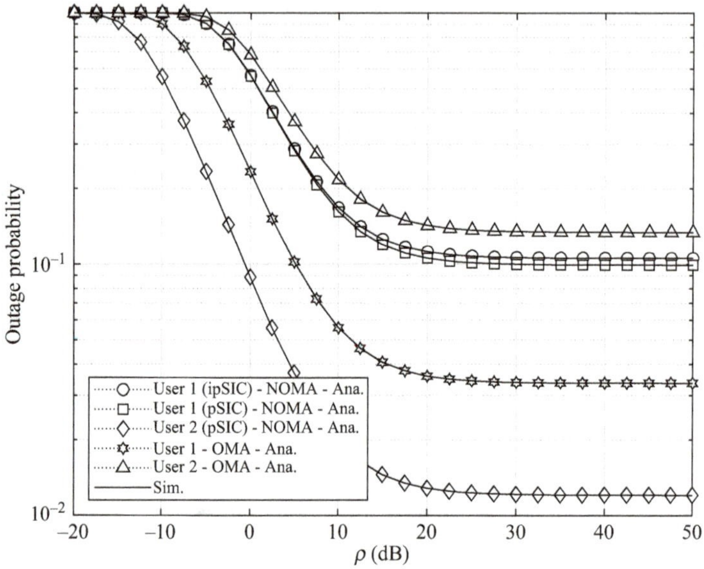

Figure 7.5 Comparison between NOMA and OMA with $N = 2$

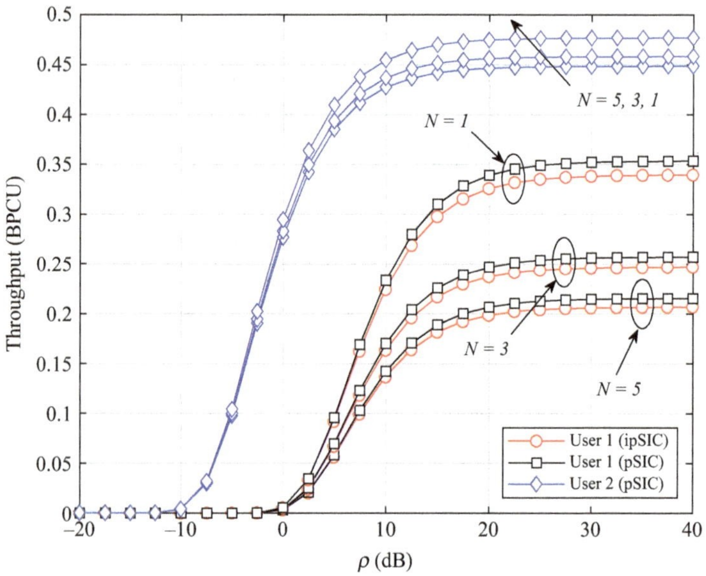

Figure 7.6 Throughput performance versus transmitting SNR

observation confirms the impact of the NOMA scheme exhibiting outage probability while evaluating the performance of two users.

In Figure 7.6, throughput performance can be determined. It plots the throughput of both user versus the transmit SNR with different N. First, it is clear that the throughput does not always increase with the increment of SNR. The reason is that the corresponding outage probability is limited by a high SNR regime as the previous simulation. Moreover, the throughput of the considered system gradually declines as N increases. This is justified that many eavesdroppers easily access confidential information with degraded performance in terms of throughput and outage behavior. In addition, we can see the tradeoff between system performance of each user and power allocation.

7.5 Conclusion

This chapter analyzed a downlink NOMA network with impacts from jamming signal. With regard to impairment of SIC, that is, imperfect SIC and perfect SIC were used in the receiver to provide a comparative study. To examine system performance, the outage probability was considered in this chapter under various parameters. Specially, we derived the closed-form expressions for the outage probability. In order to evaluate other metrics, the throughput of the system is studied and system performance remains stable at high SNRs. These results can be achieved and verified by the simulation results and they were also further evaluated under varying the number of jamming sources. In future work, more users and more pairs of the group need to be investigated to implement such scenarios in real IoT systems.

References

[1] Zhang, Q., Liang, Z., Li, Q., and Qin, J. "Buffer-aided non-orthogonal multiple access relaying systems in Rayleigh fading channels." *IEEE Transactions on Communications*. 2017; 65(1): 95–106.

[2] Cai, G., Fang, Y., Han, G., Xu, J., and Chen, G. "Design and analysis of relay-selection strategies for two-way relay network-coded DCSK systems." *IEEE Transactions on Vehicular Technology*. 2018; 67(2): 1258–1271.

[3] Kader, M. F. and Shin, S. Y. "Cooperative relaying using space-time block coded non-orthogonal multiple access." *IEEE Transactions on Vehicular Technology*. 2017; 66(7): 5894–5903.

[4] Yu, Z., Zhai, C., and Liu, J. "Non-orthogonal multiple access relaying with truncated ARQ." *IET Communications*. 2017; 11(4): 514–521.

[5] Liu, Y., Ding, Z., Elkashlan, M., and Poor, H. V. "Cooperative non-orthogonal multiple access with simultaneous wireless information and power transfer." *IEEE Journal on Selected Areas in Communications*. 2016; 34(4): 938–953.

[6] Do, D.-T. and Le, C.-B. "Application of NOMA in wireless system with wireless power transfer scheme: outage and ergodic capacity performance analysis." *Sensors*. 2018; 18(10): 3501.

[7] Nguyen, T.-L. and Do, D.-T. "Exploiting impacts of intercell interference on SWIPT-assisted non-orthogonal multiple access." *Wireless Communications and Mobile Computing*. 2018; 2008.

[8] Do, D.-T., Van Nguyen, M.-S., Hoang, T.-A., and Voznak, M. "NOMA-assisted multiple access scheme for IoT deployment: relay selection model and secrecy performance improvement." *Sensors*. 2019; 19(3): 736.

[9] Yue, X., Liu, Y., Kang, S., Nallanathan, A., and Ding, Z. "Exploiting full/halfduplex user relaying in NOMA systems." *IEEE Transactions on Communications*. 2018; 66(2): 560–575.

[10] Wei, Z., Ng, D. W. K., Yuan, J., and Wang, H. M. "Optimal resource allocation for power-efficient MC-NOMA with imperfect channel state information." *IEEE Transactions on Communications*. 2017; 65(9): 3944–3961.

[11] Do, D.-T. and Van Nguyen, M.-S. "Device-to-device transmission modes in NOMA network with and without wireless power transfer." *Computer Communications*. 2019; 139: 67–77.

[12] Wei, Z., Ng, D. W. K., and Yuan, J. "Joint pilot and payload power control for uplink MIMO-NOMA with MRC-SIC receivers," *IEEE Communications Letter*. 2018; 22(4): 692–695.

[13] Do, D.-T. and Le, A.-T. "NOMA based cognitive relaying: Transceiver hardware impairments, relay selection policies and outage performance comparison." *Computer Communications*. 2019; 146: 144–154.

[14] Do, D.-T., Vaezi, M., and Nguyen, T.-L. "Wireless powered cooperative relaying using NOMA with imperfect CSI." *Proceedings of IEEE Globecom Workshops (GC Workshops)*. Abu Dhabi, UAE. 2018. pp. 1–6.

[15] Zhang, Y., Wang, H. M., Zheng, T. X., and Yang, Q. "Energy-efficient transmission design in non-orthogonal multiple access." *IEEE Transactions Vehicular Technology*. 2016; 66(99): 2852–2857.

[16] Hui, H., Swindlehurst, A. L., Li, G., and Liang, J. "Secure relay and jammer selection for physical layer security." *IEEE Signal Processing Letters*. 2015; 22(8): 1147–1151.

[17] Park, K. H., Wang, T., and Alouini, M.-S. "On the jamming power allocation for secure amplify-and-forward relaying via cooperative jamming." *IEEE Journal on Selected Areas Communications*. 2013; 31(9): 1741–1750.

[18] Ding, Z., Xu, M., Lu, J., and Liu, F. "Improving wireless security for bidirectional communication scenarios." *IEEE Transactions on Vehicular Technology*. 2012; 61(6): 2842–2848.

[19] Goeckel, D., Vasudevan, S., Towsley, D., Adams, S., Ding, Z., and Leung, K. "Artificial noise generation from cooperative relays for everlasting secrecy in two-hop wireless networks." *IEEE Journal Selected Areas in Communications*. 2011; 29(10): 2067–2076.

[20] Huang, J. and Swindlehurst, A. L. "Cooperative jamming for secure communications in MIMO relay networks." *IEEE Transactions on Signal Processing*. 2011; 59(10): 4871–4884.

Chapter 8

QoS of communication networks using MPLS protocol

Azeddien M. Sllame[1]

This chapter describes the place of multiprotocol label switching (MPLS) in current state-of-the-art computer networking systems which emphasize the use of MPLS as a quality of service (QoS) technique in different networking environments such as data centers and multimedia networking as backbone networks. The chapter also includes practical experiments, which are applied using the OPNET network simulation tool for transmitting multimedia over IP/MPLS networks and a fat tree data center architecture employing MPLS. The examples show the feasibility of MPLS as a QoS tool compared to best-effort IP networks. In addition, this chapter describes an effort of designing a network-on-chip system engaging MPLS mechanism as on-chip communication method which has been completely developed in C++. However, all results clearly confirm that MPLS is an efficient QoS tool.

8.1 Introduction

The multiprotocol label switching (MPLS) technique allows constructing a one converged network over all currently operational networking infrastructures such as 10/100/1000/10G Ethernet and IP, ATM, and Frame Relay by simple switching and routing technique based on attaching labels to the packet headers. Therefore, one of the great features of the MPLS is the reduction of router processing requirements, since the routers will forward packets based on fixed-length labels by consulting smaller label forwarding tables unlike the IP-routing process which requires reading of the whole IP header. Consequently, this forwarding mechanism employed in the MPLS domain will deliver packets with sufficient speed and minimum end-to-end delay over the network by building specified reliable end-to-end paths over the existing topologies. Thus, this enables MPLS to provide assured delivery of services with a specified level of quality of service (QoS) guarantees to

[1]Department of Computer Network, Faculty of Information Technology, University of Tripoli, Tripoli, Libya

different flows of packets by marking packets differently using the forward equivalent class (FEC) field found in the MPLS's header, and the mapping to different classes available in IP's packet headers such as DSCP (DiffServ Code Point) to EXP's field found in MPLS header [1–3].

Moreover, MPLS is standardized by Internet Engineering Task Force (IETF) to allow counting the inserted MPLS's header into IP networks' routing information. MPLS is used to ensure that all packets in particular flow take the same route over a specified path or through the backbone network. MPLS is now counted as one of methods that provide the QoS required for reliable transfer of real-time voice and video with guaranteed end-to-end delays over the available bandwidth [4,5].

This chapter describes the role of the MPLS networking technique as a tool providing QoS in backbone networks, data centers, and network-on-chip systems. Practical examples implemented by using OPNET, a well-known network modeling and simulation tool are provided to demonstrate the efficiency and significance of MPLS as a QoS mechanism.

8.2 Definitions related to chapter context

The following terms are essential to understand chapter context [1–3,6]:

- **Packet**: It is the smallest unit of communication in packet switching networks, which has a fixed size that contains a piece of user data, addressing, sequencing, and error and flow control information in its header.
- **Flow**: It is a well-defined unidirectional stream of packets that have been produced by the same application's instance of a specific user targeting a receiver (s), which need identical QoS level. The flow can be characterized by 5-tuple (transport protocol, source address, source port number, destination address, and destination port number). What is more, an aggregate flow is two or more flows that have some shared values from the 5-tuple with other different features such as multimedia streaming and applications that can be classified into the same service class.
- **Switching:** Switching is the mechanism that defines how network resources are assigned to transfer a packet from a switch input channel to a selected output channel of the switch using the multiplexing technique implemented in the switch.
- **Routing**: It is the process of selecting a path to transfer a packet from a sending node to a receiving node over a set of connecting nodes composing a path over a network's topology based on address.
- **Flow control:** It is a technique involved to organize and synchronize the transmission process over a chosen path between a source and destination in order to avoid overwhelming the destination by the source's data. Flow control needs to outline appropriate procedures and actions to avoid filling of destination's buffers, links' congestion, busy output channels, network faults, and deadlocks.

- **Congestion**: It is defined as the deficiency of available bandwidth dynamically over a path on a network which can be controlled by means of scheduling and load management mechanisms. However, the congestion is the main cause of queues fill up and packet loss, which will incur an extra end-to-end delay to the packet's transfer time.
- **Bandwidth**: It is the competency of the network to offer a specified QoS level to different applications according to their requirements within the real capacity that is available by the existing network's resources.
- **Packet loss**: It is the ratio of the number of packets truly received by the receiver to the number of the packets sent by the sender in a specified time interval (e.g., second) of a network or application or a link. But, the packet loss happens due to heavy traffic, devices' errors, and variations of flows' nature and requirements which yield to different types of congestion and unbalanced use of the network resources by different applications running over the network.
- **Throughput:** The amount of successfully transmitted error-free data between a sender and a receiver per unit of time (bps).
- **Queuing**: The different types of queuing or buffering at all nodes of the selected transmission paths are needed for different multimedia applications. However, at the receiving end, queuing will be used to help the correct construction of the received multimedia streams, due to the out-of-order and rerouting of packets that are allowed when using real-time transmission such as the real-time transmission protocol (RTP). Of course, this has so-called queuing delay.
- **End-to-end delay (latency):** It measures the one-way delay of the packet starting from the sender and ending at the receiver. In other words, it is defined as the difference in time between the packet sending's time from the source and the receiving's time of that packet at the receiver, which includes processing delay, queuing delay, propagation delay, transmission delay, and rerouting delay.
- **Delay jitter**: It is the delay variation between two successive packets within the same flow. However, sufficient queuing can considerably reduce the delay jitter variation.
- **QoS**: It is the capability of a network to offer assured levels of sufficient services to diverse applications in order to satisfy their different requirements of delay jitter, end-to-end delay, packet loss, and throughput utilizing the manageable bandwidth, in order to maximize the exploitation of the available network resources.
- **Real-time application**: It is the application that imposes tight timing requirements on each packet of the flow, in such a way that every packet should be arrived in a specific time and, if not received by that time at the receiver, it is considered useless. Real-time applications are affected by congestion and variable queueing delays at nodes along network's paths. Thus, real-time applications impose using of different QoS techniques over networks to ensure their transmission within specific end-to-end limits with minimum delay jitter. For example, the real-time voice traffic needs short end-to-end

delay and strict delay jitter of the packets, while video streaming requests short delay jitter and accept small packet loss.

- **Tolerant/intolerant application:** It can tolerate some delay to the arrival of packets at the destination. On the other hand, intolerant "real-time" application needs packets to arrive within delay bounds (fixed offset playback delay) otherwise the delayed packets will introduce a distortion to the playback' time of the stream.
- **Elastic application:** It means the sender of traffic can adjust its transmission rate in response to any changes in the offered transfer rate, loss, or delay. The elastic application such as file transfer protocol (FTP) is always able to wait for packets to arrive and no admission control is required (even if they are delayed along the path), since it uses the best-effort service class for its transfer. It does not enforce strict delay requirements on receiving data over the network, such as FTP. On the other hand, an *inelastic "real-time" application* requires packets to be available on time at the right place on the running stream at a receiver's end or otherwise considered as lost packets.

8.3 Quality of service principles

The network that makes QoS available as a service to their customers is the network that provides a definite QoS guarantee levels for delivering different applications' packets, according to each application's requirements. Application requirements include different values of some important parameters, such as short delay jitter and minimum end-to-end delay (latency), least possible packet loss count, specific network bandwidth, and throughput. To satisfy such diverse requirements of various applications, the QoS imposes definite mechanisms and management procedures over the network's resources to control such parameters in order to satisfy the QoS level needed by those applications. In real-time multimedia, the packets of the voice application necessitate short delay and strict delay variation (jitter) of the VoIP stream, while the packets of the video application need short delay variation and small packet loss ratio [1,3,7].

On another hand, the QoS can be seen from the operational viewpoint as the capability of the network to provide a sufficient service to different applications without disturbing the steady operation of the network as well as keeping a certain level of the overall performance of the network by avoiding congestion as much as possible. However, congestion reduces the bandwidth of the network due to (1) some applications start requiring more bandwidth than the available capacity; (2) impulsive request of big bandwidth for unexpected flow transfer; (3) rerouting of some flows to other paths for consumption of the available bandwidth; and (4) different sizes of the LAN/WAN links' bandwidth. However, the links may cause congestion as a result of interconnecting high bandwidth LANs through a lower bandwidth of WAN links, aggregating multiple traffic streams from high bandwidth links into one lower bandwidth link, and transferring the traffic between two core networks over a lower bandwidth temporary network [1,6–8].

8.3.1 QoS models

Due to the importance of QoS, there are some essential mechanisms that have been widely known by network design and management communities, Internet service providers (ISP), and Internet standardization groups which will be discussed in this section.

8.3.1.1 Integrated service (IntServ) model

IntServ model offers a mechanism for the unicast or multicast applications to explicitly select and reserve some network resources by using resource reservation protocol (RSVP) with the aim of meeting the requirements of multiple applications. In IntServ model, the resource reservation and admission control are considered as the two essential building elements. In addition, IntServ model is contained inside the network service interface to be near to the application requesting mechanisms which will apply for the required services. However, such applied requirements are looking to minimize the latency of the packets sent between any pair of senders and receivers in order to satisfy the requested QoS level. The proposed IntServ framework also includes packet scheduler and packet classifier as building blocks. In IntServ model, packet scheduler, classifier, and admission control are proposed to be collected together in a so-called traffic control unit inside the router [1,3,9].

Scheduler module employs a set of queues to make proper packet forwarding according to class priorities.

Classifier module responsible for mapping packets into different prescribed classes; all packets belonging to the same class are treated equally by the scheduler served in the same queue. The classification may be done based on the protocol id, application type, IP address, or port address. Classification can be done per packet, flow, or aggregation of flows in different parts of internetworking infrastructure. Practically, the appropriate implementation of the classifier can help the routers in doing the packet forwarding process efficiently, since each classified class of packets will be treated equally when the router performs the packet forwarding.

Admission control component is in charge of deciding the possibility of the availability of requested reservations (QoS level) for the new flow without affecting the previous flows' reservations within the network's bandwidth. Admission control is operated at each router in the path to accept or reject the requested reservation [1,3,9].

RSVP protocol unit is part of endpoints and routers along the selected path and it is liable for establishing, maintaining, and managing the network resources for a specific flow along the preserved path from sender to the destination(s). However, RSVP is a network control protocol that permits receivers to apply for a specific end-to-end QoS for its data flows by setting up reservations for required network resources. Therefore, the RSVP is accountable for preserving the states of the host and the routers along the path selected to provide the requested service between the sending and receiving nodes. The routers are taking care of all of their resources such as buffers and links' bandwidth that is taken to perform the requested reservation in the soft state. RSVP is designed to handle heterogeneous receivers with different bandwidth along the paths as well as can support a wide range of applications with different resource requirements [1,3,9].

8.3.1.2 Differentiated service (DiffServ) model

DiffServ model is proposed to allocate network resources made only to aggregated traffic. The model applies policing, provisioning, and traffic classification at the edge routers. The model uses the same network infrastructure to provide different levels of QoS to different traffic flows depending on their requirements by using the code point field defined in the IP protocol's header. The model gives emphasis to the work at edge routers which is obliged to perform classification, marking, metering, shaping, and policing of packets entering the network domain [1,3,9].

The DiffServ model describes a class of service (CoS) and a set of node-based (called per hop operation) functions that control resource management for QoS purposes, which incurs minor per-node overhead. Thus, the DiffServ model provides a modular, implementable, and scalable IP QoS method. The node-based approach allows a granular control over buffers and links' bandwidth per hop behavior (PHB) on the nodes between the competing flowing streams. However, the DiffServ defines the functionality of the edge routers as follows: doing per-flow management, marking, and classification into classes defining PHB by six bit of DiffServ code point (DSCP) in DS field of IP's header, and performing traffic conditioning (policing, marking, and dropping). In addition, the functions that are performed by the core router are as follows: use the DSCP of the packet to perform PHB, traffic management referring to the class of the packet, do packet buffering and scheduling according to packet's marking, and to perform assured forwarding. The PHB is selected at a node by a mapping of the DS code point in a received packet. All of that make the DiffServ model a practical and scalable model when compared to IntServ model [1,3,9].

The DiffServ model uses six bits of the IPv4 or IPv6 to carry the DiffServ code point which decides on a PHB for the packet (two "currently unused" bits). However, the DiffServ renames the type of service (TOS) octet of the IPv4 header and traffic class octet of the IPv6 header, respectively, to the DS field. Therefore, the DSCP is used as an index into forwarding table to support the packet classification and forwarding [1,3,9].

Classification unit sorts packets according to prescribed classification rules by checking up the contents of the packet's header. In the DiffServ model, the metering, marking, shaping, and dropping functions are contained in a unit named as traffic conditioner.

Metering module is responsible for examining if the received stream of packets is matching the agreed communication profile.

Marking part is in charge of marking the flow that is consistent with the negotiated transmission profile. The traffic profile contains a collection of information about the temporal attributes of a flow such as acceptable data-rate, minimum required latency, and burst size, which can be used by the packet classifier to decide which packets agree with the in-profile properties or falls in out-profile specification. However, the packets that agree to in-profile specification do not need to pass extra conditioning, while the out-profile packets may need to be queued in order to pass through additional shaping and discarding policies.

Shaping is the process of imposing the traffic flow confirming to the prescribed profile by delaying the packets of the flow in order to become in the desired traffic shape. However, while performing traffic shaping, some packets of the stream may get discarded in the case of buffer overflow or according to specific dropping rules, and this helps shaping to match the agreed flow profile.

8.3.1.3 Traffic engineering

Traffic engineering (TE) is a method of providing efficient resource utilization by controlling the bandwidth consumption and balancing the traffic load distribution. In addition, TE is targeting to support QoS by assisting applications to satisfy their requirements in end-to-end delay, delay jitter, and packet loss. Therefore, by applying TE, the traffic streams will be distributed over all available links in the available paths between the senders and receivers making efficient use of the provided infrastructure of the network, beyond the utilizing of shortest paths found by routing protocols such as open shortest path first (OSPF) or intermediate system to intermediate system (IS-IS) [1,10].

8.3.1.4 MPLS as QoS approach

MPLS protocol offers a way for QoS support through network routing control by attaching short fixed-size labels in the packet's header. QoS within the MPLS domain is maintained using network resource reservation and TE approaches on the per flow basis. However, the MPLS is developed in such a way that some streams consist of a flow of correlated packets such as in multimedia streaming which needs to be processed continuously in the same way through all routers making up the path from sender to destination by the same FEC value so as to be treated by the same way in all routers along the path. MPLS is used to ensure that all packets in a specific flow take the same route over the backbone path of the network [3,4,10].

8.4 MPLS principles

MPLS operates in-between layer 2 and layer 3 of the OSI model; it is referred to as layer 2.5. MPLS networks are made of a contiguous set of routers called *label switch routers* (LSR) that can switch and route packets based on labels attached to all packets passing through the MPLS domain. Also, for each correlated packets composing of any specific flow, a unique FEC identifier is defined for those packets, which is used to make a label/FEC binding to identify the designated path that will be used by subsequent LSR routers to forward the packets over the *label switched path* (LSP) through the MPLS domain.

Thus, associating FEC with the label binding with the LSP makes MPLS functioning as a connection-oriented network over an IP connectionless network associated with a definite degree of QoS. However, the MPLS is an emerging technology that facilitates efficient packet forwarding, assures reliable delivery of the Internet services over backbone networks, possesses high transfer speeds, holds lower network end-to-end delay, and offers scalable networking by merging different networking technologies into one network [2,3,10].

8.4.1 MPLS related terms

The main terminologies of MPLS technology are described as follows [1,3,4,10]:

- **Label**: A short fixed-length identifier that is attached to the packet entering the MPLS domain and is used to identify an FEC, usually having local significance. However, label binding is the process that is made of two MPLS nodes to agree on the assignment of a label to an FEC. The label is used as an index into a label information base (LIB) and label forwarding information base (LFIB) tables that specify the next hop and a new label.
- **MPLS node**: A node that understands MPLS which is implementing the forwarding packets based on labels. The node also should be able to understand MPLS control protocols such as *label distribution protocol* (LDP).
- **MPLS domain**: It is an interconnected group of nodes that employ the MPLS routing and forwarding functions under a particular administrative domain. It is divided into MPLS core routers named as LSRs that are found in the center of the MPLS network and MPLS edge routers named as *label edge routers* (LERs) that operate at the border of the MPLS domain.
- **Label:** A short fixed-length identifier that is attached to the packet when arriving to pass through the MPLS domain, which is usually having local significance. However, label binding is the method made of two MPLS nodes to agree on the assignment of a label and FEC.
- **Label switch router (LSR)**: It is the router found in the heart of the MPLS domain which forwards packets based on label switching technique. When an LSR receives a packet, it checks the label information base to determine the next hop based on the attached label as a reference. Then, it removes the in label from the header and attaches the out label and forwards the packet to the next hop. In addition, LSR router is capable of routing layer 3 packets.
- **Label edge router (LER):** These routers are found in the borders of the MPLS domain, where packets enter the MPLS domain through the LER router. However, the entering LER router is called *ingress* router and the leaving LER router is named *egress* router. LER is capable of L3 routing and it is in charge of adding/removing labels and FEC to/from the packets when they enter/leave the MPLS domain.
- **Label distribution protocol (LDP)**: It is the essential MPLS signaling protocol that is responsible for forming, preserving labels/paths, and making label mapping/exchanging labels/FECs information between LSRs to build LSP paths between senders and receivers over the MPLS network before forwarding users' data. However, LDP is a control protocol that employs a set of procedures and messages which are exploited by any LSR router to inform another LSR router about the label/FEC bindings it has made. The LSR

employs LDP protocol to form LSP over the MPLS network by mapping network-layer routing information directly to data-link layer switching paths (LSP).

- **Forward equivalence class (FEC)**: All packets that have correlated characteristics are given the same FEC identifier (class) as well as they are given the same label when they are entering the MPLS domain via the ingress LER router. In addition, all packets that have the same FEC are forwarded through the similar next hop with the same priority over the same LSP path, operated equally by the router, handled on the same queue and should have the same QoS service class with equivalent drop preferences since they have the same FEC.
- **Label switched path (LSP):** LSP is a unidirectional path in the MPLS domain built over a series of LSR routers and followed by all the packets which have the same label and same FEC class. These paths are established and maintained by the LDP protocol. Several LSPs are organized in an MPLS domain, which originates at ingress LER router and traverses one or more core LSRs routers and ends at LER egress router. Usually, LDP uses TCP protocol to operate at session's setup for LSP to ensure the reliability of created LSP paths. However, the LSP path needs to be setup before data are forwarded and removed as soon as no longer needed.
- **Label forwarding information base (LFIB)**: All LSR routers keep up all learned labels in so-called LIB, in which each entry of LIB links an FEC with an LDP identifier (label) pair. However, whenever a packet needs to be forwarded to a next hop, the LSR will retrieve the label from the LIB table. This, in turn, eliminates looking for the whole header of the IP packet, which gives the speed to the MPLS forwarding technique over the IP routing, since the IP routing uses longest prefix match (slow) while the MPLS uses labels from the LIB which is an exact match (provides speed).

8.4.2 The MPLS router

8.4.2.1 Traditional router structure

Essentially, any router performs two fundamental tasks: *routing* and *packet forwarding* rely on the information exchanged between neighboring routers using different routing protocols. The routing procedure builds a routing table as a view of the network topology and computes the best paths to different destinations and stores them in the forwarding table. The packet forwarding routine moves a packet from an input interface ("ingress") of a router to the suitable output interface ("egress") depending on the information found in the forwarding table. However, the performance of the forwarding technique determines the overall performance of the router, because every packet arriving at the router requires forwarding to the next network hop. The functions of the packet forwarding process can be categorized into two classes: elementary forwarding and complex forwarding functions. Elementary forwarding defines the basic conventional functions that the router should do to transfer packets between interfaces. Complex forwarding tasks include some other additional processing functions which depend on the deployment environment and some data flow special requirements [2,3,10].

8.4.2.2 The elementary router functions

The elementary router functions include [3,10]:

- *IP header validation:* This function is performed to make sure that only regular-formed packets will be further processed while the inconsistent shaped packets are discarded. This means the header length and its checksum is correct.
- *Packet lifetime control:* This is done by decrementing TTL value of IP packet in order to avoid the packet getting in routing loops.
- *Checksum recalculation:* The header checksum needs to be updated after every hop TTL decrement.
- *Route lookup:* The packet's destination address is employed to examine the forwarding table to find the appropriate output port that the packet should be passed toward its final destination.
- *Fragmentation:* The packet may need the fragmentation/assembly at different routers to cope with different network technologies packet formats.
- *Handling IP options:* Some packets may need special processing according to different flow characterization such as applying quality of service.

8.4.2.3 The advanced router functions related to QoS

As the Internet usage grows up with different applications such as multimedia applications, different requirements have been imposed to be discussed during the design of advanced network infrastructures such as security and QoS service level or real-time flow classification. However, these introduced functions have overheads translated into extra processing inside the router during forwarding of the packet en route for its final destination. However, the most important advanced functions are [3,5,9,10]:

- *Packet classification:* It is the function of distinguishing packets into several service classes in order to operate necessary procedures or certain rules to each class of data streams such as VoIP or real-time video streaming. However, to do packet classification, the router may need to check other IP packet fields such as destination address or payload type. After classification, other traffic handling policies like policing and marking can be involved.
- *Packet translation:* The router may act as a gateway to a network so it is required to take care of the network address translation (NAT) functions. NAT assigns a public IP address into a group of private IP addresses and vice versa. This necessitates a router to keep a record of the linked hosts and their local addresses with a translation of the incoming and outgoing packets accordingly.
- *Traffic prioritization*: The router may need to apply a definite traffic prioritization mechanism in order to preserve certain QoS levels to meet the agreed different application requirements and link's bandwidth such as those required to transmit real-time streaming such as VoIP and video conferencing.

- *Policing and marking*: Policing is a technique which is applied to make sure that a traffic flow does not rise above a defined maximum rate value, that is, the process of handling out of profile traffic. However, a simple policy could be as follows: transmit the packet if it conforms to the specified bit rate and to drop the packet if it exceeds the rate limit.
- Marking is commonly used in combination with policing to impose a defined agreed rate in service level agreement (SLA) to a specific stream of traffic by allowing the marking of the traffic exceeding the specified transmission rate in order to handle it differently when the capacity accepts that or discard it and acknowledge it for retransmission.
- *Queuing and scheduling*: Due to high traffic through the Internet, the buffers are used to store packets in queues temporarily inside the routers before getting transmitted. The order of the transmission is determined by several factors such as the packet's service class or the service guarantees level associated with the packet's class. Different queues have different priority levels and every queue consumes a different amount of bandwidth on the same output link according to their priority value. However, the scheduling algorithm is used to select the next packet to be transferred from such queues subject to the bandwidth prescribed for each queue.

8.4.2.4 The difference between IP and MPLS router

The traditional TCP/IP networking model is used to facilitate data transmission among a mixture of networked systems from a definite source to a specified destination. Data are assembled as a chain of packets. All the packets are routed through a series of routers over multiple networks to reach the destination. In IP connectionless networks, every router employs a layer 3 routing protocol. Once the packet enters the IP network, each router along the path toward the final destination makes an independent forwarding decision to select the "next hop" for that packet. The forwarding decision is based on the information contained in that packet's IP header and on information stored in the routing table [10].

However, the forwarding decision is repeated at each hop by reading the whole IP header of the packet and matching routing table entries to select proper output of the router to the next hop, which may be considered as a processing-intensive task. On the other hand, in MPLS networks, forwarding is made based on looking up inside small label-based forwarding tables named as LFIB table, which is much smaller than IP forwarding tables. Therefore, the packet forwarding process inside the MPLS networks is quicker than IP networks.

Usually, the router is constructed by two parts, as seen in Figure 8.1: control plane and data (forwarding) plane. The control plane is the part in the router that holds routing information and all required control information, such as label bindings, which are exchanged between label switching routers (LSRs in MPLS environment). MPLS is a control driven plane protocol, in which the control information must be accessible and exchanged before the forwarding of the first data packet through the MPLS's router taken place in the proposed path. Therefore, in MPLS networks, the control plane uses LDP or RSVP protocol to build the paths

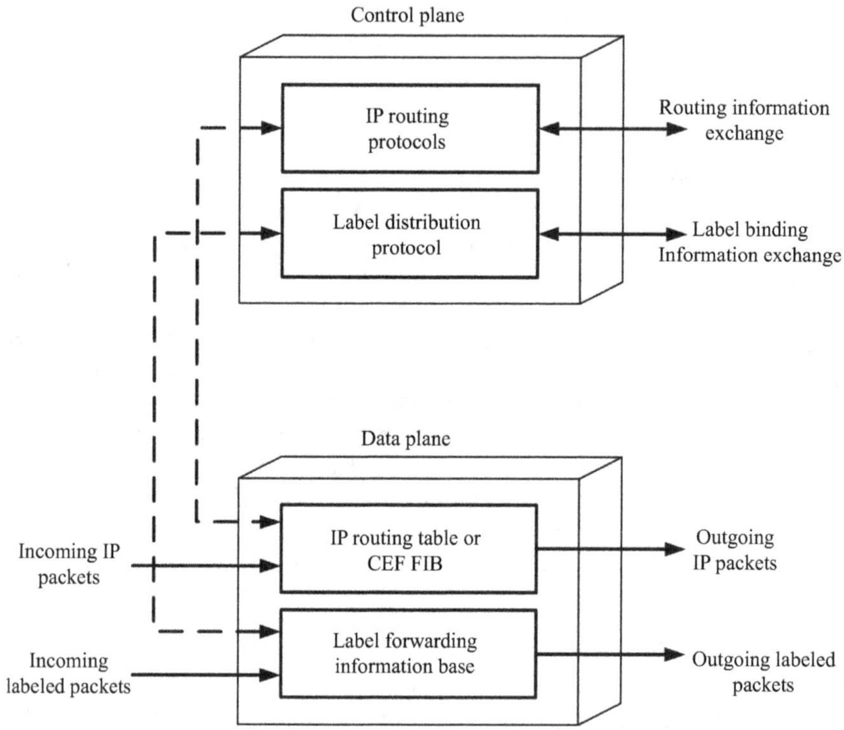

Figure 8.1 MPLS router architecture

by exchanging and making label/FEC mappings to different LSP paths to form paths that cover all source/destination pairs over the MPLS domain before any packet transfer inside the MPLS domain. Each packet that enters the MPLS domain at the ingress node will be given a label and tied to an appropriate FEC; so all packets that have the same FEC should be forwarded along the same route since they have label/FEC binding made by LDP protocol with a specific LSP path. Consequently, by applying this approach, the MPLS builds a connection-oriented network over any connectionless networks. Each MPLS node preserves two tables related to MPLS switching: the LIB and LFIB tables. The LIB contains all the labels assigned by the local MPLS node and the mappings of these labels with FECs to labels/FECs exchanged with its MPLS neighbors. The LFIB employs a subset of the labels enclosed in the LIB with the information included in the packet's attached label to make definite quick packet forwarding [2,4,9,10].

8.4.3 MPLS operation

Understanding the above-presented terms of the MPLS technique, the MPLS operation can be summarized into the following steps: label creation and distribution, LFIB table creation at each LSR router, LSP paths creation, label insertion/

table lookup, and MPLS-packet forwarding through the network, as seen in Figure 8.2. Prior to routing and delivery of packets through the MPLS domain, an interior routing protocol, such as OSPF or IS-IS protocol, is used to exchange reachability and routing information to create the traditional IP routing tables and forwarding tables over the available network topology. Then, the LDP protocol is employed to generate labels and distributed them between adjacent LSRs routers to create LSP paths as well as FEC is allocated to support QoS to different flows. Then, each LSR router forms the LIB tables and produces the LFIB tables to facilitate the forwarding based on MPLS labeling instead of IP conventional forwarding process. When a packet enters MPLS domain through ingress, the LER router is assigned a label and allocated to a particular FEC, which will follow a defined LSP path and the packet is forwarded to the next hop (first LSR router in the MPLS domain) [2,4,9,10].

Within the MPLS domain, as each LSR receives a labeled packet with its distinguished FEC as QoS means as seen in Figure 8.2, it removes the incoming label and attaches the appropriate outgoing label to the packet and forwards the packet to the next LSR router along the specified LSP. When the packet reaches the final router which is the egress LER router, it strips out the label, reads the IP-packet header, and forwards the packet to its final destination, and more details could be found in [1–3,10].

8.5 MPLS and multimedia

This section describes multimedia session description, multimedia session requirements, and a brief discussion about commonly used multimedia protocols.

8.5.1 Multimedia session

Any multimedia session as seen in Figure 8.3, such as the Skype or IPTV streaming session, includes multimedia transfer between numerous participants needs to be described by the following items: (1) IP addresses and locations of the participants; (2) status of the session's members as online or offline and their desire to take part in the session; (3) session capabilities such as the required bandwidth and payload type to define the proper codec to keep the session steady and continuous; (4) session arrangement which includes negotiating transmission parameters using proper protocols and applications; (5) session continuity that keeps the steady flow of the multimedia according to the negotiated parameters during the proposed session time; and (6) session teardown that makes the termination of the session in a smooth way between the engaged participants [1,3,8,10].

Furthermore, in order for multimedia sessions to run efficiently over Internet, the following requirements need to be fulfilled: (1) sequencing which is used by recipients to find lost packets, do reordering and decoding, and to reconstruct the stream properly each packet in its correct place; (2) buffering that is used to eliminate delay jitter and time stamping that define correct timing for each packet to do efficient playback; (3) payload type identification that is used by end devices

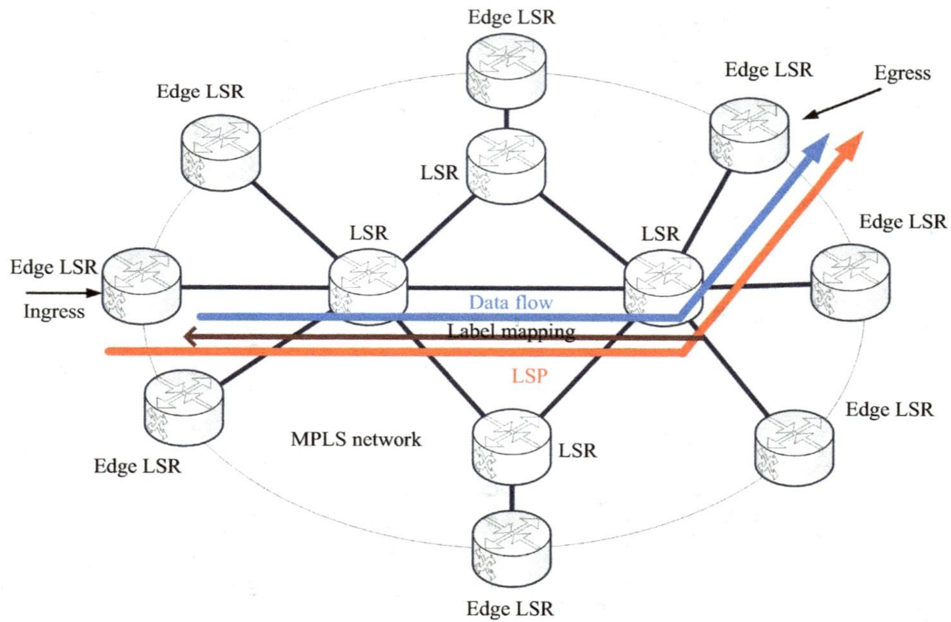

Figure 8.2 MPLS operation

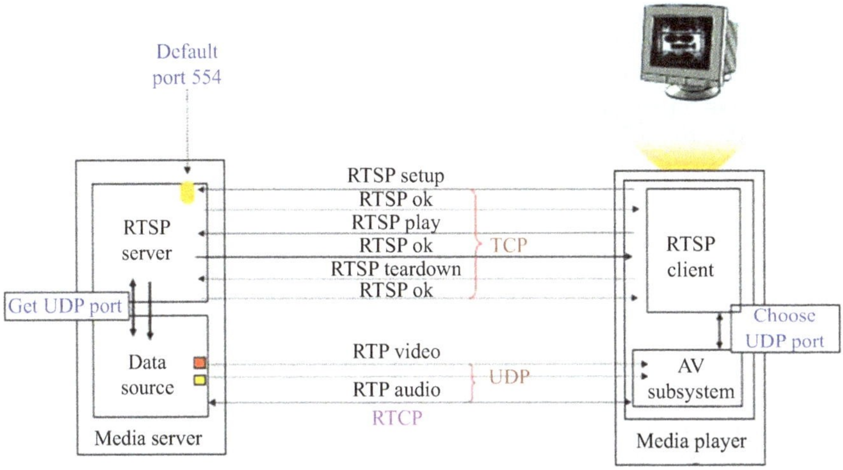

Figure 8.3 Typical multimedia session flow

and applications to do suitable multimedia interpretation; (4) error suppression by means of redundancy techniques; (5) QoS reporting along the paths between session' participants in order to keep the session's progression smooth and continuous with proper bandwidth; and (6) proper reaction to network congestion to keep the multimedia session with suitable QoS level [1,3,8,10].

8.5.2 Multimedia protocols

The minimal multimedia protocols that are involved in any multimedia session include TCP/IP, user datagram protocol (UDP), session initiation protocol (SIP), RTP, real-time control protocol (RTCP), and session description protocol (SDP). The SIP is an application layer protocol with client–server model used for signaling and controlling VoIP and Internet conferencing sessions as well as it supports instant messaging. The SIP protocol was developed by IETF and proposed to be very simple and scalable, and it uses a small number of text-based commands using much similar text messaging that are used with hypertext transfer protocol (HTTP) and the simple mail transfer protocol (SMTP). SIP works in combination with numerous other protocols such as RTP which is used to transfer the multimedia's data after establishing the session by SIP between the endpoints successfully. In addition, media payload type and associated description parameters are handled by the SDP protocol. Moreover, RTCP is usually accompanying the RTP protocol to help the senders and receivers agree on a certain QoS level and provides feedback for the session continuity purposes [1,3,8,10].

It is common in multimedia sessions to use the TCP protocol during the session setup period in order to establish a reliable session between the participants. Furthermore, after the session is created correctly, the multimedia sessions commonly employ RTP and UDP for transporting multimedia of the session. The reasons behind using UDP instead of TCP during the session's playout is that TCP is a reliable and connection-oriented protocol, makes retransmissions within unbounded delays, has no delivery of time-stamping, has slow-start for congestion, and is not a multicast protocol, and all these drawbacks are unacceptable in real-time transmission. Whereas, UDP is a simple protocol offering a minimum of transport protocol processing, port-number addressing, multiplexing for sharing the capacity of the link, and header's checksum for the correction purposes. The RTP compensates for the simplicity of UDP by providing sequencing, payload type identification, and time-stamping, and RTP is designed to support multicasting for end-to-end transfer services for real-time multimedia ensuring the preservation of sequencing to reconstruct the transferred multimedia. What is more, with RTP, there is an RTCP protocol that complements RTP in delivery monitoring and QoS control [1,3,8,10].

However, RTCP is used to provide QoS feedback by collecting end-to-end information about the QoS along with the session's path resources and contributes to all senders and receivers to aid in session's flow control by adjusting session's parameters in order to observe the session. RTCP collects all the information such as recorded loss rates to different session's paths; the last packet's sequence number; RTP timestamp; delay jitter from different senders; and numbers of packets lost [1,3,8,10].

In addition, a real-time streaming protocol (RTSP) is designed to use RTP/UDP/IP for multimedia streaming. SDP is used to describe media data carried by other protocols such as SIP and RTSP which include negotiation and capabilities. However, SDP can be used for multicast announcements about sessions by users, as well as SDP can be considered as a method to deliver necessary information to encourage participants to join the currently announced session. Therefore, the SDP message will include session name, identifier, owner and purpose, status (active, timing), media type and formats, session moderator, addressing, protocol ID, and required bandwidth [11]. More details about multimedia protocols can be found in the standard recommendation [12–16].

8.5.3 VoIP application

VoIP is a real-time inelastic application that sends packets at the rate produced by the voice codec irrespective of capacity availability. The VoIP uses codecs that are used by the sender to transform the voice analog signal to a digital form containing equivalent bit stream and return the digital bit stream into an analog voice signal when received by the recipient of the VoIP call keeping the accuracy of the received VoIP call. As mentioned earlier the VoIP is transported using RTP over UDP protocol in order to overcome the drawbacks of the TCP protocol that are related to the characteristics of the real-time multimedia transmission [1–3,11].

8.5.4 Example: multimedia over MPLS networks

The MPLS technique depends on the QoS of a multimedia session in terms of end-to-end delay and delay jitter. A simulation consists of two scenarios considering the same network topology as seen in Figure 8.4 with different router capabilities is implemented using OPNET network simulation tool version 14.5 (see [17]). There are two scenarios: scenario one is an MPLS with TE backbone network, while scenario two is a pure TCP/IP network without TE. However, OSPF is used as a routing protocol and the QoS is represented in this example by scheduling with a weighted fair queuing (WFQ) scheme [18–22].

Figure 8.5 shows the VoIP end-to-end packet delay with WFQ; the result obviously proves that the MPLS technique has registered the lowest (best) voice end-to-end packet delay over the normal TCP/IP network, which marks MPLS as an advantage technique when used in multimedia networking. Figure 8.6 displays the VoIP packet Jitter delay WFQ queuing as QoS means; the result understandably demonstrates that the MPLS technique has plotted the lowest (best) voice jitter packet delay than the ordinary TCP/IP network, which makes MPLS a useful tool for efficient multimedia networking. Figure 8.7 describes the results of the VoIP delay jitter of SIP and H.323 as signaling protocols with MPLS networking. The delay jitter of VoIP is higher (worse) with SIP than that of the jitter of H.323. Thus the H.323 protocol is better than the SIP protocol in this performance parameter because it has less jitter than SIP with MPLS networking. Figure 8.8 describes the VoIP end-to-end delay with OSPF routing protocol and MPLS networking when using SIP and H.323 voice signaling protocols. The end-to-end delay is higher (worse) with SIP than that of the end-to-end delay of H.323; thus the H.323 protocol is better in this performance parameter because it has a less end-to-end delay (less than 1.4 s) than SIP which has more than 1.5 s with MPLS networking.

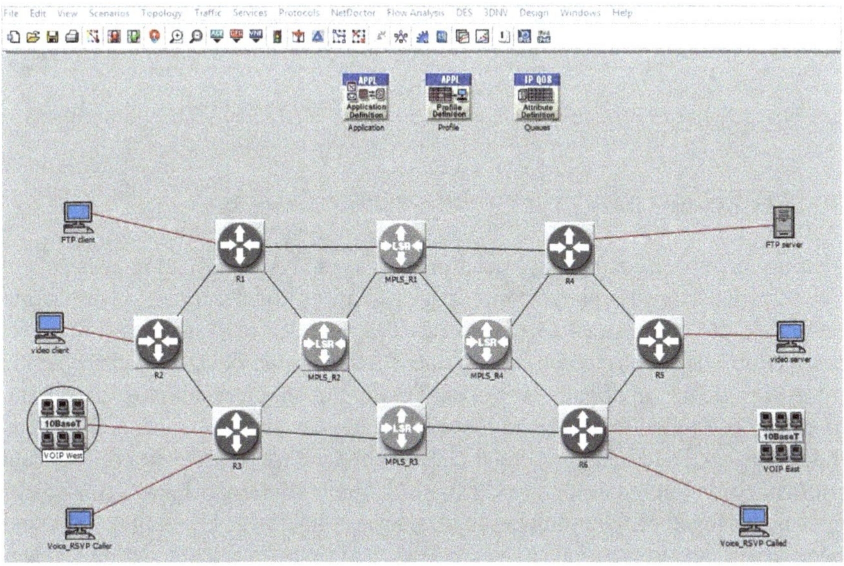

Figure 8.4 Example of MPLS multimedia backbone network

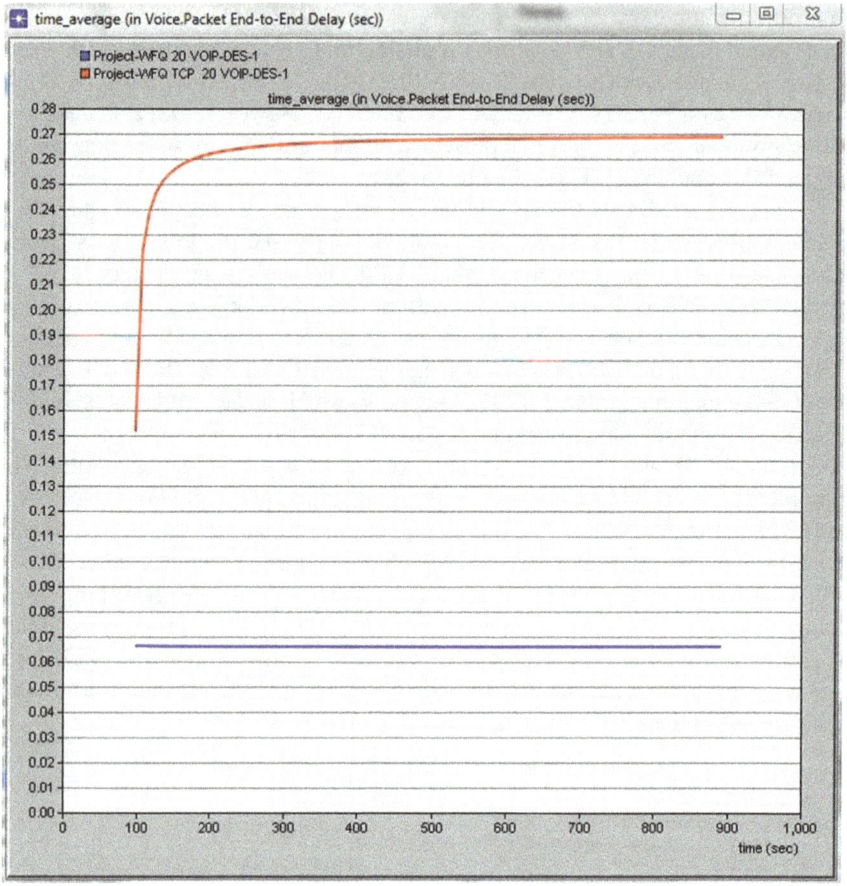

Figure 8.5 VoIP end-to-end packet delay with WFQ for IP/MPLS networks

8.6 MPLS mechanism and data center design

Networks have become the top crucial part of nowadays' information systems era. They form the essential part for information sharing in enterprises, governmental institutions, the economical sector, social sites, and scientific societies. With the extensive utilization of cloud services, data centers' networks are emerging on the way to significant, multipath networks. To support the very massive data transferred between individuals around the world, new advanced network architectures for data centers have been presented as fat tree structure in order to produce high performance on aggregate bandwidth with a simple topology. However, the data center is put together from numerous data servers housed in the same compound, where servers require assured environmental conditions, such as moderate levels of heat and humidity, as well as network connections assuring availability

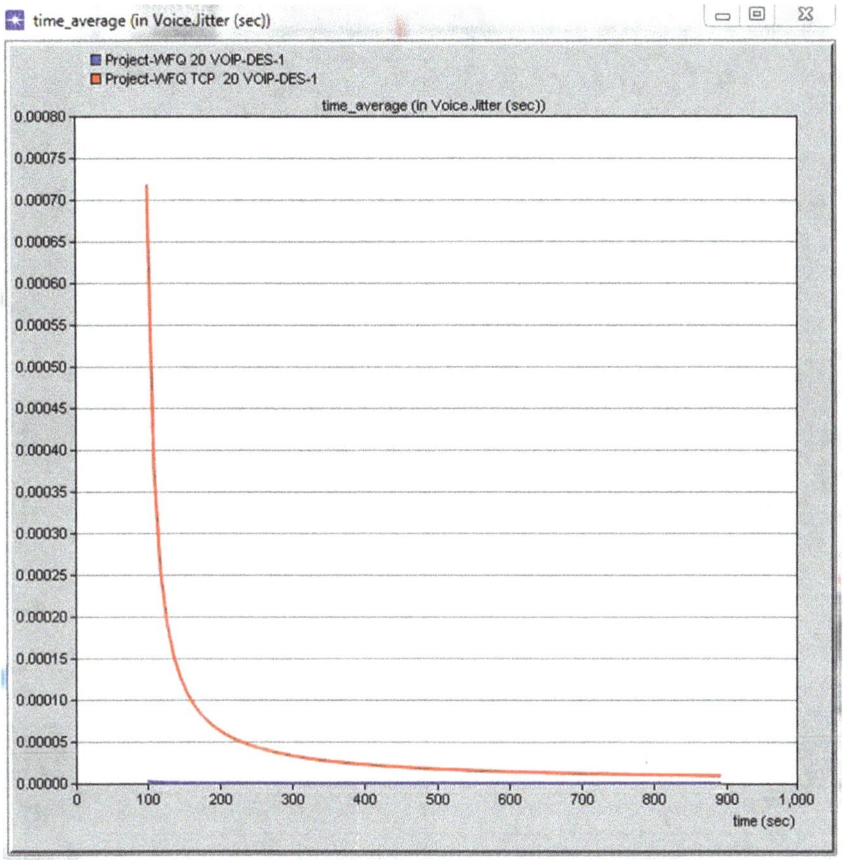

Figure 8.6 VoIP jitter delay with WFQ queuing for IP/MPLS networks

performance, resiliency, and scalability to access the Internet. Many Internet service providers (ISP) and content distribution companies have and control their own dedicated data centers such as Google, Amazon, and Akamai while some data centers offer storage services for multiple clients (see [23–26]).

The modern data center design needs to be flexible to deploy quickly and can adapt to supporting any new added services or applications in a short time frame to reach a significant competitive advantage. However, from the design point of view, the routing has become one of the most fundamental issues in the networking world's infrastructure. Routing involves selecting an efficient path from a client to a data center's server in an efficient manner to reduce latency between the user and the data center practically. In addition, MPLS has been evolved as an efficient technology that guarantees stable and steady delivery of the Internet services with high transmission speed and low end-to-end delay (latency). Therefore, applying an

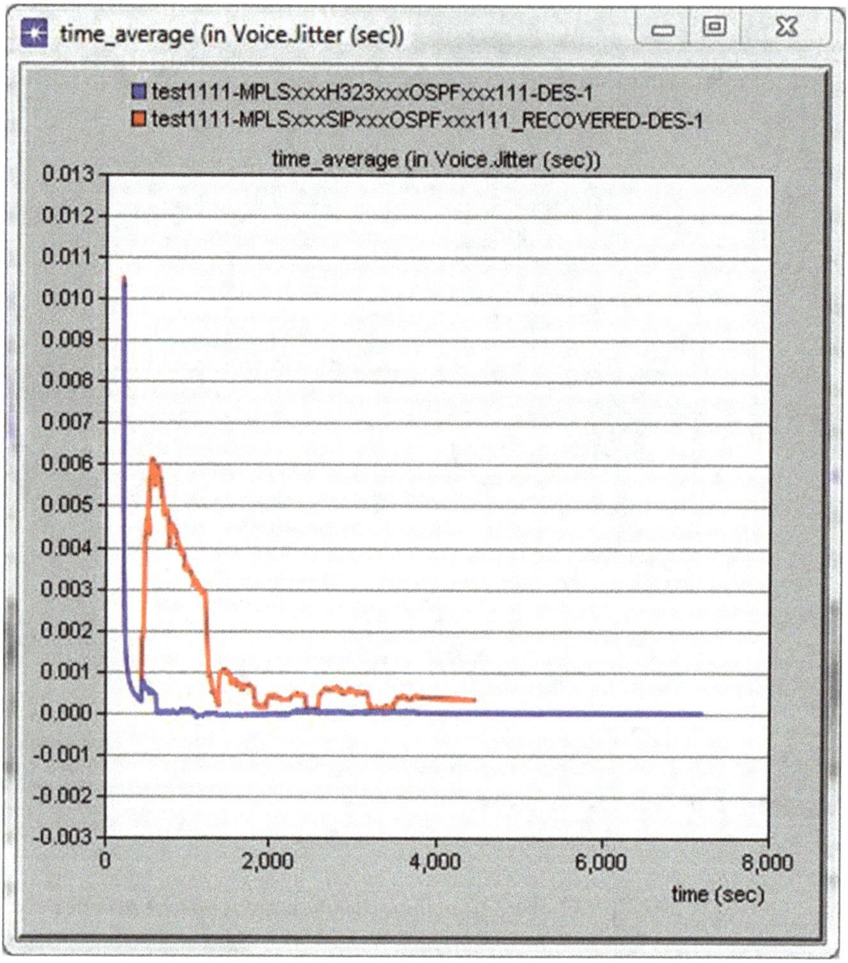

Figure 8.7 VoIP packet delay jitter of SIP and H.323 signaling protocols with MPLS networking

effective routing protocol with the MPLS forwarding technique can produce an enhanced measure to the delivery of data between clients and data centers [5].

8.6.1 Data center structure

Internet servers supplying required data to web browsers over network connections and the stored data are composed of packets for distribution and are forwarded by means of routers that decide the efficient path for that data to transfer via a sequence of networks to reach ISP and to the targeted end-user. The data center always follows the hierarchical topology, as seen in Figure 8.9 (see [23–25]) which includes the followings.

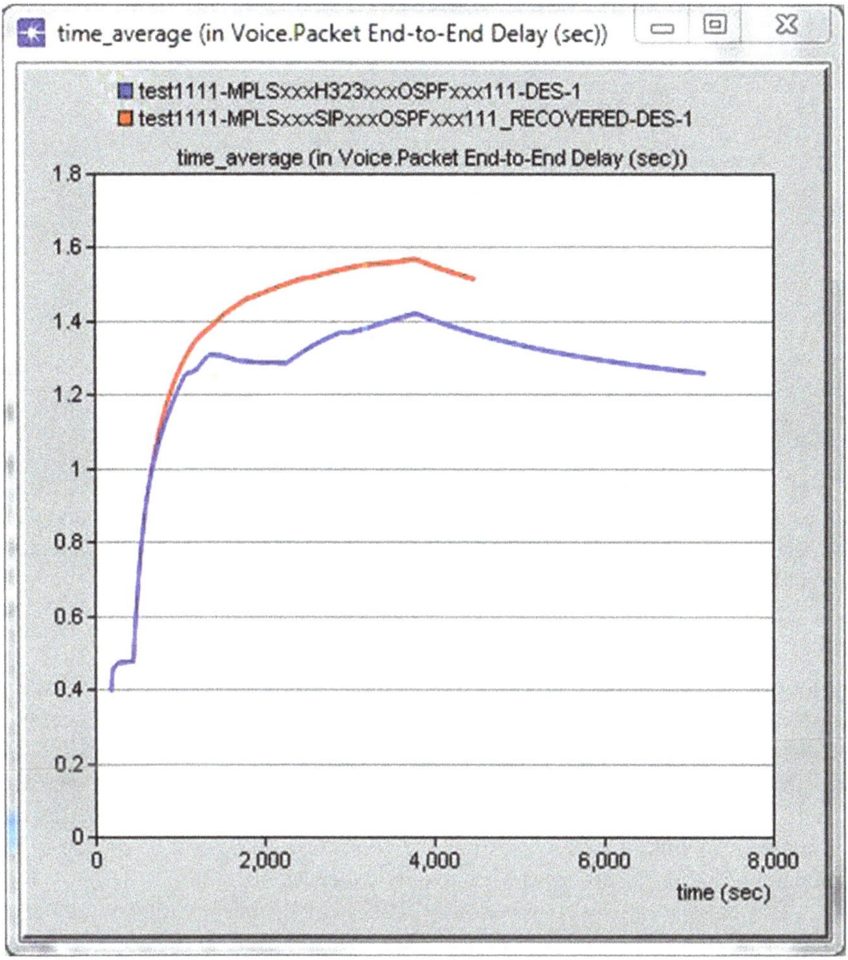

Figure 8.8 VoIP end-to-end delay with OSPF routing protocol and MPLS networking, when using SIP and H.323

8.6.1.1 Core layer

The core layer contains reliable and high-speed switches and routers which optimize all data streams flowing to and from the data center for performance, availability, and scalability. The core layer uses OSPF, enhanced interior gateway routing protocol (EIGRP), or IS-IS as routing protocols complemented with the efficiency of MPLS protocol with proper load balancing.

8.6.1.2 Aggregation layer (distribution layer)

This layer implements important functions such as policing, firewalls, redundancy, service integration and security, network analysis, and spanning tree processing.

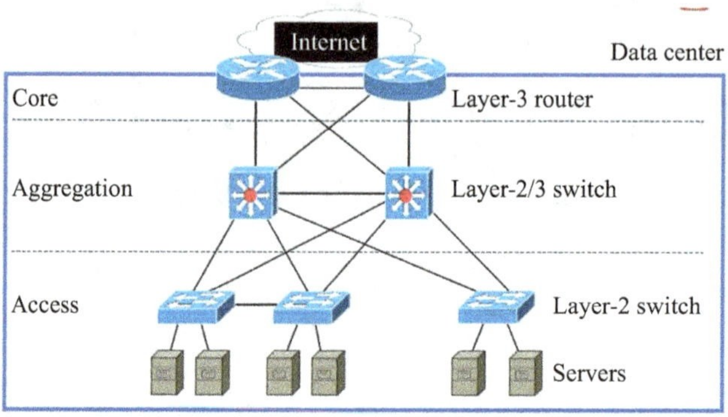

Figure 8.9 Typical data center hierarchical topology

8.6.1.3 Access layer

The access layer defines the physical positions of the servers inside the network. The server components consist of rack unit servers, blade servers, optical switching, cabling, farm, and clustered servers.

8.6.2 Example of fat tree data center structure

Fat tree is the typical example of the indirect (dynamic) interconnection networks that permit to change the organization of interconnection between the processing nodes (servers) dynamically through the use of network's switches. The fat tree is a tree structure with redundant communication links on its branches so that the packets have many routes to the destination due to structure connectivity layout. The number of links increases toward the root. The purpose is to increase the bandwidth at higher levels, where it is the most needed [23,27].

In fat tree topology, the number of switches is independent with the number of servers, while switches with different designs can be used. However, due to the redundant communication links in fat tree, the adaptive routing enables efficient rerouting and network traffic balance over the deterministic routing will result in higher throughput inside/outside the network of the data center with low end-to-end delay. Therefore, this enhances network utilization and proper congestion management while the scalability feature due to the recursion characteristic of the fat tree structure can lead to a performance that is proportional to the number of servers working in the network structure of the data center. Consequently, fat tree networking topology is considered as the most scalable topology structure than many other network structures as its bandwidth is the highest as compared to mesh architecture for example. Another advantage of a fat tree structure is that transmission distances are short for local communication patterns [28].

The simulation of a data center that employs MPLS networks is carried out by the OPNET Modeler 14.5. The simulations consist of two main architectures considering the same network topology with different router capabilities. The network topology of the data center shown in Figure 8.10 consists of MPLS routers,

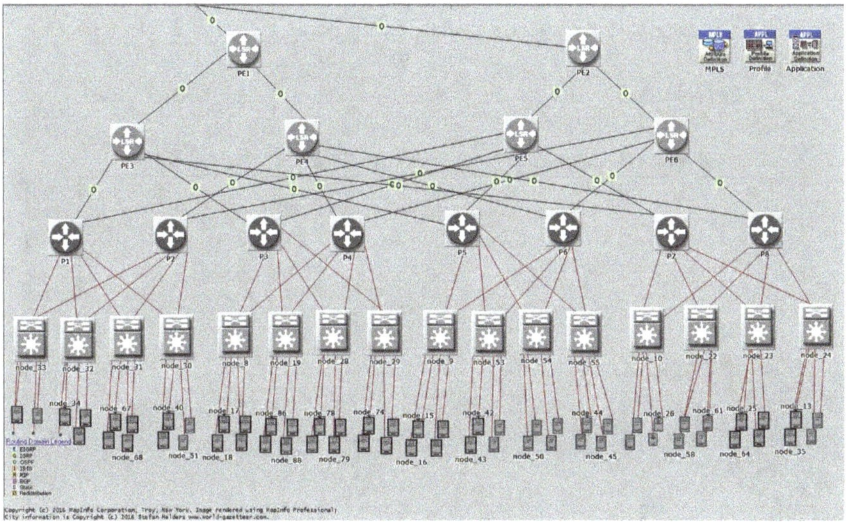

Figure 8.10 Topology of the fat tree data center example

64 servers for different applications (VoIP, FTP, DB, and HTTP), and IP cloud representing an Internet. The simulation run lasts for 60 min (3600 seconds), and the VoIP traffic has been configured between all nodes by using "create traffic flow" option in the OPNET tool with the following input parameters: call rate is 100 calls per hour, average call duration is 300 s (5 min), voice flow duration is 3600 s, the encoder scheme is G.711, traffic type is an interactive voice with delay, and throughput and reliability including overhead (bytes) of RTP/UDP/IP. Therefore, the voice traffic generated by OPNET produced a huge data inside the case study which is measured in hundreds of Terabytes data flown in the network. Besides that, FTP, database, and HTTP traffic streams are generated to flow also through the data center network in order to measure the effectiveness of the fat tree data center type efficiently. However, the simulation comparison is made of two scenarios: one applying MPLS with OSPF while the other applying OSPF only.

Figure 8.11 shows the VoIP packet end-to-end delay; the result clearly illustrates that the employing of MPLS in the data center network has recorded the lowest voice packet end-to-end delay, which makes MPLS a beneficial solution to the new data center design networks. Figure 8.12 describes the FTP download response time, which shows the lowest download response time with MPLS network (better result), while OSPF only scenario looks worse because it has produced the higher download response time value for FTP files. Figure 8.13 demonstrates the HTTP object response time, which illustrates that the MPLS data center network registers the higher (worse) value, while OSPF only network looks better because it has produced the lowest object response time value. Figure 8.14 shows the database query response time; however, when using MPLS with OSPF, it produces the higher (worse) database query response

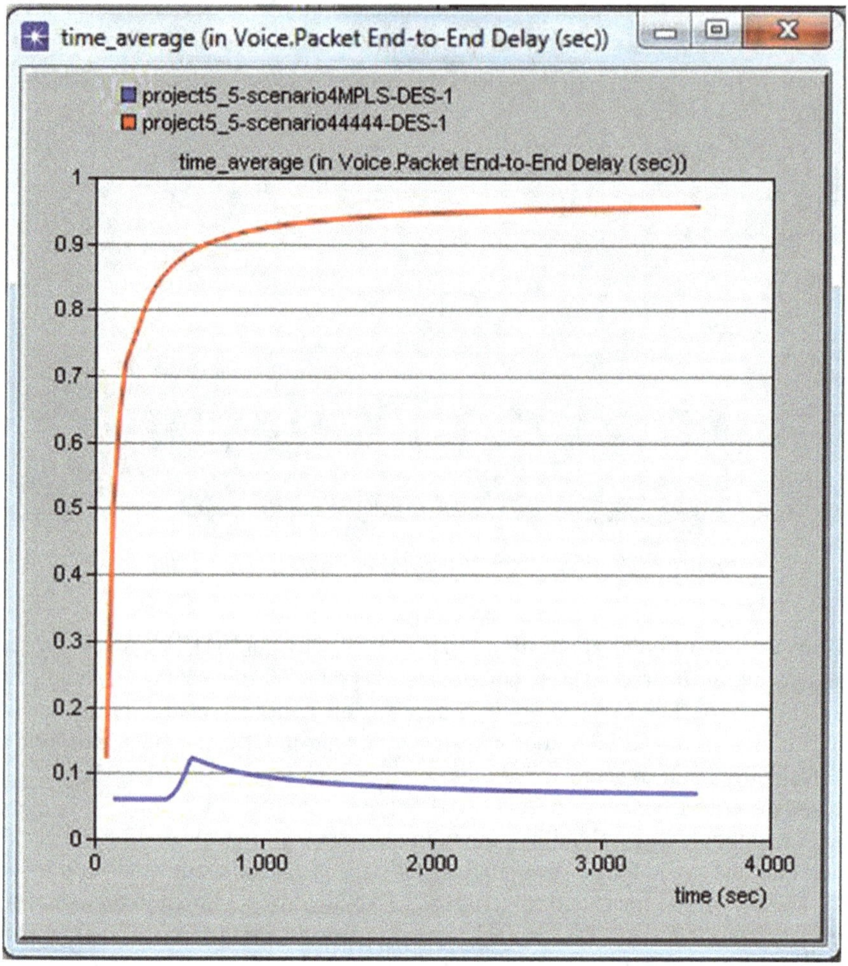

Figure 8.11 VoIP packet end-to-end delay for IP/MPLS data centers

time, while OSPF looks better because it has recorded the lowest upload response time value.

8.7 MPLS-based network-on-chip simulator

In this section, an experimental study using an MPLS-based network-on-chip (NoC) discrete-event simulator developed under the supervision of the author is described. The simulator is written in C++ programming language for modeling and simulating fat tree type of NoC systems employing MPLS as on-chip communication technique. The simulator structure includes a main component that builds butterfly fat tree interconnection network structure and supplementary

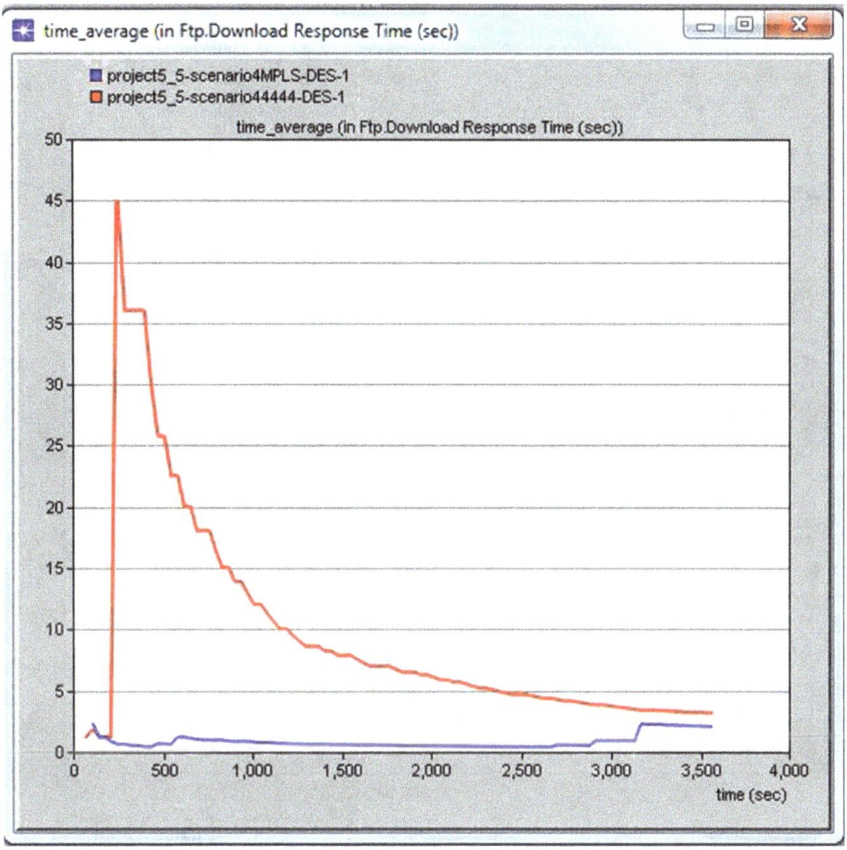

Figure 8.12 FTP download response time for IP/MPLS data centers

components that used to generate a suitable number of switches and processing nodes (blocks and cores) (see Figure 8.15). The switch component is designed to consist of input ports, output ports, input/output link controllers, multi-buffers, crossbar switch, MPLS-based routing unit, and arbitration unit [29].

The network-on-chip design paradigm is the successor of the system-on-chip (SoC) system's design approach. SoC is an integrated circuit that implements almost all the functions of a complete electronic system. The SoC is a design archetype that deals with making a system using hundreds of cores in a single chip with billions of transistors. The central feature of the SoC system is its complexity. Most SoCs are digital because it is the only way to construct such complex functions consistently. The SoC system may contains memory, general-purpose processors, digital signal processors, application-specific processors, specific glue logic, busses, radio frequency (RF) functions, digital filtering cores, and multimedia functions, as seen in Figure 8.15. The structure of the SoC chip is usually made for a specific application rather than being a general-purpose chip such as cell phones, telecommunication

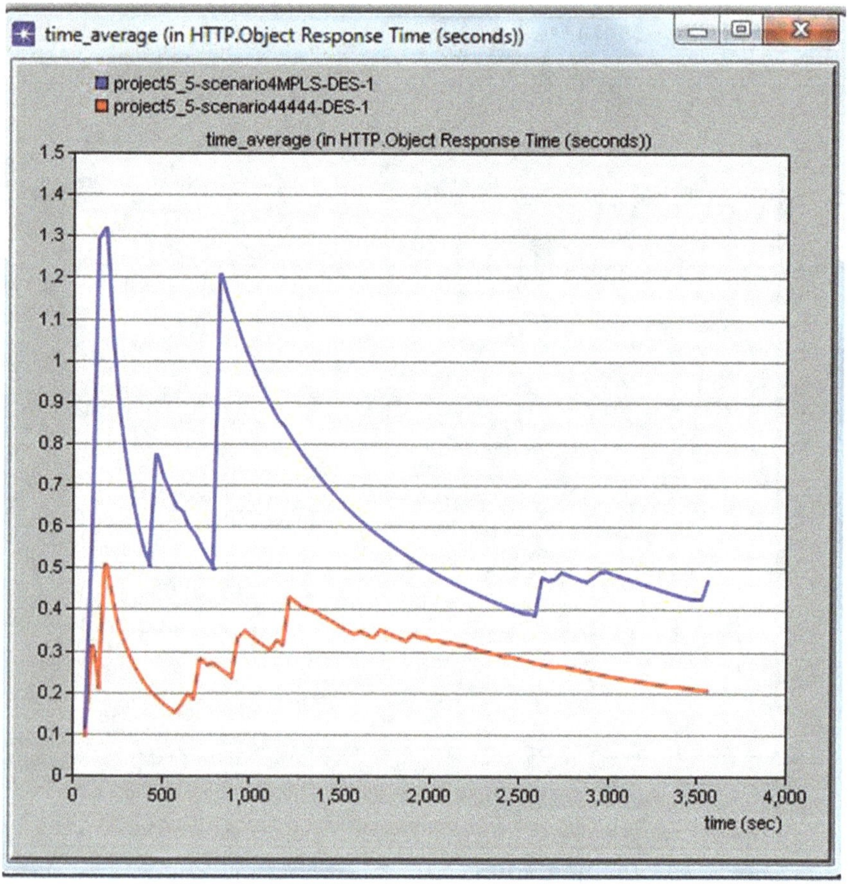

Figure 8.13 HTTP object response time for IP/MPLS data centers

devices (switches and routers), and video games. On the other hand, the growth in chip capacity due to the advances of the semiconductor technologies has allowed including multihundreds of cores into the single-chip which leads to move from SoC design principles toward NoC design style due to the limitations in communications and exchanging of data among cores using buses or multibuses communication links as the number of cores grows rapidly. Thus, that guides to shifting the on-chip communication style from electrical signals via wires into packet communication methods. Furthermore, that has led to bring computer network communication switching and routing theories to SoC design approaches [30,31].

The use of the NoC design paradigm is consequently resulted in realizing power-efficient, modular, flexible, and scalable SoC systems for up-to-date intensive multiprocessing devices in a single chip. Consequently, NoC design approach overcomes the design problems faced with SoC traditional design techniques such as a variety of dedicated interfaces, design and verification complexity,

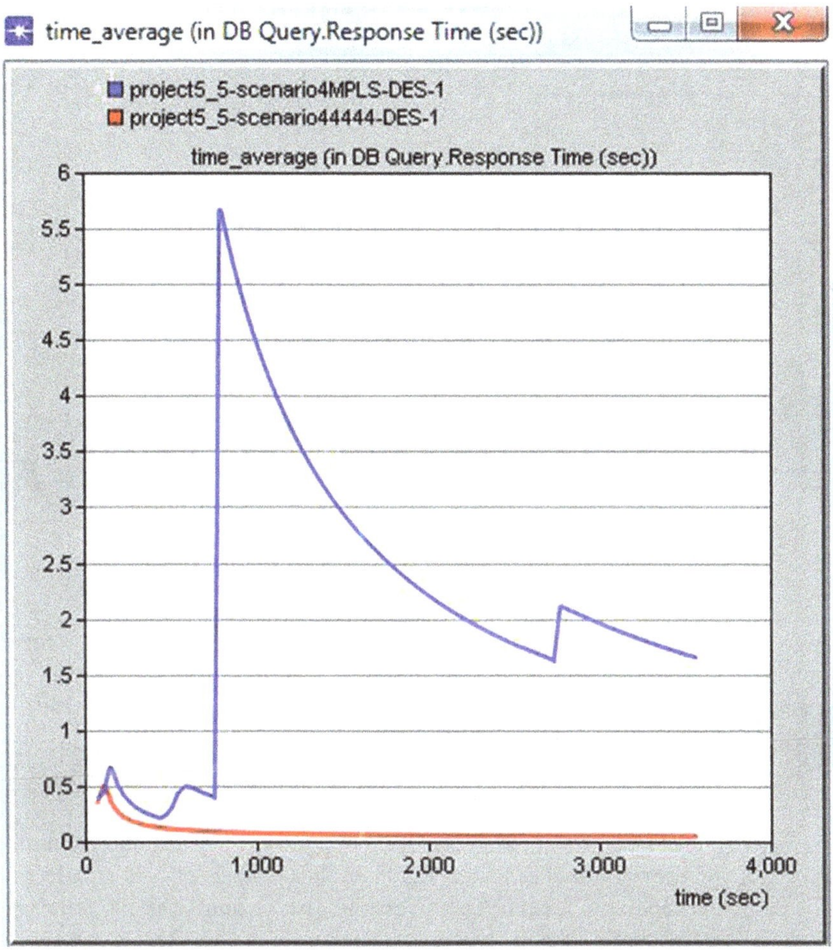

Figure 8.14 Database query response time for IP/MPLS data centers

unpredictable performance, and many underutilized wires by using buses and wires inside the chip. In addition, NoC communicates by packets of bits, routing of packets through several hops via switches, and efficient sharing of wires, and provides parallelism for intensive data processing. However, the NoC communication paradigm separates computation (cores) from communication (switches and routers), which reduces the need for global synchronization and decreases the number of global wires, and the energy consumption of cores can be individually controlled [30–33].

NoC is also reliable since fault-tolerant techniques can be applied from hardware redundancy to adaptive routing protocols that look for alternative paths for communication. The NoC design style started reusing previously developed interconnection network systems inside the chips due to the well-design of interconnection networks, efficiency, and high-performance of their structural

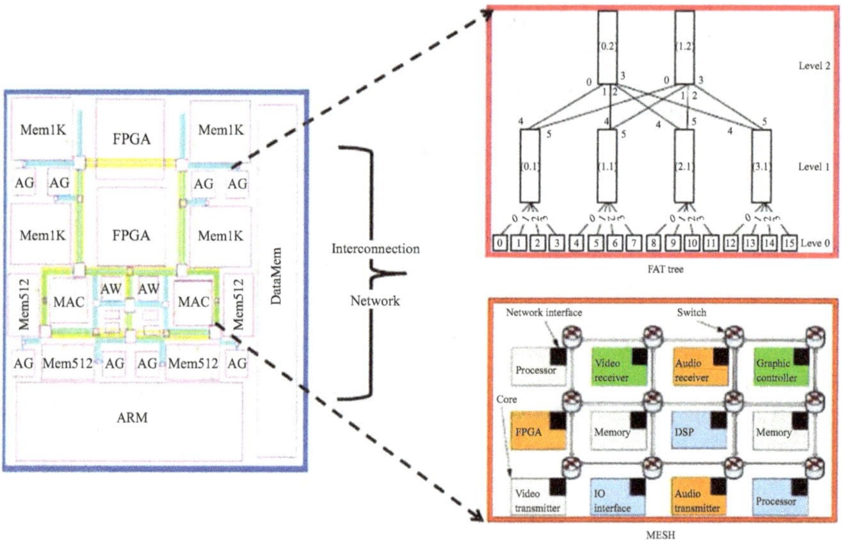

Figure 8.15 SoC and NoC structures

organization. Also, interconnection networks are already in use in many supercomputers and telecommunication switches for many decades, where they are intensively and thoroughly researched as well as practically applied with available different network topologies such as mesh and fat tree [28,32].

8.7.1 Interconnection network

Interconnection network is a structure made up of a group of nodes interconnected with suitable number of switches linked together to make up a network architecture with a particular topology. Switches are in charge of switching, routing, error, and flow control of packet's stream over the interconnection architecture. Network topology of the interconnection networks specifies the structure details such as the static organization of processing cores and the number of stages and switches. The topologies fall into two classes *static* (direct) and *dynamic* (indirect). Butterfly fat tree (BFT) and mesh architectures are typical examples of those interconnection networks. In a static interconnection network, a point to point links interconnects the network nodes in some fixed topology as a regular organization as mesh or hypercube structures. A dynamic network allows the interconnection configuration between the network nodes to be changed dynamically which is made by some methods of switching such as fat tree and multistage networks [28,32].

The fat tree is a form of interconnection network, where the cores (processing nodes) are interconnected by a tree arrangement, in which the processing cores are at the leaves of the tree, while the internal nodes are switches (or routers) placed in other levels of the tree. The advantage of a tree construction is that communication spaces are short for local communication patterns. In addition, the fat tree has redundant interconnections on its branches, thus the number of interconnections communications links increases along the path toward the root. The purpose is to increase the bandwidth of the communication at higher levels of the tree, where it is the most desired. Because it is not practicable to offer a channel between every pair of processing nodes, the network channels are shared between the nodes. Figure 8.16 shows a butterfly fat tree with 64 processing blocks (cores) placed at leaves in zero levels interconnected by a suitable number of switches in intermediate levels [27,32]. The number of levels can be calculated by the relation [34]:

$$L = \log_4 N \tag{8.1}$$

where N is the number of processing nodes (cores).

Figure 8.16 shows three levels of a fat tree, and the switches are placed in levels ranging from $l > 0$ and $l >= \mathrm{L}$. Each core (processing node) is denoted by pair $(i,0)$ where i is ranging from 0 to 63 which denotes the index of the core in level zero, where each node has two ports to connect with its parent switches and each port has two unidirectional physical links. Each level of the network has the number $\frac{1}{2^{(l-1)}} \times \frac{N}{4}$ of the switches and the total number of switches in the network is the summation of the number of switches in each level. Each switch is described by a pair of coordinates (i, l), where i represents the index of the switch in the level and l represents the level of the switch, and the pair (5,2) represents switch no. 5 in the level no. 2. Each processing node (core) at the coordinate $(i, 0)$ has the parent at coordinate $(p, 1)$ and $p = i/4$. For example, if we have the core (62, 0), it has the parent switch (15, 1). Each switch has two-parent (p) coordinates (p1,l+1) and (p2,l+1) [34].

$$\begin{aligned} p1 &= \frac{i}{2^{(l+1)}} \times 2^{l} + i \bmod 2^{(2-l)} \\ p2 &= p1 + 2^{(l-1)} \end{aligned} \tag{8.2}$$

For example, if we have the switch (15, 1), then from the above relations it has the parents:

$$P^1 = 6 \text{ and } p^2 = 7 \tag{8.3}$$

And each switch has four children (switches or nodes) as seen in Figure 8.16.

However, the previous work on this topic started by Leiserson who had described the fat tree structure and demonstrated that fat tree is the most cost-efficient structure for VLSI implementation [27]. In succession, the fat tree has been shown in many advanced parallel computing systems. Dally, in [33], had illustrated the efficiency of applying on-chip interconnection networks instead of using global ad hoc wiring way in SoC designs. Kim *et al.* in [35] have shortly displayed MPLS-NoC design with the NS-2 network simulator. Talwar *et al.* in [36] described a simulation tool developed in System-C that is able of searching designs for mesh and torus NoC systems toward hardware realization based on label switching protocol that is one of the MPLS protocol suits. Kurdy *et al.* in [37] outlined a work that compares the performance of IP and MPLS fat-tree-based NoC using high-level simulation in NS-2. However, authors in this work focused on simulating a 2D fat tree topology using MPLS with recovery technique and

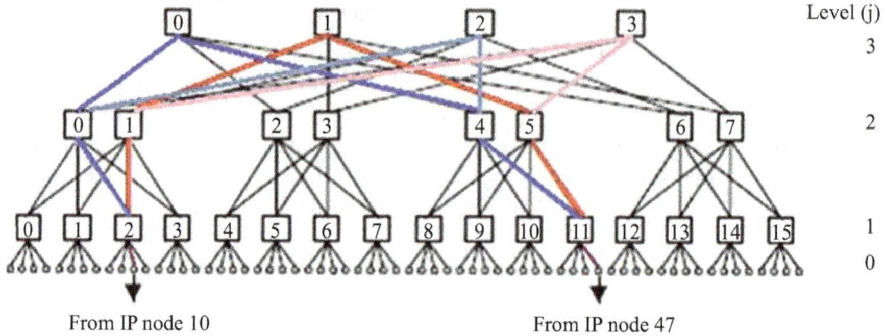

Figure 8.16 A 64 node butterfly fat tree structure with different communicating paths for a pair of processing cores

bisection routing in source switches to provide efficient bandwidth to the multimedia-like applications. But, they had not shown any design details of the NoC system's switches. However, the effort described in this section is different from the mentioned work in such a way that it describes a detailed design of fat tree NoC system based on MPLS technique with all necessary components, procedures, and functions for switching, arbitration, routing, and buffering.

8.7.2 The proposed simulator architecture

The detailed architecture of the MPLS based fat tree NoC system is described in this section. However, Figure 8.16 displays the butterfly fat tree interconnection network structure and Figure 8.17 clarifies the communication flow model between processing cores (blocks) and switches which is used in NoC design.

8.7.2.1 The fat tree generation process

As we know, fat tree is an indirect type of interconnection network organization, where the cores (or processing nodes) are interconnected by a tree structure and placed at the leaves of the tree, and the interior nodes are called communicating switches [27]. The indirect architecture is a dynamic structure with an adaptive connectivity layout that allows changing of the interconnection arrangements of the intermediate switches among the processing cores dynamically that facilitate transferring packets using many routes to destinations making it as a crucial method to produce efficient NoC systems.

In the implemented simulator of the fat tree interconnection network construction that has been established in C++ has following methods:

1. Generating fat tree network method: Instantiates all the cores and butterfly fat tree switches of fat tree NoC system and establishes the nodes connections to their parent switches [27,34,38,39]. The method works as follows:

(i) Calculate the number of switches in the first level by the relation:
Number of switches = number of processing nodes (cores) DIV number of child ports.

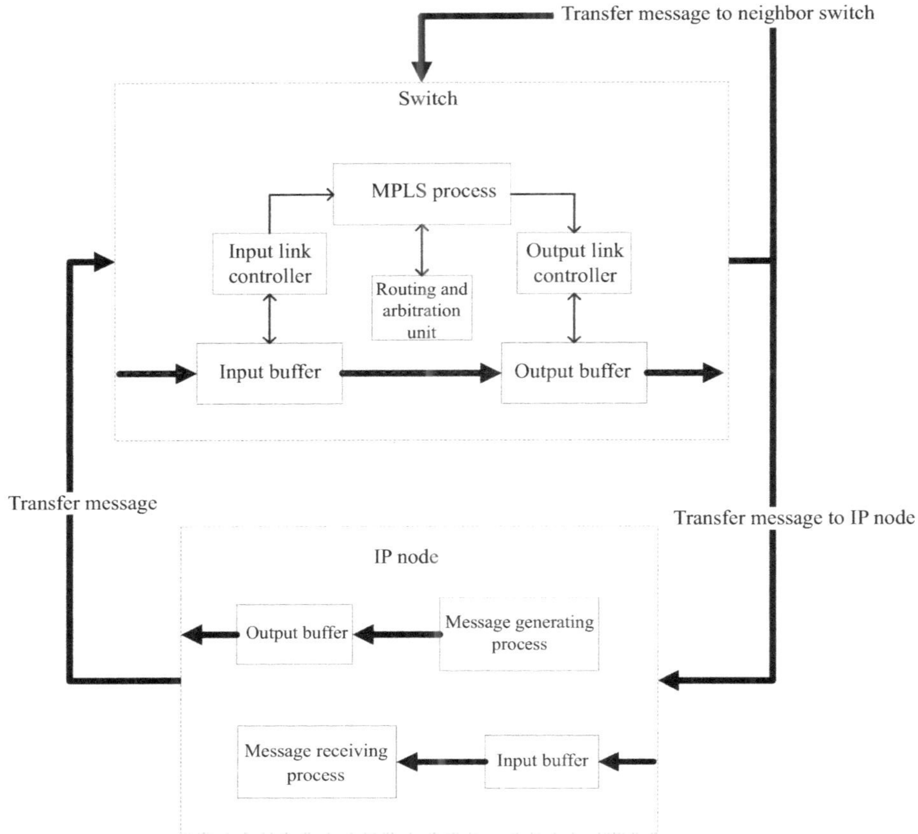

Figure 8.17 Communication flow model between cores and switches

(ii) For all levels, the network constructs a switch, adds the switch to switch list, and creates four cores associated to this switch using the equation:

Number of switches in level (i) = number of switches in previous level DIV number of child ports.

1. Setting the adjacent switches method: This method finalizes the generation of the fat tree interconnection network by setting the adjacency relationship between the switches in all different levels using the equation:

$$\begin{aligned} p1 &= \frac{i}{2^{(l+1)}} \times 2^l + i \bmod 2^{(2-l)} \\ p2 &= p1 + 2^{(l-1)} \end{aligned} \tag{8.4}$$

8.7.2.2 The simulator main process

In this fat tree NoC system simulator, the messages are generated from a source core and forwarded to its parent switch. Then, from that switch, an appropriate MPLS processing is applied, then the message is either forwarded to the adjacent switch or supplied to the destination core. This can be seen in Figure 8.17 which shows simplified switch and core structures and describes the communication model that is followed by messages after being generated by processing nodes and transferred through series of switches toward final destination (processing core) for consumption. However, as a part of the communication model, Figure 8.18 illustrates the procedure of moving the packet from node's output buffer to the parent switch's input buffer to be ready to be forwarded toward its destination core.

8.7.2.3 The node structure

Processing nodes (cores) positioned in the leaves of the fat tree NoC architecture simulate all types of processing cores of the SoC by randomly generating and consuming messages (in terms of packets). In addition, the core encompasses all the methods that are needed for message creation, receiving, and traffic configuration. However, the cores are connected to the parent switches through one input physical link and one output physical link. Each node has its address, its generated message list to save the generated message, internal buffers to keep generated messages (its size is determined by the simulator), and received message list to store the received messages from different nodes.

However, the node includes the following methods to implement:

1. Message generating process: This method is in charge of producing and reproducing messages, where it is implemented at a specified generation time which is determined by the simulator. The facts of this point are illuminated as follow:
   - Message generation method: This is used to generate a message with random length, random data for random destinations. Then this method adds these packets to the produced message list in the form of Internet

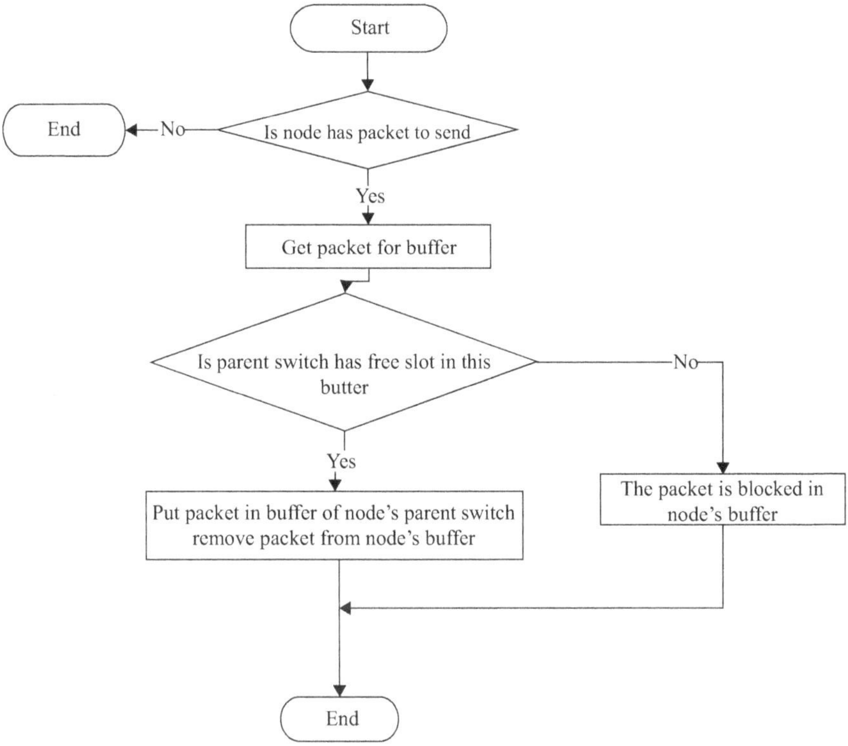

Figure 8.18 Flowchart of moving the packet from the node's output buffer to the parent switch's input buffer

Protocol (IP) packets. Still, the message is divided into many packets according to its length in bytes (where the maximum length of both the massages and packets are determined by the simulator).

- Set the next message's generation time: This method determines the next message generating time by adding a random value to a current-generation time of the current processing node. However, if the internal buffer is full, then the generated message list will be erased and the generated time of the node is set in the next cycle. Else, transfer the message from the generated message list to the internal core's buffer, then empty the generated message list.

2. Fill empty buffer: This method checks whether there is an empty slot in the output buffer, if so then transfer packets from the internal buffer to the output buffer of the node to be ready for sending to the core's parent switch. The number of transferred packets are based on the number of buffer's empty slots.
3. Move the received packets from node to the received list: This method moves packets from the input buffer of the processing node to its received message list.

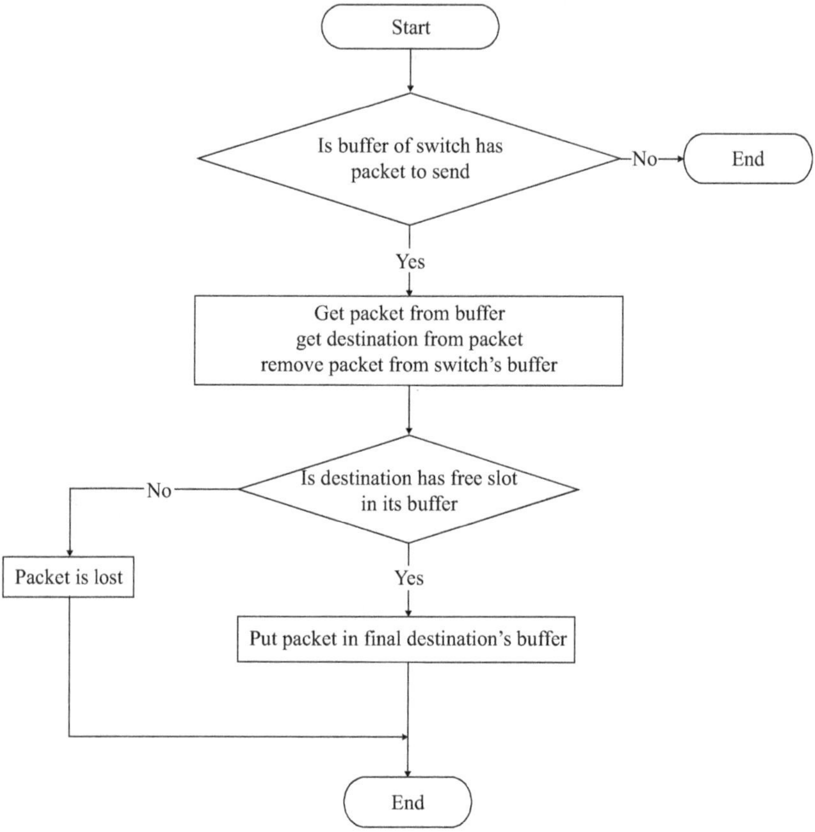

Figure 8.19 Flowchart of moving the packet from switch output buffer to the child node input buffer

4. Move packets from the node output buffer to the parent switch input buffer: This method is responsible for transferring packets from the node's output buffer to its parent switch's input buffer. The method's functionality is described in Figure 8.18. Firstly, the node checks if the output buffer has a packet to send. Secondly, the node finds whether the switch's input buffer has a free slot for that packet, if yes, then adds that packet to switch's input buffer, and removes it from node's output buffer. Else it blocks that packet in the node's output buffer until the next turn. In addition, Figure 8.19 describes the procedure of moving the packet from the switch output buffer to the child node input buffer which completes the arc of the communication model of the implemented simulator.

8.7.3 The switch structure

The designed switch consists of input ports, output ports, input/output link controllers, multi-buffers, crossbar switch, routing and arbitration components, and MPLS unit.

8.7.3.1 The arbiter

Due to the absence of virtual channels technique in the designed MPLS-based switch, the only shared resources are the buffers and switch ports, which need an arbiter to allocate the right to use the link to only one request at a time. To avoid any contention that may happen on links' usage, a fair arbiter (*round robin*) is implemented to serve all links of the switch by providing equal service to the different requesters at each simulation cycle.

8.7.3.2 The MPLS unit

The main MPLS process implemented in the simulator emphasizes on a path creation and deletion among all cores pairs that are starting to communicate to exchange packets with each other. The MPLS unit contains an LDP module that operates to establish the required LSP paths among all sending cores to receiving cores. However, the LDP message is generated by ingress switches and it contains message type, message ID, message, FEC, label, and some parameters that are needed for LSP creation and routing. However, LDP is a set of procedures by which one LSR informs another LSR of the bindings (label/FEC) it has been made. Two LSRs that use LDP as a distribution protocol to exchange label/FEC binding data are known as "label distributing peers". To complete such binding, LDP sets up a bidirectional LDP session between both sides and can request or advertise labels. However, LDP usually uses TCP for transporting session initiation messages because LDP needs reliable transport and in-order delivery (e.g. binding + tear down), LDP uses a single ACK timer for a session and piggybacking an ACK on data packets. Moreover, LDP uses UDP protocol for transporting discovery (hello) messages. However, Figure 8.20 describes the algorithm of moving the LDP-packet from the switch's output buffer to the adjacent switch input buffer.

8.7.3.3 The routing unit

This piece of switch is based on the MPLS technique which is responsible for the routing process in the switch (see Figure 8.21).

8.7.3.4 The input/output link controllers

These components control the incoming packets and support the choice of an appropriate route of the packets in a switch and they manage and control the input/output buffers. They assist in transmitting packets to the input link controllers in the adjacent switches or to the buffers of the receiving cores. The component accomplishes the following functions:

- Adds/gets/sends/removes incoming packet to/from an input buffer.
- Adds/gets/sends/removes incoming LDP packet to/from an LDP input buffer.
- Adds/gets/sends/removes incoming MPLS packet to/from an MPLS input buffer.

8.7.4 Example and simulation results

A discrete event simulator written in C++ for fat tree network-on-chip systems based on all technical details of MPLS switching and forwarding has been

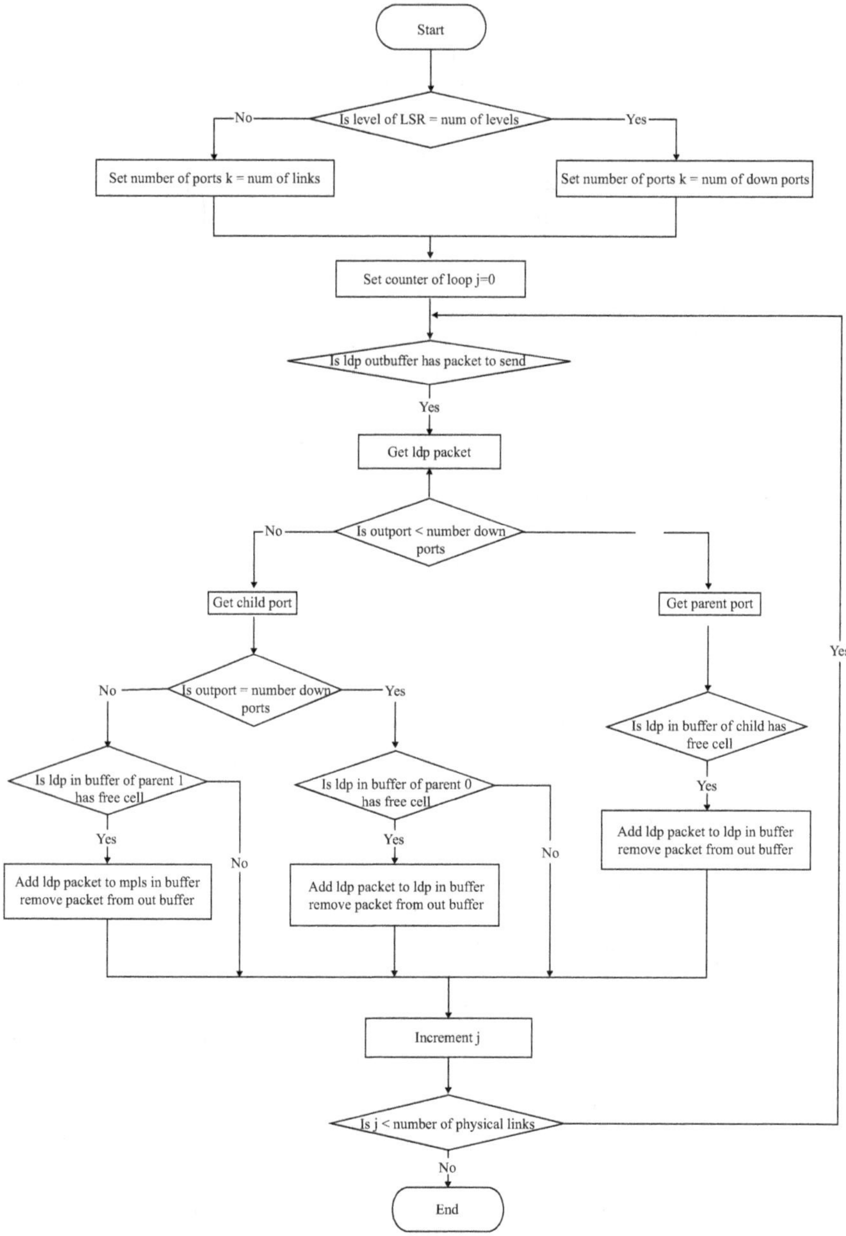

Figure 8.20 Flowchart of moving LDP-packet from switch's output buffer to the adjacent switch input buffer algorithm

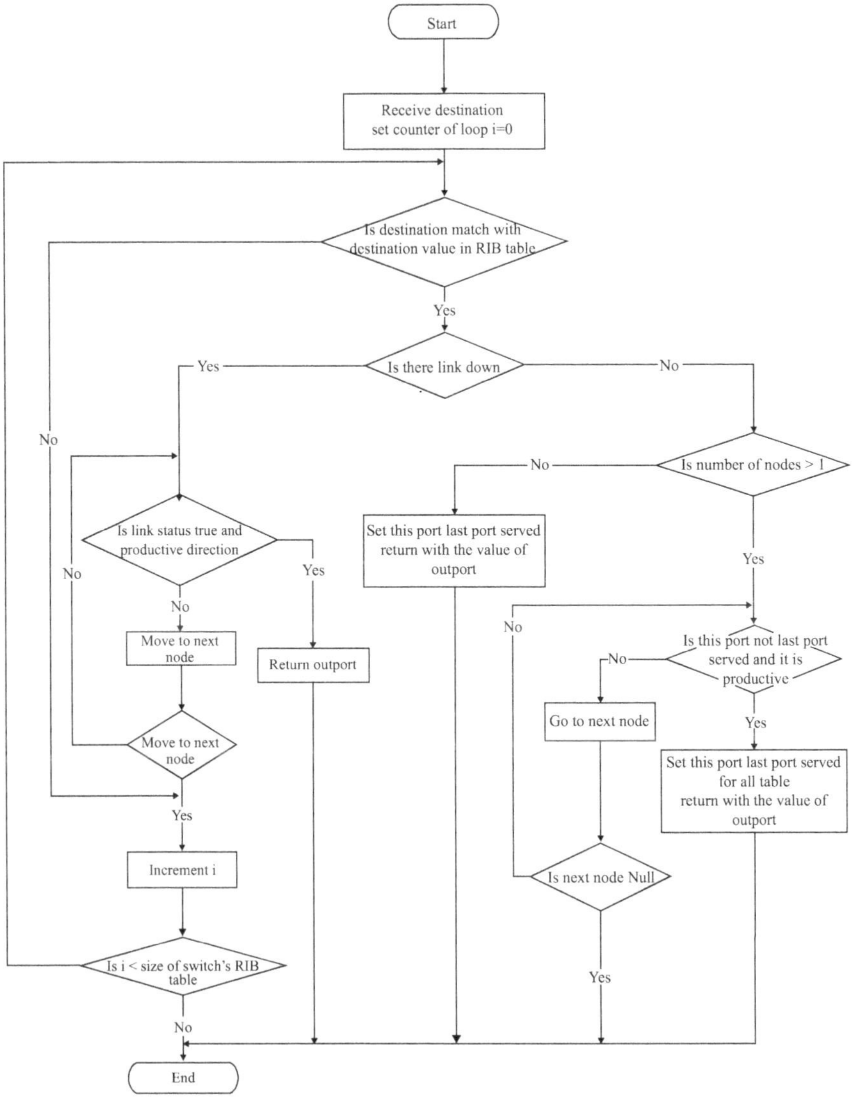

Figure 8.21 Flowchart of the implemented routing process

presented. The simulator's functionality has been verified with fat tree structure with 64 cores as seen in Figure 8.16. To validate the obtained results of the developed simulator, it is compared with fat NoC system with wormhole and virtual channels concepts that had been described in [39]. However, the simulator has the following features:

- The simulator displays the label advertising and label bindings with FECs as well as the LSP path creation process, which is done by LDP protocol before

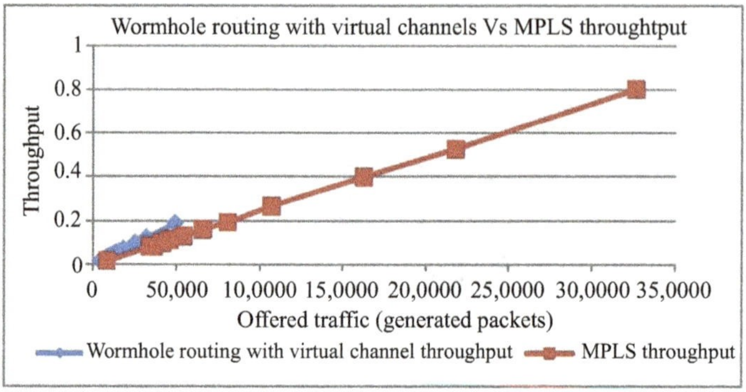

Figure 8.22 Throughput of fat tree NoC system (MPLS vs. wormhole with virtual channels)

transferring any data between any processing cores. For each packet, the simulator shows in a cycle-by-cycle approach its movement from the source core to the destination core [40–45].

- The simulator presents an output log file that records the movements of packets from output buffer of the switch to the input buffer of the adjacent switches and/or nodes;

However, the results of the comparison between MPLS-based approach with wormhole + virtual channels NoC simulator is shown in Figure 8.22 and Figure 8.23 [29]. However, Figure 8.22 shows the relation between the system throughput and the number of supplied traffic (packets). Thus, the results demonstrate that the throughput of the MPLS-based NoC fat tree system is much better than that of the wormhole with virtual channels based NoC system. However, the throughput in the MPLS case is increased linearly with the traffic increase due to the MPLS path creation and packet forwarding mechanisms which are based on routing based on labeling [46–52]. The figure shows also that congestion makes wormhole fat NoC systems losing lots of traffic after increasing the traffic further than 50,000 messages. In another hand, the MPLS case can still accept messages even after reaching 2,50,000 messages. Figure 8.23 illustrates the comparison of the average packet end-to-end delay of MPLS-based fat tree NoC system compared to the wormhole + virtual channels-based fat tree NoC system. The figure obviously says that the MPLS-based fat tree NoC system achieved better latency (much less) even with the rapid increase of the load [53–59].

The following listing is taken from the simulator's output log file:

- Some lines from the output log file of the simulator which produced in a cycle-by-cycle fashion.

————Network configuration....

Number of IP nodes–>64

Number of levels–>3

Number of physical link/LSR–> 6

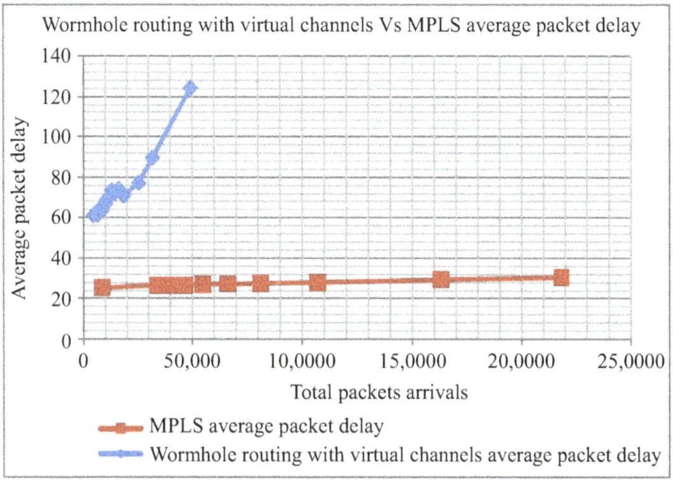

Figure 8.23 Average packet delay of fat tree NoC system (MPLS vs. wormhole with virtual channels)

Fixed message length (packets)–> false
Average message length (bytes)–> 6656
Packet length (bytes)–> 128
Average message production rate (cycle/msg)–> 4000
Number of packet/buffer –>4
Number of cycles–> 99999

Cycle #42
Moving traffic from IP node output buffer to parent switches
Packet is arriving to switch (1,0) from node [0]

- Moving LDP messages between switches............
  Request label packet of FEC 55 is transferring from switch (2,0) To switch (3,0)
  Mapping label packet of FEC 38026 is arriving to switch (2,4) to switch (1,9)
  Mapping label packet of FEC 61033 is arriving to switch (2,6) to switch (1,15)
  Mapping label message of FEC 38026 is arriving to switch (1,9)
  Mapping label message of FEC 61033 is arriving to switch (1,15)
  Request label message of FEC 55 is transferring from switch (3,0) to switch (2,6)
- Moving MPLS packets between switches

– Tag 20496 is pushing to packet # 1 in switch 9
– Tag 28688 is pushing to packet # 1 in switch 15

- MPLS packet of tag 20496 is transferring from switch (1,9) to switch (2 ,4)
- MPLS packet of tag 28688 is transferring from switch (1,15) to switch (2 ,6)

Cycle # 102366

- Moving MPLS packets between switches

- Tag 33567 is pushing to packet # 31 in switch 13
- Arrive IP packets to its destination
- Packet is transferred from switch (1,4) to node [18]
- Packet is transferred from switch (1,8) to node [35]
- Total generated messages: 6485(162897 packets)
- Total not produced messages: 0(0 packets)
- Total sent messages: 6485(162897 packets)
- Total received messages: 6485(162897 packets)
- Throughput [packets leaving switch]-> 0.257895
- Node throughput ——> 0.0254595
- Average message delay–> 37.9025
- Average packet delay–> 29.595
- Link utilization–> 0.0790183
- Buffer (node and switch) utilization–> 0.00858195
- Average message injection rate 0.0648506
- Average message not produced 0
- Average hop count–> 4

8.8 Conclusion

The key features of the MPLS technique are its efficient forwarding mechanism and TE, which are used for efficiently managing the networks for effective deployment of network resources. In addition, due to lower network delay, scalability and predictable performance of the services provided by MPLS technology makes the MPLS the most appropriate tool for fulfilling real-time multimedia applications requirements, when used in backbone networks. Also, MPLS is now introduced efficiently as an on-chip communication tool for NoC systems.

References

[1] Kurose, J. F., and Ross, K. W. *Computer networking: A top-down approach featuring the internet*. USA: Addison Wesley Publishers; 2012.
[2] Stallings, W. *Computer networking with internet protocols and technology*. USA: Prentice Hall (Pearson Education); 2004.
[3] Evans, J., and Filsfils, C. *Deploying IP and MPLS QOS for multiservice networks: Theory and practice*. USA: Elsevier, Morgan Kaufmann Publishers; 2007.

[4] Rosen, E., Viswanathan, A., and Callon, R. *Multiprotocol label switching architecture: RFC 3031*. Internet Engineering Task Force. 2001. p. 61.
[5] Perros, H. G. *Connection-oriented networks SONET/SDH, ATM, MPLS and optical network*. UK: John Wiley & Sons Ltd.; 2005.
[6] Peterson, L. L., and Davie, B. S. *Computer networks: A systems approach* (4th ed.). San Francisco, CA: Morgan Kaufmann Publishers; 2007.
[7] Bhatti, S. N., and Crowcroft, J. "QoS-sensitive flows: Issues in IP packet handling." *IEEE Internet Computing*. 2000; 48–57.
[8] Perros, H. *Networking services: QoS, signaling processes*. USA: Create Space Independent Publishing Platform; 2010.
[9] Forouzan, B. A. *Data communications and networking* (4th ed.). New York, NY: McGraw-Hill Publisher; 2007.
[10] Medhi, D., and Ramasamy, K. *Network routing: Algorithms, protocols, and architectures*. CA, USA: Morgan Kaufmann (Elsevier); 2007.
[11] Forouzan, B. A. *TCP/IP protocol suite*. 4th ed. New York, NY: McGraw-Hill publisher; 2010.
[12] Rosenberg, J., Schulzrinne, H., Camarillo, G., *et al. Session initiation protocol (SIP): RFC 2361*. Internet Engineering Task Force (IETF)." Available from https://tools.ietf.org/html/rfc2361 [Accessed June 2002].
[13] Schulzrinne, H., Casner, S., Frederick, R., and Jacobson, V. *A transport protocol for real-time applications (RTP): RFC 3550*. Internet Engineering Task Force (IETF). Available from https://tools.ietf.org/html/rfc3550 [Accessed July 2003].
[14] Schulzrinne, H., Rao, A., and Lanphier, R. *Real time streaming protocol (RTSP): RFC 2368*. Internet Engineering Task Force. (IETF). Available from https://tools.ietf.org/html/rfc2368 [Accessed April 1998].
[15] Handley, M., and Jacobson, V. *Session description protocol (SDP): RFC 4566*. Internet Engineering Task Force (IETF). Available from https://tools.ietf.org/html/rfc4566 [Accessed July 2006].
[16] ITU-T Recommendation G.114. "International telephone connections and circuits – General recommendations on the transmission quality for an entire international telephone connection One-way transmission time." 2003.
[17] OPNET Product Documentation v.11.0.A. OPNET Technologies, Inc., Bethesda, MD, 2004.
[18] Sllame, A. M., and Aljafry, M. "Performance evaluation of multimedia over IP/MPLS networks". *International Journal of Computer Theory and Engineering*. 2015; 7(4): 283–291.
[19] Jannu, K. P., and Deekonda, R "OPNET simulation of voice over MPLS with considering traffic engineering." Master's thesis on electrical engineering. Thesis No: MSE-2010-5311. Sweden: School of Engineering, Blekinge Institute of Technology; 2010.
[20] Đogatović, V. R., Stojanović, S., Kostić-Ljubisavljević, A., and Mikavica, B. *Quality of service considerations for two diffServ scenarios in IP networks. ICEST2015*. 2015.
[21] Awduche, D. O. "MPLS and traffic engineering in IP networks." *IEEE Communications Magazine*. 1999; 12: 42–47.

[22] Marzo, J. L., Calle, E., Scoglio, C., and Anjali, T. "QoS online routing and MPLS multilevel protection: A survey." *IEEE Communications* Magazine. 2003; 126–132.

[23] Vahdat, A., Al-Fares, M., Farrington, N., *et al.* "Scale-out networking in the data center." *IEEE Micro*. 2010; 30(4): 29–41.

[24] Oppenheimer, P. *Top-down network design*. 3rd ed. USA: Cisco Systems Inc., Cisco Press; 2011.

[25] Alessandro, G., and Santana, A. *Data center virtualization fundamentals*. 2nd ed. USA: Cisco Systems Inc, Cisco Press; 2014.

[26] Benltayef, N., El-Neffaty, A., and Al-Alewy, A. "Datacenter virtualization using VMware vsphere." *First Libyan International Conference on Electrical Engineering and Technologies*. Tripoli. Libya. 2018.

[27] Leiserson, C., "Fat trees: Universal networks for hardware-efficient supercomputing." *IEEE Transactions on Computers*. 1985; C-34(10): 892–901.

[28] Duato, J., Yalamanchili, S., and Ni, L., *Interconnection networks: An engineering approach*. Morgan Kaufmann; 2002.

[29] Salama, N., and Sllame, A. M. "Designing an efficient MPLS-based switch for fat tree network-on-chip systems". *ACM Proceedings of the 1st International Workshop on Advanced Interconnect Solutions and Technologies for Emerging Computing Systems (AISTECS Workshop)*. Prague, Czech Republic. 2016.

[30] Jantsch, A., and Tenhunen, H. *Networks on chip*. USA: Kluwer Academic Publishers; 2003.

[31] De Michelli, G., and Benini, L. *Network on chips*. Berlin: Morgan Kaufmann; 2006.

[32] Dally, W. J., and Towles, B. *Principles and practices of interconnection networks*. San Francisco: Morgan Kaufmann Publishers; 2004.

[33] Dally, W. J., and Towles, B. "Route packets, not wires: On-chip interconnection networks." *Proceedings of the DAC'38 Conference*. Las Vegas. 2001.

[34] Sllame, A. M., and Elasar, A. "Modeling and simulating network-on-chip designs: A case study of fat tree interconnection architecture." *International Journal of Computer Theory and Engineering*. 2013; 5(5): 823–829.

[35] Kim, M., Kim, D., and Sobelman, G. E. "Network-on-chip quality-of-service through multi protocol label switching." *IEEE International Symposium on Circuits and Systems (ISCAS 2006)*. 2006. pp. 1843

[36] Talwar, B., Kulkarni, S., and Amrutur, B. "Latency, power and performance trade-offs in network-on-chips by link microarchitecture exploration." *22nd International Conference on VLSI Design*. 2009.

[37] Kourdy, R., Rad, M. R. N., Pooyan, M., and Nasab, M. R. "Improvement Mpls-NoC bandwidth by dividing bandwidth in Fat-tree topology." *The 2nd International Conference on Computer and Automation Engineering (ICCAE). IEEE Conference. Singapore,* February 2010, Vol. 3. pp. 470–474.

[38] Pande, P. P., Grecu, C., Ivanov, A., and Saleh, R. "Design of a switch for network on chip applications", *Proceedings of ISCAS, Bangkok*. 2003; V: 217–220.

[39] Hemayet, H. M., Ahmed, A., Al-ayeemn, T. I., and Mostofa, A. M. "GPNoCSIM: A general purpose simulator for network-on-chip." *Proceedings of the International Conference on Information and Communication Technology IEEE ICICT07*. 2007.

[40] Bourdena, A., Pallis, E., Kormentzas, G., Skianis, C., and Mastorakis, G. "QoS provisioning and policy management in a broker-based CR network architecture." *IEEE Globecom 2012*. Anaheim, CA, USA. 2012.

[41] Bourdena, A., Pallis, E., Kormentzas, G., and Mastorakis, G., "Radio resource management algorithms for efficient QoS provisioning over cognitive radio networks." *IEEE ICC*. 2013.

[42] Mavromoustakis, C. X., Dimitriou, C. D. Mastorakis, G., and Pallis, E. "Real-time performance evaluation of F-BTD scheme for optimized QoS energy conservation in wireless devices." *IEEE Globecom. Workshop*. 2013. pp. 1151–1156.

[43] Mastorakis, G., Pallis, E., and Kormentzas, G. "A DVB/IPQoS aware backhaul networking environment." *Wireless Personal Communications*. 2010; 52(3): 637–649.

[44] Sundarasekar, R., Shakeel, P. M., Baskar, S., Kadry, S., and Mastorakis, G. "Adaptive energy aware quality of service for reliable data transfer in under water acoustic sensor networks." *IEEE Access*. 2019; 7: 80093–80103.

[45] Mavromoustakis, C. X., Mastorakis, G., Pallis, E., Mysirlidis, C., and Dagiuklas, T. "On the perceived quality evaluation of opportunistic mobile P2P scalable video streaming." *2015 International Wireless Communications and Mobile Computing Conference*.

[46] Mavromoustakis, C. X., Mastorakis, G., and Batalla, J. M. *Internet of things (IoT) in 5G mobile technologies*. Cham: Springer. 2016.

[47] Bourdena, A., Mavromoustakis, C., Kormentzas, G., Pallis, E., and Mastorakis, G. "A resource intensive traffic-aware scheme using energy-aware routing in cognitive radio networks." *Future Generation Computer System*. 2014; 39: 16–28.

[48] Batalla, J. M., Mastorakis, G., Mavromoustakis, C. X., and Zurek, J. "On cohabitating networking technologies with common wireless access for home automation system purposes." *IEEE Wireless Communications*. 2016; 23(5): 76–83.

[49] Dimitriou, C., Mavromoustakis, C., Mastorakis, G., and Pallis, E. "On the performance response of delay-bounded energy-aware bandwidth allocation scheme in wireless networks." *IEEE ICC*. 2013.

[50] Bourdena, A., Pallis, E., Kormentzas, G., and Mastorakis, G. "Efficient radio resource management algorithms in opportunistic cognitive radio networks." *Transactions on Emerging Telecommunications Technologies*. 2014; 25(8): 785–797.

[51] Mavromoustakis, C. X., Bourdena, A., Mastorakis, G., Pallis, E., and Kormentzas, G. "An energy-aware scheme for efficient spectrum utilization in a 5G mobile cognitive radio network architecture." *Telecommunication Systems*. 2015; 59: 63–75.

[52] Skourletopoulos, G., Mavromoustakis, C. X., Mastorakis, G., Batalla, J. M., Sahalos, J. N. "An evaluation of cloud-based mobile services with limited capacity: A linear approach." *Soft Computing*. 2017; 21(16): 4523–4530.

[53] Mavromoustakis, C., Mastorakis, G., Bourdena, A., and Pallis, E. "Energy efficient resource sharing using a traffic-oriented routing scheme for cognitive radio networks." *IET Networks*. 2014; 3(1): 54–63.

[54] Batalla, J. M., Kantor, M, Mavromoustakis, C. X., and Skourletopoulos, G. "A novel methodology for efficient throughput evaluation in virtualized routers." *IEEE International Conference on Communications (ICC)*. 2015. pp. 6899–6905.

[55] Mavromoustakis, C. X., Kormentzas, G., Mastorakis, G., Bourdena, A., and Pallis, E. "Context-oriented opportunistic cloud offload processing for energy conservation in wireless devices." *IEEE Globecom Workshops (GC Wkshps)*. 2014. pp. 24–30.

[56] Markakis, E. K., Karras, K., Zotos, N., Sideris, A., Moysiadis, T, and Corsaro, A. "EXEGESIS: Extreme edge resource harvesting for a virtualized fog environment." *IEEE Communications Magazine*. 2017; 55(7): 173–179.

[57] Mavromoustakis, C., Dimitriou, C. D., and Mastorakis, G. "Using real-time backward traffic difference estimation for energy conservation in wireless devices." *The Fourth International Conference on Advances in P2P Systems. AP2PS* 2012.

[58] Markakis, E. K., Politis, I., Lykourgiotis, A., Rebahi, Y., and Mastorakis, G. "Efficient next generation emergency communications over multi-access edge computing." *IEEE Communications Magazine*. 2017; 55(11): 92–97.

[59] Mastorakis, G., Mavromoustakis, C. X., and Pallis, E. *Beyond the internet of things*. Cham: Springer. 2017.

Chapter 9

Damaged critical infrastructure for VANETs

Grace Khayat[1], Constandinos X. Mavromoustakis[1], George Mastorakis[2], Jordi Mongay Batalla[3], Hoda Maalouf[4], Evangelos Pallis[5], Evangelos K. Markakis[5], Imran Khan[6] and Naercio Magaia[7]

Vehicular ad hoc network (VANET) is a spontaneous creation of wireless network among vehicles to exchange data. Breakdown or interruption of communication is one of the most immediate and significant impacts of natural disasters. Clustering is one of the most common networking protocols for data propagation in those networks. The application of the clustering algorithm is effective in VANET because the algorithm makes it a more robust and scalable network. For stable cluster formation in VANET, some constraints such as vehicles' velocity and vehicles' separation distance must be considered while selecting the cluster head which is the node responsible for data propagation to the infrastructure. In damaged VANET, the cluster head might be unable to successfully transmit the packets to the infrastructure. In such cases, overlapping clusters with double cluster heads is one of the best solutions to guarantee successful packet delivery to the destination. One of the major challenges in clustering is the cluster head (CH) election since the CH has a critical role in data routing.

This chapter discusses a new clustering algorithm based on a weighted formula for cluster head election. The weighted formula is based on three parameters: the trust, the distance, and the velocity. This protocol will elect two cluster heads to ensure an end-to-end communication path between source and destination.

[1]Department of Computer Science, University of Nicosia, Nicosia, Cyprus
[2]Department of Management Science and Technology, Hellenic Mediterranean University, Agios Nikolaos, Greece
[3]National Institute of Telecommunications and Warsaw University of Technology, Warsaw, Poland
[4]Computer Science Department, Faculty of Natural and Applied Science, Notre Dame University–Louaize, Zouk Mosbeh, Lebanon
[5]Department of Electrical and Computer Engineering, Hellenic Mediterranean University, Heraklion, Greece
[6]Department of Electrical Engineering, University of Engineering and Technology Peshawar, Peshawar, Pakistan
[7]Departamento de Informática, Faculdade de Ciências, Portugal, Universidade de Lisboa

9.1 Introduction

In 1998, the Intelligent Transportation Society of America adopted a definition of intelligent transportation systems (ITS) that exploits all new technologies in terms of electronics, information, and telecommunication. Mobile ad hoc network (MANET) is a network made of stations that communicate via the wireless medium (WM) whenever they are within the communication range of each other. Vehicular ad hoc network (VANET) is a specialized type of MANET consisting of stations being cars equipped with short-range radios, rapid change in topology, frequent link failure, high vehicle mobility, etc. In VANET, the nodes exchange information aimed at offering safety and nonsafety applications such as parking or multimedia services [1].

VANET communication strategies can be classified as flooding or forwarding strategies. In flooding, each node, acting as a relay, delivers multiple copies of each packet to all other nodes. In forwarding strategies, the node selects the best path to the destination using the distance metric to estimate the cost of message delivery [2]. Those networks have many challenges which include vehicles' mobility, variable vehicles' speed, restricted vehicle movements on the road, time-varying vehicle density signal degradation, security, privacy, and lack of an online centralized management and coordination entity [3]. The most important challenge is to ensure the communication path between the source and the destination with low communication delay and overhead. Buildings are an important factor that affects communication in VANET since these obstacles attenuate the communication signal [4].

9.2 VANET architectures

VANET architecture can be divided into three different types:

1. Pure cellular wireless local area network

   The vehicles connect to the Internet through cellular gateways and wireless local area network access points. This type of architecture has a high cost due to the cellular towers and wireless access points. In addition, this architecture is geographically limited [5].
2. Pure ad hoc network

   Also known as inter-vehicles ad hoc network, establishes communication between nearby vehicles with no deployment cost. This network architecture is fragmented due to the rapid topology changes and high vehicular mobility thus being very challenging for routing [6].
3. Hybrid network

   The hybrid architecture is a combination of ad hoc and infrastructure domains providing richer contents and improved flexibility [7].

9.3 VANET hybrid architecture

This architecture consists of two types of communication: vehicle to vehicle communication (V2V) and vehicle to infrastructure communication (V2I) which is also known as vehicle-to-roadside unit (V2R). In 1999, the Federal Communication Commission (FCC) allocated a frequency spectrum for vehicle to vehicle (V2V) and vehicle to roadside (V2R) wireless communication. In 2003, dedicated short-range communications (DSRC) uses the 5.850–5.925 GHz frequency band (5.9 GHz band) for the use of public safety and private applications for wireless communication within a 1000-m range [8,9].

In V2V, each vehicle is equipped with global positioning system (GPS), sensors, networking devices, digital map which has the road segment information, and computing devices. V2V communication is mainly used in safety applications such as safety warning, traffic information, road obstacle warning, intersection collision warning, etc. Vehicles are the only type of nodes available in this type of communication. Therefore, the vehicles must act as a source, as a destination, or as a router. The intermediate vehicles are responsible for packets transportation to the destination. Thus, V2V communication might be one-hop communication which is a direct source to destination communication or multihop communication where vehicle relies on other nodes to retransmit the packet until being delivered to the destination. V2V communication has link breakdown due to high nodes' mobility yet it is efficient and cost effective. Neighbor vehicles exchange packets using short-range wireless technologies like Wi-Fi and WAVE.

In V2I or V2R, nodes are either infrastructure devices named roadside units (RSUs) or vehicles known as onboard units (OBUs). RSU is a wireless transceiver and receiver having data storage, computing power, and routing characteristics. RSUs are considered as anchors in VANET communication that exchange information concerning the network and balance the traffic to avoid congestion. RSU represents wireless LAN access point with a range of communication up to 1000 m. RSU periodically broadcasts beacon messages to vehicles within its coverage area. Using these beacon messages, vehicles get the current state and request services. Whenever a vehicle moves from one RSU coverage area to another, it must pay for a handoff cost. RSU's major role is to authenticate the messages exchanged and ensure the confidentiality of communication. In V2R, it is difficult to sustain an end-to-end communication between the vehicles and RSU due to the high mobility of vehicles. RSU has limited bandwidth which will give rise to multiple RSUs having overlap area environment where RSU access becomes crucial for increasing vehicles' throughput [10]. Traffic lights or road signs could be equipped with an RSU to actively inform vehicles on traffic flow, road conditions, or construction sites ahead. Another example for RSU is in static hazard areas such as construction sites, where RSU will warn surrounding vehicles of the construction [11].

In a damaged network, the RSUs might not be functional, thus resulting in a disconnected network. In this case, the vehicle must obtain the message authentication by themselves. The vehicles must communicate confidentially, authenticate messages, and resist malicious attacks without the need for RSUs.

9.4 VANET routing

VANET has no entity able to synchronize and manage the communication leading to high packet collisions and low channel efficient usage. The major key requirement in VANET applications is the availability of efficient and effective routing protocols. This field has been an emerging research area which resulted in several routing protocols to overcome the VANET network challenges [12]. Routing protocols are divided into four classes based on the number of the receivers:

- **Unicast routing**: packets are sent from the source to one destination.
- **Multicast routing**: packets are sent from the source to several receivers using multihop communication.
- **Geocast routing**: packets are sent from the source to nodes in a specific geographic region.
- **Broadcast routing**: packets are sent from the source to all nodes in the network.

V2V uses both unicast and multicast packet forwarding techniques using IEEE 802.11p standard. VANET routing is very challenging in urban environments due to obstacles. Due to RSU's nonfunctionality caused by an unpleasant crisis, vehicles might encounter a disconnected communication where the RSU coverage area does not overlap anymore. VANET routing protocols target to find the optimal routing path to ensure a reliable packet transmission to the destination [13].

9.5 VANET routing protocols

The main target of routing protocols is to find the best path for packet routing from source to destination with minimum communication time and network resources. Due to the special characteristics of VANET, several parameters such as location, scope, speed, security, amount of information, etc. must be taken into consideration whenever designing a new VANET routing protocol.

Several routing protocols which are classified into several groups have been designed for VANET. The grouping of these protocols is based on different aspects such as protocol's characteristics, used techniques, routing information, quality of services, network structures, routing algorithms, and so on.

Based on routing characteristics and techniques VANET protocols are grouped into five classes:

1. Topology based routing protocols

   Data packets are transferred using link's information based on nodes' topology information. Each node builds up a routing table to store the link information based on which the routing decision is to be taken [14].
2. Position based routing protocols

   This group includes one of the most promising protocols since they support the vehicle's geographical position. These protocols are very effective for highly mobile and dynamic networks, but they perform poorly in networks with obstacles. The routing decision is made based on destination and neighbor nodes' positions which require a global positioning system (GPS) [15].

3. Broadcast based routing protocols

   Those protocols are used for applications such as weather, traffic, road situation, etc. due to their simplicity. Broadcasting routing floods the packets to all available nodes within the broadcast domain which will increase network bandwidth consumption [16].

4. Geocast based routing protocols

   This group of protocols multicast packets from source to destination. In a specified geographic region, a directed flooding mechanism takes place to broadcast the packets. Similar to position-based routing protocols, position determining services such as GPS are required. Thus, every node must know its position and neighbors' positions [17].

5. Cluster-based routing protocols

   This group of protocols are known for their effectiveness in scalable networks, but with limitations from vehicles' mobility. They split the network into clusters based on different criteria such as the same direction or velocity. These protocols have low cost and delay in packet delivery [18].

9.6 Cluster routing protocol

Cluster routing protocol is one of the most promising VANET routing protocols due to its scalability. It is a dynamic mechanism that builds a hierarchical infrastructure-like overlay on top of the ad hoc network. In these protocols, several nodes are grouped to form groups known as clusters. The grouping is based on several predefined metrics such as the vehicles' density, velocity, direction, geographical location, etc. Each cluster is formed of several vehicles where one of them will be elected based on predefined parameters to be the cluster head (CH). The other vehicles will be known as cluster members (CM). The CH is responsible for coordinating the CMs and routing the packets between the cluster members and neighboring clusters [19]. Therefore, the CH will be acting as an infrastructure wireless access point in order to ensure an end-to-end communication across the VANET (Figure 9.1).

9.7 Clustering objectives

This section discusses the prime objectives highlighting clustering as one of the best protocols to be followed in VANET.

1. Scalability: Hierarchical architectures are known to increase scalability in networks having high number of nodes. In clustering protocols, those hierarchical layers are known as clusters where a group of nodes, vehicles in case of VANET, form a cluster as discussed earlier.

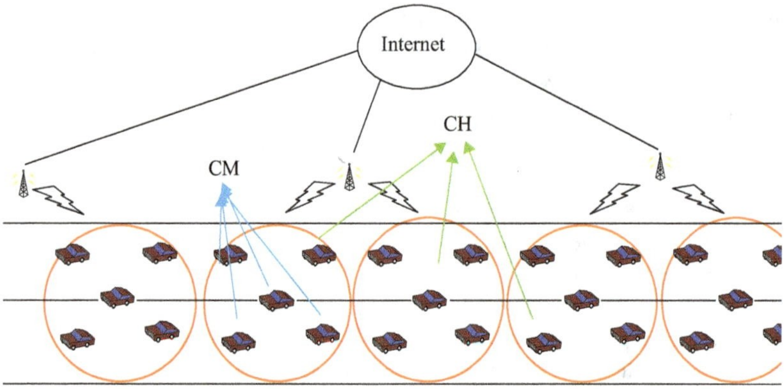

Figure 9.1 Cluster routing protocol

2. Fault tolerance: The failure of any non-CH node will not affect the performance of other cluster members since the CH is the only node responsible for routing the packets. In case the failure node is a CH, this will affect the nodes associated with it until clustering takes place and a new CH is elected.
3. Data aggregation: In networks with high number of nodes, the probability of sending repetitive information increases. Clustering protocols can aggregate all received packets into one output packet. In clusters, data aggregation will result in more energy-efficient networks by reducing the total network's load.
4. Stability: Topology changes of the nodes are reported to the CH. Therefore, the CH is responsible for tracking the nodes' status, link, and other changes.
5. Decrease routing delay: Clustering protocols relies on multihop routing approach. The cluster members will forward the packets to the CH which will in turn be responsible for routing them to the destination. The routing delay will be decreased since the number of hopes will be reduced due to the role of the CH.
6. Collision avoidance: Collision will result in packet loss which will in turn increase the retransmission rate. Clustering will reduce the collision rate due to the routing roles handled by the CH as discussed earlier.

9.8 Clustering protocols

Clustering protocol splits the networks into subgroups. Each group is formed of a predefined number of maximum vehicles. This group is known as cluster. Each cluster protocol has an algorithm for cluster head election. In most of the clustering protocols, each cluster has one cluster head. The cluster head is responsible for routing the packet outside its cluster in addition to routing any incoming packet to its destination. All noncluster head vehicles are known as cluster members.

The below flowchart displays the general procedural flow for clustering protocol in VANET (Figure 9.2).

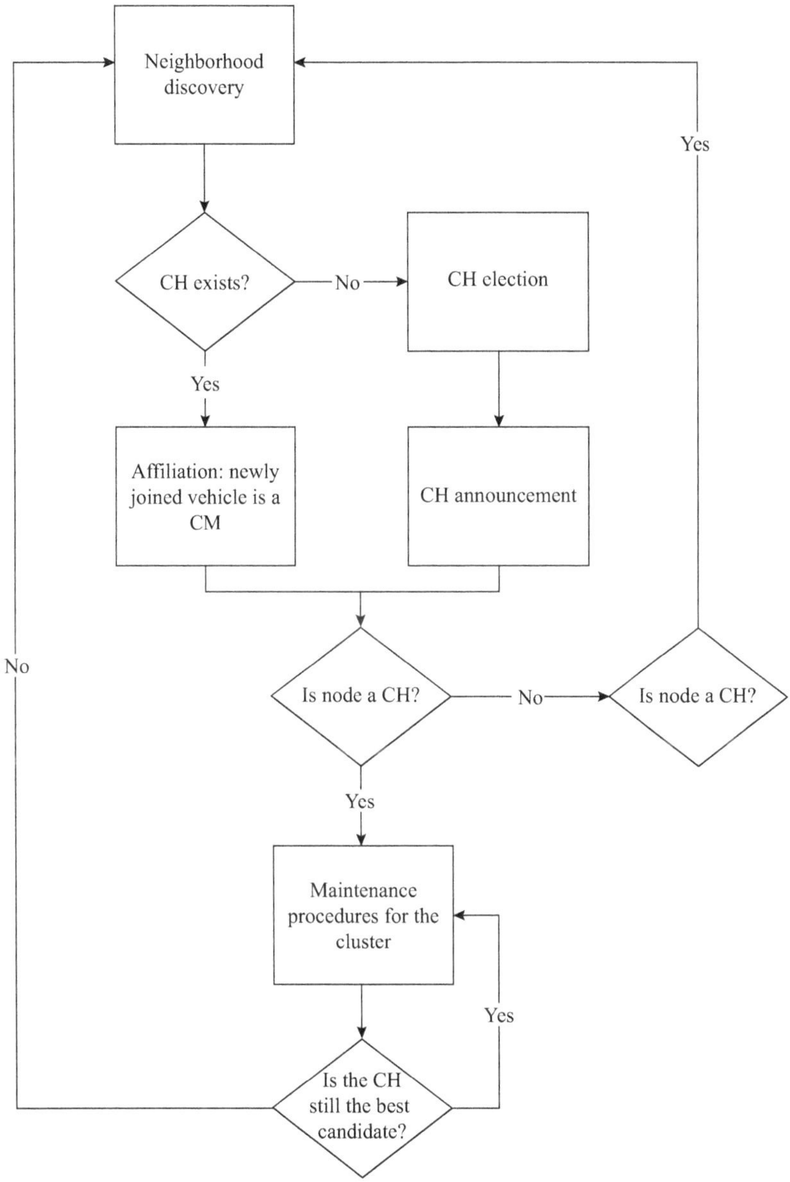

Figure 9.2 Cluster head selection flowchart

1. Neighborhood discovery: Every vehicle upon joining a network will start by broadcasting periodic beacon messages. Meanwhile, the vehicle will also be discovering its *n*-hop neighbors either by receiving broadcast messages or through neighbor solicitation requests. The newly joined vehicle will store the position for all neighboring nodes in a table to be used in the clustering algorithm.

2. Cluster head election: The newly joined vehicle will check its neighborhood table for CH existence. In case the CH does not exist, the newly joined vehicle might be a possible candidate and skip the next step.
3. Affiliation: This step is only reached in case the CH exists in the neighborhood table. The newly joined vehicle will send a request to the CH to join its cluster.
4. Cluster head election: This step varies depending on the clustering algorithm being followed. At the end of this step, only one vehicle will be elected to be a CH for a cluster consisting of *n* neighborhood vehicles.
5. Announcement: The node being elected to be the CH will announce its election to all neighborhood vehicles.
6. Maintenance: This step varies depending on whether the node is a CH or a CM.
   - Node being a CH: In this case, the maintenance depends on the clustering algorithm being followed. Some algorithms change CH, merge with neighboring clusters, and track lost links as maintenance procedures for the formed clusters. The CH and CM will go back to step 1 whenever the CH is not the candidate to be reelected after *n* time elapses.
   - Node being a CM: The node will periodically evaluate its link to the CH. In case the link is lost the CM will go back to step 1.

9.9 Cluster head selection protocols

The best CH election algorithms have always been an important field of research as the stability of the clustering protocol is based on an effective and stable CH election. CH election methods can be classified as competition or comparison-based. In the competition-based CH election algorithms, a competition is taking place between the members of the cluster to be announced as a CH. The first vehicle to announce itself as a CH wins the competition. Whereas in the comparative-based CH election algorithms, each vehicle compares a calculated index to elect the most suitable vehicle to be the CH. Below are the three main assessment mechanisms which are followed for CH election: weighted network metrics, precedence, and timers. Timers is the only assessment used in the competitive-based CH election whereas weighted metrics and precedence are used in comparative-based CH election.

1. Weighted network metrics: Each node calculates an index quantifying its fitness to be elected as a CH. This index known as fitness index or score is a weighted sum of several network metrics, such as the degree of connectivity, link stability, node uptime, ... After calculation, each node advertises this index within its one-hop neighborhood. Each node will rank all its neighbors in its neighbor table. The most highly ranked node will receive a request for CH association which will make it elected to be the CH. The most highly ranked node might be the node having the highest index or the node having the lowest index depending on the algorithm being followed.
2. Precedence: Based on a weighted sum of metrics, every node seeking an affiliation with a CH will make its own assessment of the suitability of

potential CH. In this type of election, the nodes might assess a CH which suits its own requirements. Precedence-based algorithms have better stability thus being the best suited to form hierarchical structures.

3. Timer: The nodes remain in an unclustered state for a period while waiting for a CH to be announced. For any node in the unclustered state if a CH is announced within the timeout period it will affiliate with it; otherwise, it declares itself a CH. This mechanism is fast and simple. A neighborhood discovery phase is not required therefor channel bandwidth will only be used for CH announcement. On the other hand, in case multiple nodes simultaneously reached the end of their timers, the nodes within range of multiple CH will suffer from packet collisions.

9.10 Cluster head election criteria

In this section, we will discuss the most used criteria in the weighted sum for cluster head election.

1. Mobility: This criterion is based on the relative mobility between the node and the node it is calculating its weighted sum for. This is based on the fact that the link is maintained between any two nodes moving with small relative velocity. This metric is used since the potential CH will then have the most reliable link to route the node's packets as compared to a highly moving node based on the inverse square law of electromagnetic signal propagation.
2. Signal quality: The node's location and velocity are based on GPS. In many regions, the GPS suffers inaccuracies due to poor satellite reception. Therefore, this metric which is calculated based on signal to noise ratio (SNR) or received signal strength (RSS) was added by many clustering algorithms.
3. Link expiration time: This metric, also known as expected link lifespan, is the expected link expiration between the node calculating the weighted sum and the potential CH. This metric is highly important to prevent fast-moving nodes to be elected as CH which in turn will result in a fast rate of reclustering need. In addition, this metric will also prevent nodes moving in opposite direction to be elected as a CH otherwise the routed packet will never reach its destination. This metric is calculated based on the relative position, velocity, and an assumed maximum communication range.
4. Platoon leadership: This metric is based on the node's position relative to its neighbors. This will result in CH elected to be moving in the same direction as the node electing it. In case the distance between the available CH and the node is above a certain threshold, then the node will not affiliate with this CH.
5. Vehicle class: This criterion will exclude private cars from being elected as CH; therefore, only buses can be elected as CH. This is based on the fact that public buses usually have a well-known path with a limited speed as compared to private cars which have no control for both of these criteria.
6. Driver intention: This criteria targets to elect the CH moving in the same direction and expected to continue in the same path as the node electing it. This

criterion is based on the relative position, relative velocity, number of neighbors, and the lane upon which the nodes are traveling. The traffic is considered as well in this criterion as it has a direct effect on the number of lanes available on a certain road.

7. Security reputation: This criterion is essential to prevent malicious nodes from being elected as CH. Many algorithms have been used for this criterion calculation such as the reliability of the node's packet relaying, certificate authority trust including traffic obedience and reputation for packet forwarding, and public key infrastructure.
8. Mean speed: The speed of vehicles is assumed to be of normal distribution in a free flow traffic road. Therefore, each vehicle can calculate the separating distance of the neighbor vehicles. The CH will be elected to be the vehicle closest to the mean speed of the cluster.

9.11 VANET cluster routing protocols review

Cluster-based routing protocols are known for their efficiency in data routing. Therefore, this section will review several VANET clustering protocols.

1. Clustering for open IVC network (COIN)

   COIN is based on MANET protocol for purpose of enhancing stability and scalability. Cluster head election is based on driver's behavior, vehicle's ID, relative mobility, and distance between the vehicles. This protocol is effective in a network with low relative mobility which allows the vehicles to be within the radio range for the longest period.
2. Cluster-based directional routing protocol (CBDRP)

   CBDRP proposed by Song *et al.* is based on clustering of the vehicles based on the direction. As any other clustering protocol, CBDRP has two main steps: CH election and packet routing. The CH election is based on the vehicle location and velocity. The packets are forwarded from the CH to all neighboring CHs. The CHs will forward the packets until reaching the destination. A main drawback of this protocol is that the packet forwarding mechanism will only take place if another CH is within the radio coverage range of the CH carrying the packet.
3. Traffic Infrastructure based cluster routing protocol with handoff

   This protocol improves network performance by adapting handoff cellular network due to clusters' overlapping. Each CH in this protocol has one CH from the backbone infrastructure which are supposed to be covering the roads. Other assumptions are taken in this protocol which include bidirectional links, location services (GPS) on board of the vehicle, and two-dimension deployment.
4. Stability based clustering algorithm (SBCA)

   Like most other clustering protocols, SBCA has two stages: first stage is the setup where the nodes are split into clusters with a primary CH election, and the second stage is the maintenance where a secondary CH is elected. The role of the secondary CH is to take over the CH roles whenever the primary CH is no longer available.

5. Passive clustering aided routing (PassCAR)

   This protocol is based on passive clustering approach with low overhead control in packet forwarding. This mechanism is inefficient in highly mobile and dynamic networks which resulted in network overhead and network delays. In PassCAR, each vehicle is a potential candidate for CH role. The CH election is based on weighted metric calculation which includes the link lifetime, transmission count, and node degree.

6. Intelligent based clustering algorithm (IBCAV)

   IBCAV is taken into consideration several criteria: traffic flow and density, cluster size, and vehicles' velocity. The CH election is based on a series of computations, time, and network resource consumption. The roadside unit (RSU) is elected as a CH due to its high processing capability. In case of absence of the RSU, the vehicle with the lowest speed will be elected as the CH. As other protocols, the CH is elected to be the one with the lowest speed in order to guarantee the maximum time within the cluster. This protocol uses the Hello messages for cluster formation and store/carry forwarding approach for packet routing.

9.12 Clusters with double cluster head

In a crisis scenario where infrastructure is destroyed and nonfunctional, clustering protocols with double cluster heads are a possible solution to overcome the absence of RSUs. The only requirement of these protocols is the existence of a GPS and a dedicated short-range communication (DSRC). The GPS is required for knowing the real time location and velocity of the vehicles. Whereas, the DSRC is required for wireless communication between the nodes.

Each vehicle will have a unique identifier (ID), a primary CH, and a secondary CH (SCH). Each vehicle will be in one of the following four states:

- Unclustered vehicle (UV): the vehicle just joined the network (initial state)
- Cluster head (CH): the vehicle is elected to be CH
- Secondary cluster head (SCH): the vehicle is elected to be the SCH
- Cluster member (CM): the vehicle is affiliated with a CH and SCH

Whenever a vehicle joins the network, it is in the UV state. Thus, it starts by broadcasting Hello messages to advertise its existence. The Hello message carries different information based on the protocol followed. In general, all Hello messages states the location and velocity of the sender vehicle.

After broadcasting and receiving the Hello messages, each vehicle calculates its eligibility for being elected as a CH based on the protocol followed. Another type of Hello message will be broadcasted carrying the eligibility index of being elected as a CH. Each vehicle receiving this Hello message will compare the received eligibility index to its value. The vehicle with the highest eligibility index will announce itself as a CH using a third type of Hello message denoted as

CHHello. Each vehicle receiving the CHHello will check if the announced CH is within one-hop neighbor and with an eligible CH election higher than its value. If so, the vehicle will send a JoinReq message to the announced CH. Upon receiving the JoinReq, the CH will check if the maximum number of the cluster members is not reached. In case the maximum number is not reached, the CH will send a JoinReply message to the sender vehicle while adding the vehicle's information to the CH routing table. Once the JoinReply is received, the vehicle will change state to be a cluster member (CM) affiliated with that CH. The CH will elect the secondary CH (SCH) based on the eligibility value E_c of the cluster members. The CH will announce the SCH in the next Hello message denoted as CHHello.

In case the link with the CH is lost, the CM forwards its packets to the SCH to be carried to the destination. The SCH will keep on providing the services to the CM as if it is the CH. The CH, SCH, and the CM will keep on exchanging Hello messages to ensure the link between them. Once the Hello message is not received, the link is down. In case the member does not receive any Hello message from the CH or the SCH, then the CM becomes inconsistent with the current CH and restarts the clustering protocol

9.13 Cluster merging

Cluster merging is the combination of two clusters to form a new larger cluster which will eventually decrease the number of clusters and cluster heads in the network. The process of merging is initiated whenever two cluster heads are moving within the coverage range of each other. The two cluster heads must stay within the transmission range of each other for a period of time known as merge interval before starting the merge. There are some rules that the two clusters must fulfill before initiating the cluster merging. The clusters should have all cluster members consistent with the current cluster heads. In addition, the total number of cluster members after the merge should be less than the total allowed cluster members according to the protocol followed. There are several ways for the election of the new CH some of which are as follows:

- The CH with the lower eligibility CH election value should resign from its role and affiliate as a CM to the other CH which will be in turn be elected as the CH of the newly formed cluster.
- The CH with a higher number of cluster members will be elected as the new CH.
- The CH with higher stability will be elected as the new CH where the stability is based on the average relative speed. The vehicle with a lower average speed thus having higher stability will be elected as the new CH.

9.14 VANET in crisis

In the ideal case scenario, any cluster member will forward its packets to its cluster head which in turn uploads it to the RSU. Once the packet has been uploaded to the

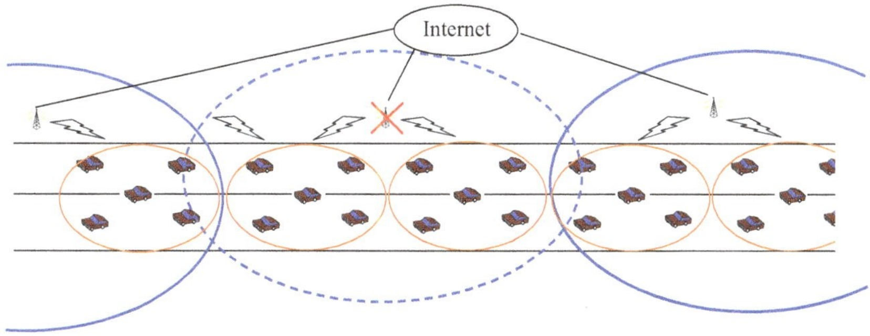

Figure 9.3 Disconnected clustered network

RSU, then it will be routed to its destination from the Internet. The destination will thus be able to download its packet from the RSU it is associated with. Note that the RSU and vehicle association is conditioned that the vehicle must be within the coverage area of the RSU [20].

In the crisis scenario, one or more of the RSUs might be nonfunctional which will result in packet failure to be uploaded to the Internet as shown below (Figure 9.3).

Note that in case the sender and the destination are cluster members within the same cluster then the nonfunctionality of the RSU will have no effect on their communication as they are within the direct communication range of each other. Clustering with double head association for each cluster is one of the best solutions to overcome the RSU nonfunctionality due to its zero cost and simplicity.

9.15 Weighted double cluster head selection

This section will discuss a cluster routing protocol with weighted double cluster head election algorithm as a solution for the disconnected network in crisis VANET networks.

As weighted cluster election is one of the most widely used mechanisms due to its flexibility, a weighted formula based on three parameters is used to elect the double heads. A wide range of criteria might be considered in the weighted formula for cluster election, but we considered the below three criteria due to their direct and noticeable effect in the designed protocol [21].

The distance is the first parameter in the weighted cluster head election. This criterion had been used by several cluster head protocols such as the fuzzy-based cluster selection and triple cluster-based routing protocol (TCRP) [22,23]. This criterion had a direct effect on the cluster head election. The cluster with the smallest distance to the largest number of neighbors will be elected as the cluster head. This is based on the theory of the centroid as being equidistant from the vertices.

The second parameter in the weighted cluster head election is the velocity which is used by several protocols such as angle-based clustering algorithm [24]. It is highly important to mention that whenever velocity is used in this chapter, in reality, it is the respective velocity of every vehicle with respect to the others.

The third parameter considered in the suggested weighted cluster head election is the trust. The trust is a highly important criterion to be considered whenever dealing with autonomous networks such as the VANET. The trust is added for two reasons: the first is to rule out malicious nodes which might not perform at their best to deliver the packets to the destination and the second reason is to elect the best candidate for the CH. Whenever having two or more nodes equally satisfying the first two criteria, the trust comes into play to elect the node with better performance. We will discuss in detail how is the trust being calculated as it is not a straightforward criterion as the distance and the velocity.

Trust is of two types: direct trust and indirect trust. The direct trust is the ratio of packets successfully sent by the node with respect to the total packets requested to be sent by the node as shown below [25]:

$$\text{Direct trust } (n, m) = \frac{\Delta f_{nm}}{\Delta r_{nm}} \tag{9.1}$$

where Δr_{nm} is the total number of packets requested to be forwarded by vehicle n to vehicle m such that $n \neq m$ and m is a neighbor of n and Δf_{nm} is the effective number of packets forwarded by vehicle n to vehicle m based on the total delay of the packet.

The indirect trust is simply the summation of all direct trust values broadcasted by neighboring nodes. The below formula shows the indirect trust calculated by node n with respect to node m:

$$\text{Indirect trust } (n, m) = \sum\nolimits_{i \in Q} \text{direct trust } (n, m) \tag{9.2}$$

where n is the trustor node, m is the trustee, and Q is an element of the set n vehicles.

A weighted summation between the direct and indirect trust will result in the total trust as in formula (9.3):

$$\text{Trust } (n, m) = \lambda\, x \text{ direct trust } (n, m) + (1 - \lambda) x \text{ in direct trust } (n, m) \tag{9.3}$$

where λ is a constant threshold which varies between 0 and 1.

Now that we have the three parameters, the weighted cluster head selection formula is

$$P_{CH} = \mu D_{(n,m)} + \varnothing T_{(n,m)} + \theta V_{(n,m)} \tag{9.4}$$

where P_{CH} is the probability of the node to be elected as a cluster head; μ, $\varnothing$. and θ are constant values to be predefined according to the system requirement such that $\mu + \varnothing + \theta = 1$; $D_{(n,m)}$ is the distance between the two vehicles n and m; $T_{(n,m)}$ is the trust of vehicle n with respect to vehicle m; and $V_{(n,m)}$ is the velocity of m with respect to n.

```
Result: VANET Clustering Protocol for crisis cases with nonfunctional infrastructure
        based on weighted probability for Double Cluster Head Selection
Calculate the packet delivery delay for each vehicle with respect to others;
Calculate the rate of delivered packets with respect to the total packets sent;
Calculate the Direct, Indirect and Total Trust;
Calculate the Probability of CH selection based on the weighted formula;
while vehicle is not CH or CM do
    Select the highest probability to be primary CH;
    while CM count less than Maximum CM do
        if Next highest probability difference is less than threshold then
            Select the highest probability to be CM;
        else
            Move to the next cluster;
        end
    end
    while Cluster has 1 CH do
        while count less than Maximum CM do
            if distance between CM and CM of other clusters is min then
                select those nodes to be secondary CHs
            else
                Keep looping to find the min distance between the two nodes connecting
                 the clusters
            end
        end
    end
end
```

Figure 9.4 Pseudo code

The below pseudo code shows the double cluster head selection based on the weighted formula discussed above (Figure 9.4).

Starting by packet ratio calculation the direct, indirect, and total trust will be calculated. Having the distance and the speed, the probability of cluster head election can be calculated taking into consideration the weights as required by the network. The calculated probability will be broadcasted among neighbor nodes. The node having the highest probability will be selected as the primary cluster head. A certain number of cluster members will be associated with this primary cluster head. Each cluster member will check its distance to neighboring cluster members. The two clusters with minimal distance will be selected as secondary cluster heads. The role of the secondary cluster heads is to act as connecting bridges between the clusters. The connecting bridges will overcome the nonfunctionally of the RSU in a crisis scenario thus ensuring an end-to-end communication between any two vehicles within this VANET network.

The below schematic shows 20 vehicles numbered with IDs 1 to 20. The black circles show the primary clusters with the primary cluster head. The grouping shown in blue represents the secondary clusters with the secondary cluster heads. As shown in the schematic, there is an end-to-end communication throughout the network (Figure 9.5).

Let us suppose vehicle 5 requires to send a packet to vehicle 4. Vehicle 5 will forward the packet to its cluster heads vehicles 8, 11, and 15. Vehicle 11 is the primary cluster head whereas vehicles 8 and 15 are the secondary cluster heads. Vehicle 11 will eventually drop the packet since no connection to any RSU.

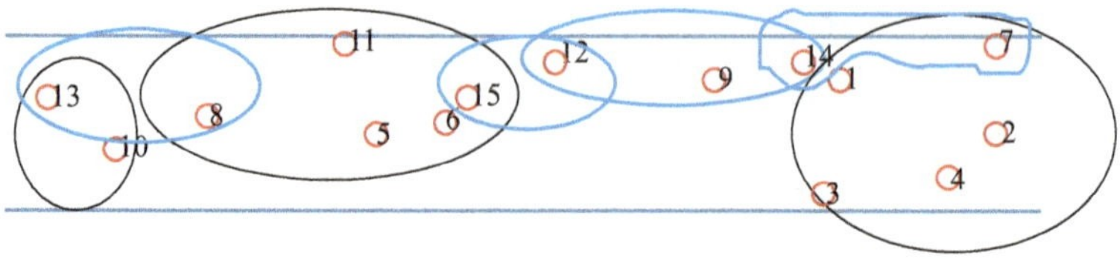

Figure 9.5 Double cluster head selection

Vehicles 8 and 15 will forward the packet to their neighboring cluster members 13 and 12, respectively. Vehicle 13 will also drop the packet since the destination is not within its communication range and no other cluster heads to forward the packet to. On the other hand, vehicle 12 will forward the packet to the neighboring cluster heads, that is, vehicles 9 and 14. Vehicle 9 being the primary cluster head will also drop the packet due to the RSU lost connection. Vehicle 14 being the secondary cluster head will forward the packet again to its neighboring cluster heads. In our considered schematic, vehicle 7 is the only neighboring cluster head for vehicle 14. Once vehicle 7 receives the packet, checking the destination address vehicle 7 realizes that the destination is within its cluster, that is, within its direct communication range. Finally, the packet will be forwarded from vehicle 7 to its destination vehicle 4.

9.16 Conclusion

In disaster scenarios, infrastructure is at the highest risk of getting destroyed. VANET communication relies on infrastructure known as RSUs. In crisis cases as mentioned earlier, one or more roadsides might be nonfunctional which results in a disconnected VANET network. The main target of the RSU is to upload the packets received from the vehicles within their coverage region to the Internet. This chapter discussed a specific type of VANET routing protocol known as cluster routing protocols. This type of protocol divides the network into subgroups known as clusters. Each cluster usually has one cluster head which is responsible for uploading the packets to the roadside unit. In addition to the cluster, each cluster has several cluster members. This chapter studies the possibility of having overlapping clusters associated with double cluster heads. The target of this newly proposed protocol is to ensure an end-to-end communication without relying on the RSUs. This protocol is a very promising protocol due to the no cost requirement, efficiency, and reliability.

References

[1] Yein, A. D., Lin, C.-H., Huang, Y.-H., Hsieh, W. S., Lee, C.-N., and Kuo, P.-C. "A novel message authentication method for VANET without RSU." *Engineering Computations*. 2016.

[2] Santa, J., Pereñíguez, F., Cano, J. C., Skarmeta, A. F., Calafate, C. T., and Manzoni, P. "Comprehensive vehicular networking platform for V2I and V2V communications within the walkie-talkie project." *International Journal of Distributed Sensor Networks*. 2013.

[3] Faiza, O., Lila, B., and Mourad, G. "The involvement of RSUs in VANETs: Survey and perspectives." *International Journal of Computational Engineering Research*. 2013; 3(7).

[4] Jiménez, F. "Connected vehicles, V2V communications and VANET." *Electronics*. 2015; 538–540.

[5] Liu, J., Wan, J., Wang, Q., Deng, P., Zhou, K., and Qiao, Y. "A survey on position-based routing for vehicular ad hoc networks." *Telecommunication Systems*. 2015; 15–30.

[6] Lin, Y.-W., Chen, Y.-S., and Lee, S.-L. "Routing protocols in vehicular ad hoc networks: A survey and future perspectives." *Journal of Information Science and Engineering*. 2010; 913–932.

[7] Senouci, O., Aliouat, Z., and Harous, S. "Survery: Routing protocols in vehicular ad hoc networks." *Second International Conference on Advanced Wireless Information, Data, and Communication (AWICT)*. 2017.

[8] Balapgol, S. and Deshukh, P. K. "Broadcast protocol for V2V and V2RSU in VANET." *International Journal of Advanced Research in Computer and Communication Engineering*. 2015; 4(7): 38–43.

[9] Raw, R. S. and Das, S. "Performance comparison of position-based routing protocols in vehicle-to-vehicle (v2v) communication." 2011; 3(1).

[10] Wu, D., Ling, Y., Zhu, H., and Liang, J. "The RSU access problem based on evolutionary game theory for VANET." *International Journal of Distributed Sensor Networks*. 9(7): 2103.

[11] Kiran, P. S., Rao, B. T., and Reddy, L. S. S. "Architectural crises in vehicular ad hoc networks." *Global Journal of Computer Science and Technology*. 2010; 10(2): 31–35.

[12] Lin, D., Kang, J., Squicciarini, A., Wu, Y., Gurung, S., and Tonguz O. "MoZo: A moving zone based routing protocol using pure V2V communication in VANETs." *IEEE Transactions on Mobile Computing*. 2017; 16(5): 1357–1370.

[13] Bilal, S. M., Khan, A. U. R., and Ali, S. "Review and performance analysis of position-based routing protocols." *Wireless Personal Communications*. 2017; 559–578.

[14] Qureshi, K. N. and Abdullah, A. H. "Topology based routing protocols for VANET and their comparison with MANET." *Journal of Theoretical and Applied Information Technology*. 2013; 58(3): 707–715.

[15] Kaur, S. and Gupta, A. K. "Position based routing in mobile ad hoc networks: An overview." *International Journal of Computer Science and Technology*. 2014; 3(4): 792–796.

[16] Misra, S. C., Woungang, I., Zhang, I., and Misra, S. *Guide to wireless ad hoc networks*. Springer; 2009.

[17] Hamza, T. and Nsiri, B. "A hybrid routing protocol for VANET using ontology." *Procedia Computer Science*. 2015; 73: 94–101.

[18] Yein, A. D., Lin, C.-H., Huang, Y.-H., Hsieh, W.-S., Lee, C.-N., and Kuo P.-C. "A novel message authentication method for VANET without RSU." *International Journal for Computer Aided Engineering and Software*. 2016; 33(8): 2288–2301.

[19] Hadded, M., Zagrouba, R., Laouiti, A., Muhlethaler, P., and Saidane, L. A. "A multi-objective genetic algorithm-based adaptive weighted clustering protocol in VANET." *IEEE Congress on Evolutionary Computation (CEC)*. Japan. 2015.

[20] Mavromoustakis, C. X. "Mitigating file-sharing misbehavior with movement synchronization to increase end-to-end availability for delay sensitive streams in vehicular P2P devices." *International Journal of Communication Systems, Wiley*. 2013; 26(12): 1599–1616.

[21] Khayat, G., Mavromoustakis, C. X., Mastorakis, G., Batalla, J. M., Maalouf, H., and Pallis, E. "Successf VANET clustering based on weighted trusted cluster head selection." *IWCMC*. 2020.

[22] Sanhaji, F., Satori, H., and Satori, K. "Cluster head selection based on neural networks in wireless sensor networks." *International Conference on Wireless Technologies, Embedded and Intelligent Systems (WITS)*. 2019

[23] Khan, Z. and Fan, P. "A novel triple cluster based routing protocol (TCRP) for VANETs." *IEEE 83rd Vehicular Technology Conference (VTC Spring)*. 2016.

[24] Mahmood, D., Javaid, N., Mahmood, S., Qureshi, S., Memon, A. M., and Zaman, T. "MODLEACH: A variant of LEACH for WSNs." *Eighth International Conference on Broadband, Wireless Computing, Communication and Applications*. 2013

[25] Khayat, G. and Maalouf, H. "Trust in real-time distributed database systems." *IEEE International Conference on Industrial Technology*. Amman, Jordan; 2017.

Chapter 10

Artificial intelligence-enabled optical wireless communication links: a revolutionary approach toward smart communication model

Rajan Miglani[1,2], *Jagjit Singh Malhotra*[3], *Sushank Chaudhary*[4] *and Gurjot Singh Gaba*[2]

10.1 Introduction

Projected to be a new direction of research and innovation, machine learning (ML) and artificial intelligence (AI) techniques will bring significant transformation in the field of optical wireless communication (OWC) links. Hailed as a promising last-mile solution in delivering optical access to end-users, the OWC systems fail miserably under severe channel conditions. AI and ML can help alleviate various signal transmission and detection issues that otherwise cannot be solved using conventional techniques and algorithms. The chapter has been broadly divided into five sections. Section 10.1 of this chapter has been dedicated to fundamental building blocks of an optical wireless link factors that influence the quality of service (QoS) in OWC, and various factors that can enhance or inhibit link performance on any given day. In Section 10.2, we undertake a brief overview of the underlying technology and transmission characteristics for optical signal propagation in a wireless channel. Section 10.3 focuses on different AI and ML related techniques that can be used to replace the existing fading mitigation strategies that are otherwise used in conventional OWC systems. Section 10.4 focuses on narrowing down to a holistic conclusion to understand the challenges and opportunities which may come along this journey of integrating AI and ML with OWC systems. Section 10.5 summarizes the chapter.

[1]Department of Electronics and Communication Engineering, I.K.G Punjab Technical University, Punjab, India
[2]School of Electronics and Electrical Engineering, Lovely Professional University, Punjab, India
[3]Department of Electronics and Communication Engineering, DAV Institute of Engineering and Technology, Punjab, India
[4]Quanzhou Institute of Equipment Manufacturing, Chinese Academy of Sciences, China

10.2 Overview of optical wireless communication systems

Optical wireless communication, also referred loosely as free-space optical (OWC) communication, is essentially a communication system that entails the use of a light source for transmission of information. The examples of OWC as communication strategy can be traced back in history to 800 B.C. when ancient Romans and Greeks used fire as a coded beacon signal for conveying different symbolic messages to their far-flung military posts. However, the real breakthrough came in 1880 when Alexander Graham Bell demonstrated the modulation of voice signal over sun rays and transmitted the modulated signal over 200 m [1,2]. Named as photophone, this device was undoubtedly a technological achievement that revolutionized the idea of using unguided media as a propagation channel. Bell described photophone as one of the greatest achievements of his life, even greater than the telephone. A little less than a century later after the invention of photophone, things started changing dramatically with significant improvement and refinement of optical devices. The invention of optical fibers, development of high powered lasers, and receiver optics propelled the idea of using light as a possible means of communication through wireless medium [3]. Some of the notable works of this time include: transmission of television signal using GaAs LED at MIT Lincolns laboratory, commercial data transmission using laser links by Nippon Electric Company (NEC) and US air force program to establish satellite to ground link, often known as space flight test system (SFTS), to name a few [4].

10.2.1 Resurgence of OWC/OWC systems

It is interesting here to note that despite having sufficient knowledge and resources, the use of light-based wireless communication systems for terrestrial applications did not garner much interest from researchers around the world. This scenario can broadly be attributed to the fact that communication means available during that time were sufficient to suit the human needs of data transmission. Secondly, rapid research in the field of high-speed optoelectronics has been one of the key reasons behind the revival of finding ways by which free space optical systems can be used for civil/commercial applications. Also, with the penetration of data-hungry devices in our daily lives, the demand for higher bandwidth for end-users has seen an exponential increase in the last couple of decades, and this trend only seems to intensify in coming years [5,6]. The radiofrequency (RF) carriers that currently form the major part of the network backbone have limited operational bandwidth and capacity; hence, it is highly unlikely that the existing RF regime will be able to cater to this bandwidth hunger [7–13]. Therefore, the OWC systems, which were initially thought to be useful for deep space and military applications only are now being tweaked to suit the needs of urban areas where the per capita data consumption is very high [14–17]. In a modified form of application, OWC technology can be used for interchip, indoor, and outdoor applications.

10.2.2 Pros and cons of OWC links over RF systems

The renewed interest in looking out possibilities for the commercialization of OWC is not merely a coincidence but an urgently needed solution for the prevailing situation. OWC systems have certain inherent features that make them favorites in delivering high-end data services to end-users. The following features will help us to understand why the future belongs to light-wave systems and in particular to OWC [18–20]:

1. **Bandwidth:** The fundamental difference between OWC and RF systems is the operating frequency. While the former uses the IR and visible band of the spectrum (10^{12}–10^{16} Hz), the latter generally operates in the VHF band (~10^{8} Hz). Therefore, the usable bandwidth in OWC links is approximately 10^{5} Hz greater than conventional RF systems. This leverage of higher usable bandwidth translates into higher data capacity in OWC systems. Although higher RF bands are capable of delivering higher data rates on the flip side, at such frequencies, not only are the operational costs too high, but general traits of RF communication like mobility, coverage area, and receiver sensitivity also become irrelevant. Moreover, it is imperative that in the near future, all-optical networks will form the backbone of high-speed communication systems. Hence, OWC links can easily integrate with these access networks [16,19]. Figure 10.1 highlights the bandwidth capabilities of different RF technologies in contrast to optical communications systems.
2. **Unlicensed spectrum:** Radiofrequency spectrum is a limited resource; hence, spectrum congestion is a noticeable effect that the telecom operators are facing today. Limited bandwidth means that the licensing fee paid by operators will become enormously gigantic, which in turn leads to higher prices being paid by the end-users [20]. Optical bandwidth, on the other hand, uses an unregulated band of the spectrum, which will significantly reduce operating costs along with relieving telecom operators from the troublesome process of leasing spectrum from government agencies [21,22]. The IR and visible band used in OWC links will help in delivering cost-effective services to end-users as operators will incur lower deployment costs. Therefore, it is being argued that free-space optical links will play a comprehensive role in future communication generations (5G) by its virtue of cost-effectiveness, reduced interference, and robust carrier reuse capability [23,24].
3. **Reconfigurable and redeployable:** Many of the commercial OWC systems available today are reconfigurable devices that can be installed and redeployed in a short period. Also, these OWC systems operate on the physical layer of the communication model; hence, there is no need for complex algorithms to streamline a full-duplex link. In fact, OWC systems can be very conveniently used as plug and play devices. Such character of ease of set up can be instrumental in events such as disaster management and natural calamities where conventional communication links often suffer large scale losses. For example, back in 2001, during the unfortunate attack on the world trade center in America, the entire communication network in that area got snapped. As a

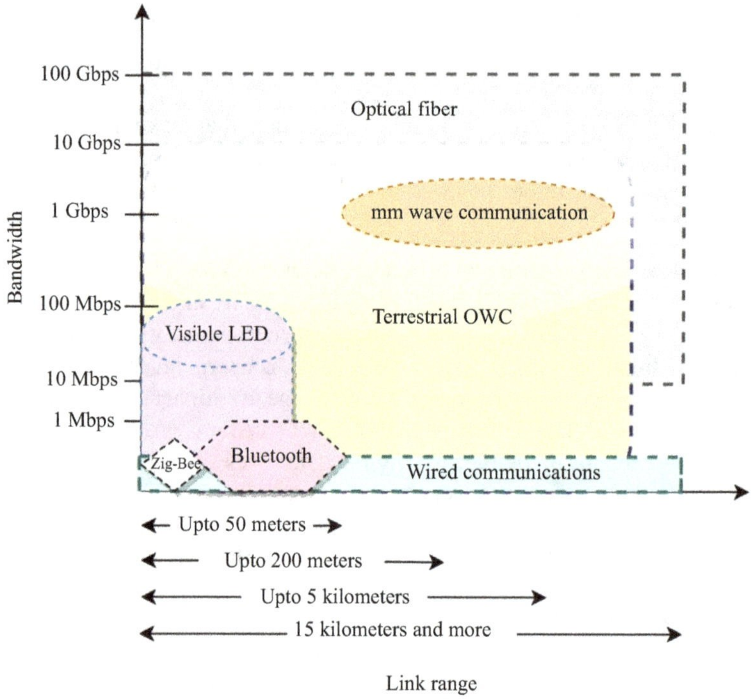

Figure 10.1 Bandwidth versus link range scenario of various communication technologies

quick alternative, OWC links were installed to facilitate communication between various rescue and search teams engaged on ground zero.

4. **Inherent security features:** OWC systems are a line of sight-based communication links that use highly directional beams for transmission of information. For any unintended user to spoof the signal, physical obstruction of the signal will be required. This is in sharp contrast to RF systems where beams are not directional but instead spread over a wide area, hence making them vulnerable to unauthorized detection.
5. **Narrow beam size:** Highly directional character of the laser beam allows frequency reuse and isolation of interference; thus, multiple OWC links can operate simultaneously without any significant threat of interception by the potential of hackers/unintended users. However, a narrow beam in OWC links mandates stringent alignment condition between transceivers and beam sway due to any channel adversities, and geometrical or geographical reasons may lead to complete link breakdown.
6. **Channel dependency:** Since the propagation medium in the case of OWC links is the unguided medium, that is, atmosphere, hence unpredictable atmospheric effects can leave a dampening impact on the link efficiency. Accordingly, the OWC link may not suit those geographical locations where heavy

rains, snow, and fog occur very often. Moreover, even in the case of clear sky conditions when there are comparatively very low attenuations, the loss of signal quality due to atmospheric turbulence cannot be ruled out [25,26].

10.2.3 Block diagram of a free-space optical communication link

Optical wireless communication, a term often synonymously used with free-space optical links, refers to the use of unguided media for transmission of an optically modulated carrier signal. The carrier can be either be part of a visible, infrared, or ultraviolet (UV) band, but generally, the visible band is the preferred choice for terrestrial and indoor links. Figure 10.2 illustrates the fundamental block diagram of the OWC system, which like any other communication system can be classified to be composed of a transmitter, channel, and receiver, which have described as follows.

1. **Transmitter section:** This section of the system is concerned with modulating the information onto an optical carrier and then transmitting the same over the channel, that is, atmosphere. But before the modulated signal propagates over the channel, it first passes through a series of collimating lenses, also known as transmission optics. Since OWC links are essentially line of sight (LoS)-based links, hence such an arrangement of lenses minimizes the beam wander as the signal propagates through the medium. Meanwhile, the laser driver circuit provides the constant current required for the operation of the light sources, which are often susceptible to performance fluctuations on account of thermal effects. Although various higher modulation schemes are now compatible with

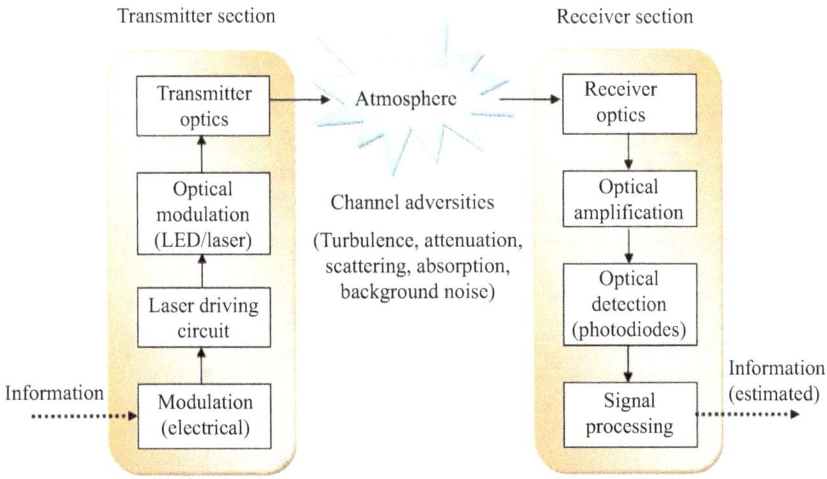

Figure 10.2 Block diagram of the optical wireless communication link

OWC systems, yet intensity modulation (IM) scheme due to its simplicity and cost-effectiveness is the most popular amongst all [20].

2. **Channel:** Channel in OWC link is an unguided medium. The modulated signal which is transmitted over the atmosphere is subjected to various meteorological effects such as rain, fog, and snow. The atmospheric phenomena are highly unpredictable in terms of occurrence and the impact they leave on the channel. Atmospheric adversities can, therefore, cause significant loss of information due to absorption, scattering, and attenuation of the information-bearing signal.
3. **Receiver:** This section is responsible for recovering the modulated information received over the channel. The receiver optics, in contrast to the transmitter section, collect the spread-out beam and focuses it on a highly sensitive receiver, that is, a photodiode. Some OWC links also employ an additional optical amplification stage before signal detection is accomplished. This is done to strengthen the signal quality, which usually deteriorates due to channel adversities. After photodiodes have performed optical-electrical (O-E) conversion, post signal processing techniques such as filtering and regeneration might be used to enhance correct signal estimation.

10.2.4 Atmospheric effects on OWC channel

Since the propagation medium in the case of OWC links is the atmosphere, hence different meteorological phenomena such as rain, fog, snow, and aerosols have performance-limiting impacts on the link performance. Based on meteorological visibility, atmospheric conditions can be broadly classified into three categories: clear, moderate, and weak which result in low, heavy, and severe atmospheric attenuation, respectively. For example, a cloudy day characterized by rain will lead to poor visibility and absorption of light-bearing signals by water droplets, thus leading to severe atmospheric attenuation [27–30]. Losses in OWC links due to the adverse atmosphere can be associated with one or more of the reasons shown in Figure 10.3 and explained below as well.

1. **Absorption and scattering losses**: The atmospheric channel consists of various inhomogeneous particles such as water droplets, fog, smoke, dust, smog, and haze. These constituent particles of varying sizes tend to absorb and scatter information-bearing photons, resulting in loss of transmission power. The attenuation of signal due to atmospheric particles is given by [31]:

$$\gamma(\lambda) = \alpha_m(\lambda) + \alpha_a(\lambda) + \beta_m(\lambda) + \beta_a(\lambda) \tag{10.1}$$

The first two terms $\alpha_m(\lambda)$ and $\alpha_a(\lambda)$ of (10.1) represent molecular and aerosol absorption coefficients, respectively, while the other two terms $\beta_m(\lambda)$ and $\beta_a(\lambda)$ represent molecular and aerosol scattering coefficients, respectively. It must be noted here that both absorption and scattering in the OWC channel are wavelength (λ) dependent. While absorption by channel particles leads to degradation of transmission power levels, on the one hand, scattering, on the other hand, causes the angular redistribution of energy without any shift in

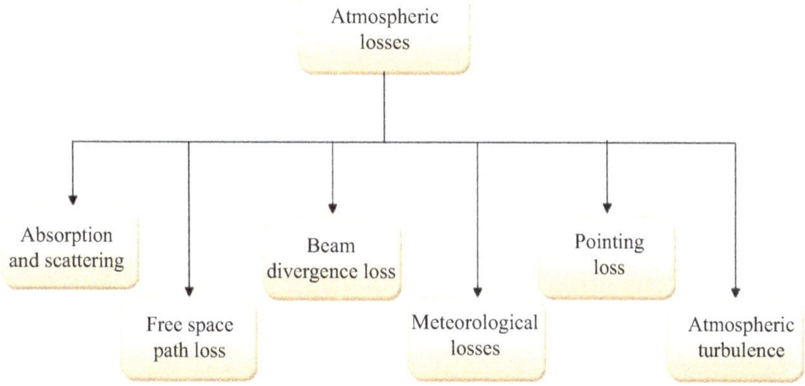

Figure 10.3 Different atmospheric losses that effect OWC link performance

wavelength. The extent of scattering depends upon the ratio of operating wavelength (λ) to the size of the scattering particle (r). For $\lambda/r > 1$, Rayleigh scattering affects the link, while for $\lambda/r \approx 1$, Mie scattering takes place. The atmosphere is composed of subparticles whose size varies from nanosized particles to the order of centimeters, resulting in absorption and (or) scattering. Commercial OWC links around the world operate in safe wavelength windows, that is, the wavelength range for which the specific attenuation (dB/km) is least. These transmission windows are generally in the wavelength range of 780–850 nm and 1500–1600 nm.

2. **Free space loss**: This is the most significant loss affecting the OWC link performance. It is the loss experienced by the optical modulated signal while propagating through the link of range L and is expressed as [31]:

$$L_s = (4\pi L/\lambda)^2 \tag{10.2}$$

As seen from (10.2), the loss factor L_s is inversely proportional to the squared value of operating wavelength. Thus OWC links in contrast to RF links suffer more link loss due to shorter operating wavelength.

3. **Beam divergence losses:** Apart from absorption and scattering of the beam, OWC signals also experience beam spreading while transversing through the channel. As a result, the size (field view) of the received signal is much greater than the actual receiver aperture size. This means that the receiver aperture will now collect only a part of transmission signal power, thus leading to divergence losses, as illustrated in Figure 10.4. The optical power collected by the receiver is mathematically expressed as:

$$P_R = P_T G_T G_R L_S \tag{10.3}$$

where P_R is received optical power, P_T is transmitted optical power, G_R and G_T are receiver and transmitter antenna gain, respectively, and L_S is free space

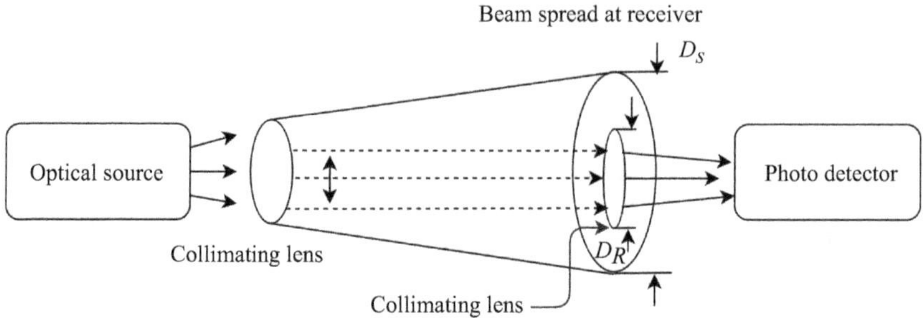

Figure 10.4 Beam divergence losses in OWC channel

path loss. It must be observed here that as per Figure 10.4 and (10.3), one may deduce that the larger the receiver aperture, the better is the detected power (gain of an antenna, in general, is $\approx (D/\lambda)^2$, where D is antenna aperture). However, unlike in conventional communication systems where the signal to noise ratio (SNR) is proportional to signal power levels, the SNR in optical links is proportional to the square of the physical size of the receiver (D_R). This implies higher SNR can be achieved using detectors of a larger area; however, it also increases the effect of shot noise in the receiver due to enhanced capacitance, which in turn limits the operational receiver bandwidth.

4. **Meteorological losses**: These losses can broadly be attributed to weather activities such as rain, fog, snow, smog, and mist. As discussed previously, the presence of inhomogeneous particles in the channel leads to power dissipation, which may cause partial or complete loss of information. Several models have been proposed in the past to characterize the relationship between atmospheric visibility and the optical attenuation of the signal will incur while propagating through the medium. Out of various models, Kim and Kruse's models are the most popular ones that are used to determine specific attenuation, that is, attenuation per unit link length (dB/km), using link visibility and operating wavelength as functional parameters. Based on various simulations and numerical analysis, it has advocated that operating wavelengths of the order of 1550 nm are best suited to minimize the impact of channel attenuation due to fog, rain, and snow. However, in general, amongst different weather phenomena, snow, in particular, has a severe impact on the link as compared to fog and rain [32].
5. **Pointing losses**: Since OWC links studied in this chapter are pointed ones that require line of sight (LOS) arrangement hence it becomes necessary to discuss misalignment losses as well. LOS links require strict alignment between the transmitter and receiver, and any loss of alignment leads to signal fading, which is called a pointing loss. Building sway, earthquakes (no matter how minor they are), loose platform fittings on which transceivers are mounted are some of the causes that may lead to loss of alignment [33,34]. To ensure that pointing losses do not exceed tolerable levels, a system known as automatic acquisition, tracking, and pointing (ATP) mechanism is used in commercial

OWC systems. The fundamental principle of ATP systems involves the use of a separate optical beam, which is only responsible for detecting and correcting misalignment losses [34]. This beam does not carry any modulated information but uses a share of total transmission power.

6. **Atmospheric turbulence:** As explained previously, scattering and absorption are wavelength-dependent phenomena in which the attenuation occurs due to collision between information-bearing photons and atmospheric molecules. Thus, on a relatively clear sky day with sunny conditions, the link attenuation will be minimal. However, even on a clear day when visibility is as good as up to 20 km, atmospheric adversities can still disable the OWC link, and this adversity is known as turbulence. Radiations from the sun cause the earth's surface and the air around to heat up. The heated air, being light, rises up and mixes up with relatively cooler masses of air. This turbulent interaction results formation of pockets of air masses of varying sizes. These pockets of air masses, also known as eddies, have nonuniform temperatures amongst themselves, and it is this character of eddies that leads to the variation of the refractive index of the medium [26].

The air pockets, as shown in Figure 10.5(a) and (b), whose size can vary from 0.1 to 10 cm, start behaving as independent mediums of different refractive indices. As a result, the optically modulated signal passing through these eddies, as

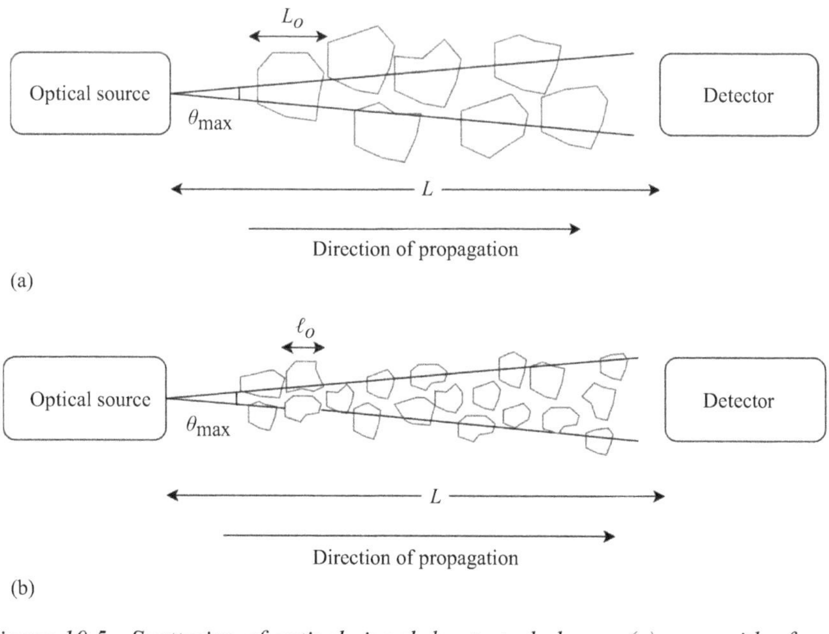

Figure 10.5 Scattering of optical signal due to turbulence: (a) cone with of propagating beam less than cell width and (b) cone with of propagating beam more than the cell width

illustrated in Figure 10.6(a) and (b) experience a scintillation effect, which means that the propagating signal undergoes amplitude and phase fluctuations. Atmospheric turbulence at any given location depends on (1) altitude of site, (2) atmospheric pressure, (3) temperature, and (4) wind speed. Apart from scintillation, turbulence may also lead to adverse effects such as beam steering, image dancing, loss of spatial coherence, beam divergence, and polarization fluctuations.

10.2.5 Strategies to limit the impact of atmospheric turbulence

As per previous discussions, we may conclude that atmospheric turbulence can play potential havoc by degrading the received signal beyond repair. Following different techniques have been proposed over the years to limit the impact of atmospheric turbulence on OWC link:

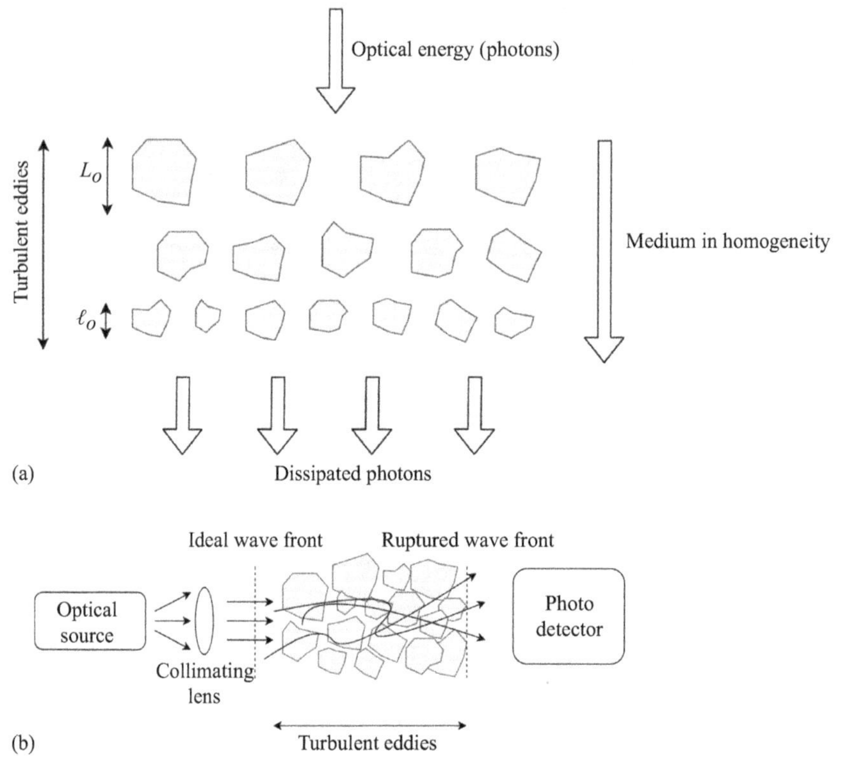

Figure 10.6 (a) Formation of eddies of different sizes (l_o–L_o) and dissipation of energy as the beam propagates through them and (b) symbolic representation of ruptured wavefront (scintillation) due to the presence of turbulence causing eddies

1. **Aperture averaging**: In an event where the receiver size is small as compared to beam diameter or alternatively, if the beam divergence is very high, then the fraction of optical signal concentrated on the optical detector is very less. Figure 10.6(a) and (b) illustrates the redistribution of optical power due to the presence of turbulence, causing eddies in the medium. The acceptance cone ($\theta_{\max}$) formed during propagation, as shown in Figure 10.6(a) and (b) is related to eddy size by [31]:

   $$\theta_{\max} = \lambda / l_o \tag{10.4}$$

   where λ operational wavelength and l_o is eddy size. Assuming that the transmitter and receiver are separated by distance R, the maximum possible cone width will be $R\,\theta_{\max}$. As per the aperture theory of turbulence, till the time the relation $\sqrt{\lambda R} < l_o$ holds good, the beam passes through a lesser number of eddies. However, $\sqrt{\lambda R} \geq l_o$ will mean that beam now passes through a large number of smaller eddies. Such a situation leads to large scale intensity fluctuations as the receiver aperture is smaller than the beam diameter. Aperture averaging, therefore, employs increased receiver diameter, large enough to average out intensity fluctuations in the received signal. Increasing aperture size is like walking a tight rope because increasing receiver aperture certainly assists in increasing received power levels, but at the same time, it may also lead to an increase in background noise and limited detector bandwidth.
2. **Spatial diversity**: Increasing receiver aperture to improve received signal detection may not always be an optimum solution. However, similar performance can still be extracted from the link if multiple smaller receivers replace a single large aperture. These receivers serve the cause of improving receiver detection and at the same contribute lesser noise when compared to one large receiver. Commonly known as MIMO (multiple input multiple output), spatial diversity techniques have already been used with a great success in conventional RF systems to improve channel capacity. Figure 10.7 illustrates different forms of spatial diversity that can be incorporated in the OWC systems. Receiver diversity uses multiple receiver segments but only one transmitter. The presence of turbulence in the channel causes the transmitted beam to split into multiple beams of smaller intensities. The receiver segments then add up these various signals such that in-phase signals will lead to a surge in power levels while out of phase signals will cause destructive interference. The use of multiple transmitter beams can reduce the occurrence of signal fading due to destructive interference. This is because in the case of MIMO, spatially uncorrelated multiple transmitter beams are used and whose overlapping only results in in-phase power addition.
3. **Error coding:** Error coding in communication involves the addition of redundant bits to the original information that can be used for forward error correction (FEC). These overhead bits help the receiver in detecting and correcting the erroneous bits that are introduced during signal propagation through the fading channel. Various coding techniques such as Reed-Solomon coding,

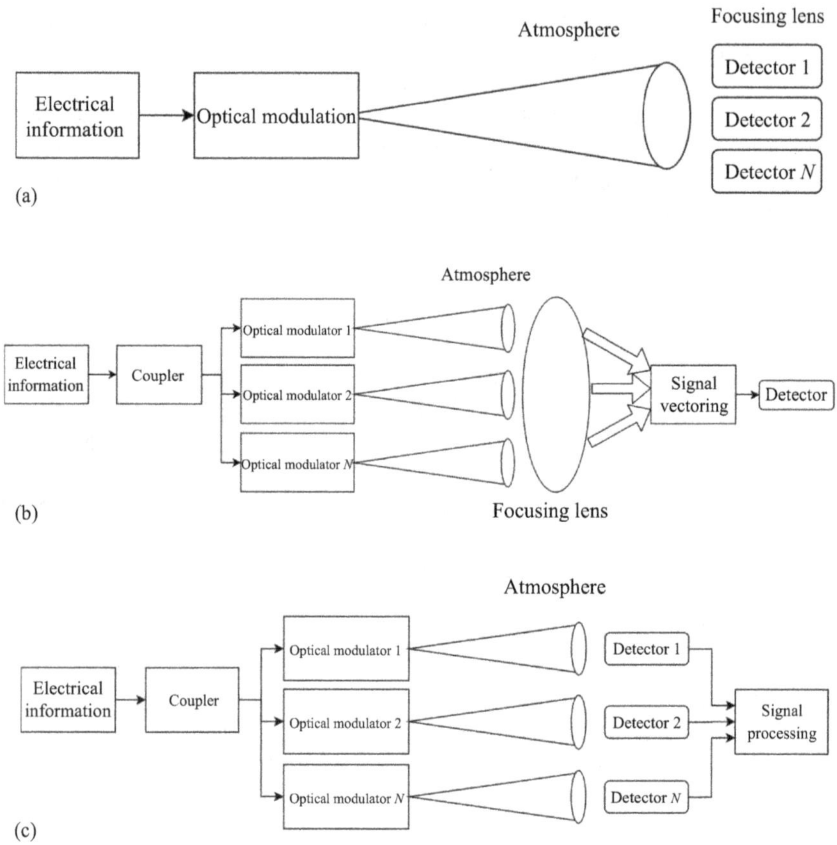

Figure 10.7 Spatial diversity techniques: (a) receiver diversity, (b) transmitter diversity, and (c) MIMO

Convolution coding, and Hamming codes can be used to reduce the bit error rate of communication links. Channel coding not only enhances the information-carrying capacity of the link but also dramatically reduces the minimum signal power required at the receiver for information recovery. This reduction in signal power or otherwise signal to noise ratio (SNR) is commonly known as the coding gain of the link and as a thumb rule, the higher the gain, the better the performance [29].

4. **Relay nodes in OWC:** In cases of extremely adverse weather conditions, strategies such as aperture averaging, spatial diversity, and channel coding may not fetch the desired results, especially if the OWC link range is over 1 km. In such conditions, the use of relay nodes can be extremely useful as these nodes facilitate forwarding of the optical beam through smaller segments in contrast to direct links. Depending upon the functionality, these nodes can be either

active or passive. The active relay nodes receive OWC signal, and condition and amplify it before forwarding it to the next node, and this chain repeats until the destination node arrives. Passive nodes, on the other hand, simply relay the information from one node to another.

5. **Modulation techniques:** Due to its simplicity and cost sufficient circuitry, baseband modulation is the most popular technique out of different methods available for optical modulation. In this technique, the information signal (electrical) directly modulates the optical source through the LED/laser driving circuit to produce an optical carrier. Among different baseband modulation schemes, on-off keying (OOK) and variants of pulse position modulation (PPM) are widely used for signal modulation in OWC systems. OOK scheme, although simple suffers from disadvantages such as weak noise immunity and low power efficiency. To overcome this limitation, various higher modulation techniques such as M-ary PSK, M-ary QAM, differential modulation, and special modulation schemes like OFDM have been proposed. Such modulation techniques offer an array of advantages such as spectral efficiency, power efficiency, and superior noise immunity.
6. **Detection theory:** Direct detection is the most simplified form of signal detection, which depends upon the number of photons that physically hit the surface of the photo-detector. Therefore, direct detection is related to the estimation energy of the received signal than its phase. The absorption of photons by the detector surface releases an equivalent number of electrons (ideally), which is known as photocurrent. Figure 10.8 illustrates the mechanism that can be used for signal reconstruction.

The incoming optical signal is initially collected and concentrated using receiver optics (telescope) from where an optical filter, preferably a bandpass filter, removes background noise. A post detection amplification stage may be added after photo-detection to strengthen the electrical signal produced by photo-detector. Detection is followed by the signal processing stage where a low pass filter and 3R regenerator can be used to improve signal reconstruction. However, it must be understood here that the direct detection systems are insensitive to frequency or phase variations of the received signal. Hence, direct detection is best suited to detect intensity-modulated

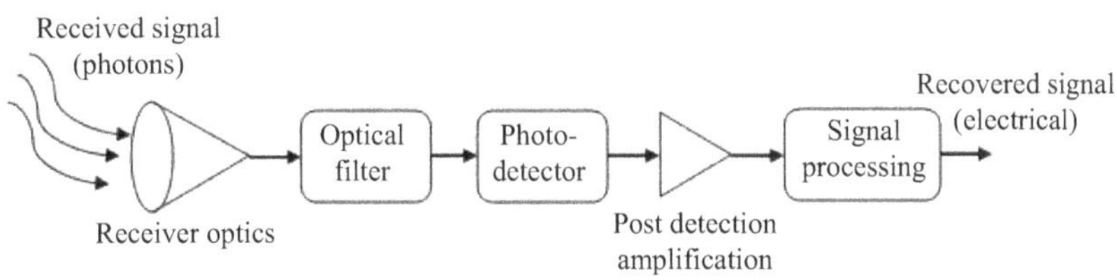

Figure 10.8 Direct detection in OWC systems

signals only. Mathematically, the received signal is related to the recovered signal as

$$i(t) = \frac{\eta q \lambda}{hc} RP(t) \tag{10.5}$$

where $i(t)$ is the instantaneous value of the recovered electrical signal, $P(t)$ is the power of the incoming optical signal, η is the detector efficiency, q is the charge on an electron, λ is the operating wavelength, h is the planks constant, and R is the photodetector responsivity (usually assumed as unity).

In sharp contrast to direct detection systems which are limited by the presence of background noise, coherent detection exhibits superior signal reconstruction due to its excellent noise margin. Although expensive and complex to operate, coherent systems can be advantageous when OWC links are used for applications that require high data rate transfers. Additionally, coherent receivers can be used to detect all kinds of signals ranging from intensity modulation to phase and frequency modulation. As shown in Figure 10.9, coherent detection involves the mixing of the incoming signal with another optical signal from a local oscillator (LO). Depending upon the choice of the frequency of LO, coherent detection can be further classified as homodyne and heterodyne detection. However, in both of these cases, the mixing of signals results in the formation of a new optical signal whose signal strength dwarfs the noise levels. Coherent detection has an exceptional quality to detect received signals irrespective of their phase, frequency, and polarization. Therefore, in the event of strong turbulence, which can introduce random fluctuations in phase and frequency of the signal, coherent detection can still produce a received signal of reasonable quality as compared to direct detection.

Table 10.1 enumerates various characteristic attributes of different mitigation strategies used to limit channel adversities in optical wireless communication links. From this table, one can figure out that the introduction of mitigation mechanisms in the link does bring out benefits which can be in terms of improved signal to noise ratio, channel capacity, and link range. However, at the same time, none of these measures is available without compromise, which may include design complexities, escalated costs, and higher transmission power.

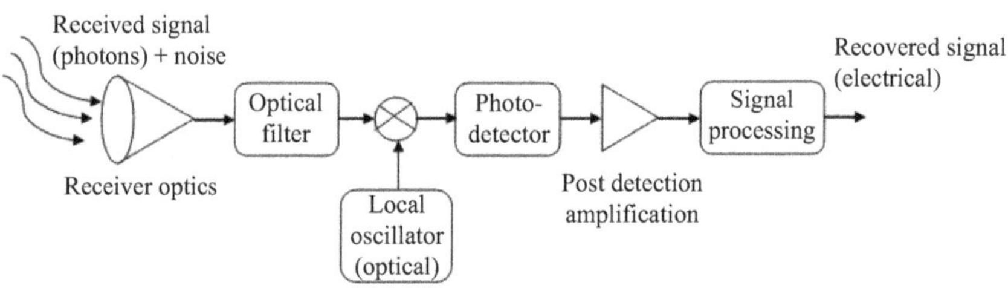

Figure 10.9 Block diagram of a coherent detection system for OWC link

10.3 Introduction to artificial intelligence

Used very commonly these days, the terminologies AI and ML are closely associated with the use of extensive computer-enabled intelligence to solve some of the routine issues faced by mankind. In a simple scenario, AI and ML can be considered as creating mechanisms which can learn, rectify mistakes, and optimize the performance of an assigned task just as the way biological learning systems do. As pointed out in Table 10.1, the use of mitigation strategies brings in benefits but is limited by constraints; in such a scenario, AI and ML can be used to optimize the performance of OWC links.

10.3.1 Role of AI in shaping our lifestyles

Artificial intelligence has already been making great headlines through applications such as healthcare, finance, data security, and astronomy to solve intricate mathematical patterns and make adaptive predictions based on observed patterns [67]. AI is helping to simplify human lives in many lesser-known ways, for example, every single minute hundreds and thousands of YouTube users upload massive video content of numerous genres. Using AI-based algorithms, YouTube chooses the most appropriate video content which matches the interests of the user, and all this is based on an array of factors such as past viewership history, likes, and subscriptions. The same is true for numerous other applications offering a host of services ranging from shopping to online music and online movie rentals.

For the past couple of decades, there has been significant thrust in the field of telecommunications to incorporate AI and ML techniques to optimize the link performance. The same is true for optical fiber-based links as well, where this technology has been used at various hierarchical levels for network management, transmission, operational management, and switching. Since OWC links share considerable similarity with optical fiber networks, hence, it is entirely appropriate and reasonable to say the AI induced smartness should help OWC links in overcoming operational and channel-based limitations discussed in Section 10.2.4.

10.3.2 Use of AI techniques in network management

This section surveys different AI techniques that are being used around the world for achieving desired goals. Irrespective of various categories of AI techniques, the common thing among them is the identification of entities that can recognize and react accordingly to optimize the desired parameter(s). In this section, we briefly examine different AI-related techniques that have been successfully used in lightwave networks for network management and enhancing transmission efficiency. To begin with, the simplest form of communication links can be assumed to have static characteristics with deterministic features. For such links, *search algorithms* and *optimization theories* have been substantially used in network planning and routing management [68,69]. Search algorithms can be beneficial in complimenting the goodness of relay-based multihop OWC links by establishing light paths through the most mesh that promises maximum probability of link connectivity [70,71].

Table 10.1 Various performance enhancement features used in entry-level OWC links

Mitigation strategies	Advantages	Limitations
Aperture averaging	• Can help improve SNR by limiting background noise by using a larger aperture at the receiver • Beam spreading and beam divergence can effectively be tackled using this scheme [26,35–38] • For moderate turbulence conditions, aperture averaging is highly recommended	• Aperture averaging can also lead to enhanced background noise and parasitic capacitance • Can be used effectively only for the plane and Gaussian beam prorogation • Useful only when receiver aperture is greater than fading correlation denoted as $\sqrt{\lambda L}$ • Alternatively, adaptive optics can also be used but it leaves OWC links economically unviable
Spatial diversity	• Compared to aperture averaging, spatial diversity can be used in strong turbulence conditions • Spatial diversity techniques used in conventional RF systems can be easily modified for use in OWC links [39–42] • Spatial diversity can be used even in the absence of channel state information (CSI) [43–45]	• Low SNR gains in moderate turbulence conditions • Considerable transmitter and receiver complexity as multiple apertures are used • The hardware is bulky due to mandatory spacing between antennas
Multihop transmission	• Increases the link coverage, especially during adverse channel conditions • Significant improvement in outage probability has been experimentally shown over direct links [46–53]	• No specific experimental information on its ability to mitigate fading effects • Can be useful only when the signal is large enough to ensure error-free transmission • Relay nodes require signal processing, which introduces cost, complexity, and time delay

(Continues)

Higher-order modulation techniques	• Modulation schemes like pulse position modulation (PPM) and multi-PPM are advantageous as they can achieve near-optimum channel capacity while performing hard detection [54–59] • SIM modulation techniques are most suited for integration with backbone optical fiber networks [60–64] • Polarization-based modulation schemes are also suited for transmission in turbulent channels [65,66]	• High design complexity and moderate spectral efficiency • Poor power (optical) efficiency and necessity to incorporate DC bias • Enhanced transmitter and receiver complexity, reduced link capacity in case of phase aberrations
Coherent detection	• Helps to increase receiver sensitivity and reduction in background noise levels • Significant improvement in the spectral efficiency of the link • Combination of spatial diversity and coherent detection is quite useful [29]	• The cost of implementation is quite high • Mandates the use of digital signal processing to reap maximum benefits of coherent detection • Phase noise estimation and rejection is a big issue in coherent receivers

A revolutionary step in the direction of the intelligent network is in the form of *game theory* wherein multiple smart agents are used, and all of these agents communicate and impact each other. Agents can be defined as entities that work toward achieving desired goals through their roles of environmental observation. *Game theory*-based intelligence has been used for optical networks [72], and it is anticipated that this theory will also be useful in the case of hybrid RF/FSO links.

Intelligence in systems can be further refined by using agents that are capable of maintaining a knowledge base of the environment and understanding its impact on overall link performance. Knowledge base (KB) intelligence agents follow a three-pronged strategy, which includes (1) *knowledge collection*, (2) *developing reasoning*, and (3) *planning*. Mechanisms based on such plans are although complicated but can be very helpful in developing a holistic framework of learning, adapting, and acting. This intelligence can be instrumental in developing cognitive OWC links that can perceive and anticipate possible environmental conditions and press of remedial actions.

Based on the discussions mentioned above, one may conclude that intelligence in the network must incorporate the ability to:

1. cope up with uncertainties,
2. make the right decisions at the right time, and
3. learn from the environment and past choices.

Complete information about the network and factors affecting its performance is something quite unusual and ideal. Network operations are full of unexpected events and uncertainties; hence, it is paramount that agents must act on laws of probability rather than depending upon apparent outcomes. A Bayesian network is one such example that creates a robust and conditional dependence model, which includes a cause–effect situation, thus streamlining predictions even under an environment of uncertainty [73].

The second critical stage of intelligence is making a sequence of correct decisions based on uncertainties detected. The principle of decision making is based on utility functions, which are defined in the form of numerical values, which are further ordered in a manner that the desired state remains at the top of the priority list. One such example is the *Markov decision process* (MDP) wherein the agent's action depends on current circumstances, and each of these circumstances is mapped discreetly with an outcome. MDP's transitional models in which actions of agents are associated with probabilistic outcomes and the agents work toward maximizing the outcome, that is, achieving the best possible state within each stage [74–76].

Learning allows the agents to improvise their performance and their utility in achieving desired outcomes. The learning is essentially the acquisition of experience gathered through previous tasks and this experience is stored in the knowledge base to decide the future course of action. Learning allows the evolution of agent intelligence by making them capable of anticipating unexpected scenarios. Bayesian learning methods, for example, incorporate probabilistic reasoning by using prior updated observations to create a hypothesis. Mainly used in optical detectors, these hypotheses select the best probable outcome by using a data set which maximizes the likelihood of that event [77–79].

Machine learning is also another subset of AI which has found wide acceptance in network intelligence. Machine learning can be broadly classified into:

1. supervised learning,
2. unsupervised learning, and
3. reinforced learning.

Supervised learning entails agent intelligence to map the relation between different input and output conditions and thus creating a mathematical function between these pairs. Linear regression, artificial neural networks, and support vector machines (SVMs) are some of the common examples of supervised learning techniques. The use of supervised learning has been suggested in [78] to determine the quality of service (QoS) and overall link performance of fiber optical networks [79–82]. Unsupervised learning, on the other hand, makes agents learn about input patterns even when there is no specific relationship between input and output. Such types of learning techniques include principal component methods that are used for recognition of modulation formats and types of impairments affecting optical networks [83–85]. Agents in reinforced learning follow optimum or near to optimum decision making, which is based on the policy of reinforced learning gathered over time from interaction with the environment. Learning over the prior data set includes the identification of rewards and punishments, where the former refers to an optimum decision. In contrast, the latter refers to a decision that did not pay off. The temporal difference is one such type of reinforced learning which has been used in optical networks for selecting optimum wavelength and optimum path during the burst switching [86].

10.3.3 Applications of AI techniques for OWC links

Since terrestrial OWC can be readily operated as plug and play devices, hence, in this section, we will discuss various physical layer issues where intelligence help optimize the system performance.

10.3.3.1 Optical transmitters

AI-based intelligent learning systems can be effectively used to generate deterministic statics of optical transmission components, which otherwise might seem impractical for human intelligence. Striving to position itself as last-mile technology that can offer high bandwidth optical access to end-users, the OWC links are being equipped with complex modulation formats such as M-ary QAM and M-ary PSK. The higher modulation transmission requires a very accurate characterization of phase, frequency, and amplitude of the carrier at the receiver to recover the modulated information. Although digital signal processing techniques [54] have also been suggested to enhance carrier and phase synchronization, their operational efficiency reduces when higher-order modulations are used or when channel induced impairments are very severe [87,88]. Coherent receivers, which have been discussed in Section 10.2.5, can assist in limiting the phase noise induced by the local oscillators and mitigate the effect of atmospheric adversities. In such cases, the use of AI can complement the advantages of coherent reception and higher modulation formats. Zibar *et al.* in [89] suggested a Bayesian filtering-based framework to precisely characterize amplitude and phase noise caused due to an optical source, which could either be due to transmitter or local oscillator. Similarly, authors in [90] have suggested the use of the genetic algorithm and machine learning to induce self-control and tune mechanism for mode-locked lasers. Speaking about transmitter power optimization, it will be relevant to discuss the work of Hraghi *et al.* [91] in which authors have proposed AI techniques to determine optimum parameters to generate flat gain characteristics for ultra-dense wavelength division multiplexing (WDM) systems.

10.3.3.2 Optical amplification using erbium-doped fiber amplifier

Erbium-doped fiber amplifiers (EDFA) have become an integral part of long haul optical fiber networks and, more recently, OWC links as well. EDFAs have become immensely popular due to features such as all-optical amplification, almost flat gain characteristics in 1530–1610 nm, WDM amplification, and ease of integration with optical networks. The usefulness of EDFA's links in handling WDM-OWC links has also been demonstrated in [51] and [54]. However, on the flipside, EDFAs also suffer from some fundamental issues predominantly which are amplified stimulated emission or commonly known as ASE noise and postamplification power excursion in WDM channels. In their work, Huang *et al.* [92] suggested supervised machine learning models to characterize the dependence of multichannel signals on power excursions in EDFA links. This model precisely determines strategy on channel add/drop in WDM links to minimize power excursions. Additionally, automatic adjustment of operating point in EDFA has been suggested in [93] using neural networks to reduce the impact of noise accumulation while assuring predefined output optical power levels.

10.3.3.3 Performance observation

OWC transmitter includes a laser driving circuit, while photo-detectors, filters, and amplifiers are included in the receiver section. In the case of multihop OWC links, relay nodes are also used, which may include amplifiers, equalizers, and regenerators. The significant success of incorporating intelligence in network management will depend upon the ability of agents to monitor performance parameters such as signal to noise ratio (SNR), bit error rate (BER), and eye patterns at different sections of the network. These qualitative parameters provide a quick assessment of the quality of network management and service being delivered to end-users [94–98]. Timely detection of various physical parameters related to transmission and detection helps launch corrective actions much before the actual damage in the form of loss of communication takes place [99–101].

Wu *et al.,* for example, in their work [78] proposed the use of artificial neural networks to quantitatively measure parameters such as SNR, polarization dispersion, and link linearity from eye patterns. The generated statistics are used to launch network diagnosis and, thus, corrective actions in relevant section/s of the link. Similarly, the use of Kalman filter to approximate carrier phase and polarization state has been suggested in [102]. Detection of potential network impairments using qualitative parameters without prior knowledge of transmission rate, choice of modulation has been suggested in [84]. Based on pattern recognition, networking monitoring in [84] uses asynchronous delay-plots to generate accurate results. Optical SNR (OSNR) and modulation format detection have been studied by Thrane *et al.* in [103]. The authors here have proposed the use of neural networks to estimate the OSNR and support vector machine for classifying and detecting modulation formats. However, the authors here have considered only white Gaussian noise effects while ignoring other channel impairments.

10.3.3.4 Intelligent receiver for loss mitigation

The channel capacity and link range in OWC links depend on the intensity of atmospheric adversities. Severe channel losses have a direct bearing on limiting the useful link range and achievable data rates. Recently published works have suggested the use of digital signal processing (DSP) [54] techniques at the OWC receiver to overcome intersymbol interference, carrier frequency, and phase offsets. However, with increasing transmission rates and multiple modulation formats, the DSP algorithms become computationally heavy and practically unviable. To overcome this situation, Gonzalez *et al.* [83] presented machine learning techniques in DSP modules to enhance accuracy in symbol detection at the receiver. This has been achieved using clustering algorithms that can identify the modulation format of an incoming signal without any need for prior control messages. This form of autonomous identification can be used to design multimodulation compatible OWC links wherein the choice of modulation will be based on the severity of channel turbulence. Additionally, carrier synchronization, polarization distortion mitigation, and expectation-maximization of optical elements in DSP module units using state-space models have been presented in [104]. Recently, Giacoumidis *et al.*, in their study [105], highlighted the use of nonlinear SVM with reduced complexity to bring out higher spectral efficiency in advanced modulated links such as that of coherent orthogonal frequency division multiplexing (CO-OFDM). The proposed model has been shown to reduce the effect of inter-channel cross talks with an ability to withstand higher transmission of optical power.

10.4 AI for optical links: opportunities and challenges

AI tools have been successfully used in conventional optical links to solve various issues related to transmission, reception, and network management. With its natural tendency to optimize and maximize the assigned tasks, the vision for future AI technologies will go far across the present horizons of applications. Since OWC links are plug and play devices, hence they can be used during wartime or any other emergencies at short notice of time. In either case, it is vital to protect OWC links from network attacks and disruptions enforced by adversaries. Unexplored primarily to date, AI-based monitoring agents can be used to detect unusual activities by entities that may lead to data breaches and violations. Being a physical layer technology, OWC links are highly susceptible to disruptive attacks, which are mainly aimed at degrading transmission quality or even blocking the communication links. Communication networks and infrastructure are new forms of national assets that play a crucial role in our day to social and economic life. Any attempts to hijack or eavesdrop are treated as nothing less than an attack on the national and financial sovereignty of the country. Using the measurement of optical signals as a statistical tool, the detection of unsolicited intrusions in optical networks has been suggested in [106,107]. However, the inclusion of such micro agents to monitor the system will add to the massive cost, complexity, and enhanced delays in data transmission on account of stringent screening.

AI-based techniques such as machine learning (deep learning) are expected to play a major role in network planning and resource allocation in next-generation OWC links. The resource allocation may include choosing the optimum pathway in the case of multihop OWC links, choosing transmission type in case of hybrid RF/OWC and bandwidth allocation to suit the traffic load requirements. With the onset of the Internet of things (IoT), it is anticipated that traffic will increase exponentially in the coming years. In such a scenario, intelligence can be used to decide bandwidth allocation to different data types and even prioritize the transmission of tactile traffic over the other.

10.5 Summary

With evident bandwidth crunch staring at us, alternatives must fill the space that will be left behind by RF networks once they are outdated. Optical wireless communication links with transmission and bandwidth properties vary the same as that of conventional fiber links that can be installed at 1/10th of the cost of the latter. This means OWC links can be the new face of cost-effective network infrastructure, which can conveniently converge between backbone networks and end-users. OWC links are, however, extremely vulnerable to atmospheric adversities, which can deplete the transmission quality and operational link range. Although various mitigation strategies like spatial diversity, relay transmission, and coherent detection have been proposed in the past, each of them possesses certain inherent limitations, which ultimately prohibits large-scale deployment.

Artificial Intelligence uses multiobjective agents which can be trained to achieve desired objectives. In this chapter, we have discussed various literature sources that have demonstrated the use of AI to optimize multiple functional characteristics of optical networks. Since the majority of these characteristics such as optical transmission and detection obey similar principles of physics irrespective of optical network type, hence, it is highly encouraging to incorporate the proven intelligence mechanisms in OWC links as well. We have also described future aspects of AI in OWC technology and it is expected that the inclusion of these features will go a long way in commercializing OWC systems.

References

[1] Killinger, D. "Free space optics for laser communication through the air." *Optics and Photonics News*. 2002; 13: 36–42.

[2] Bell, A. G. "On the production and reproduction of sound by light." *American Journal of Sciences*. 1880; 3: 305–324.

[3] Dennis, R., Mark, L., Ganesh, G., Bruce, P., and Gerald, N. "Optical wireless propagation, theory vs. experiment." *Proceedings of SPIE: Optical Wireless Communications III*. 2001; 4214: 38–45.

[4] Goodwin, F. E. "A review of operational laser communication systems." *Proceedings of IEEE*. 1970; 58: 1746–1752.

[5] Nadeem, F., Geiger, B., Leitgeb, E., Muhammad, S. S., Loeschnig, M., and Kandus, G. "Comparison of link selection algorithms for free space optics/radio frequency hybrid network." *IET Communications*. 2011; 5: 2751–2759.

[6] Fehske, A., Fettweis, G., Malmodin, J., and Biczok, G. "The global footprint of mobile communications: The ecological and economic perspective." *IEEE Communications Magazine*. 2011; 49: 55–62.

[7] Nikoloudakis, Y., Panagiotakis, S., Markakis, E., *et al.* "A fog-based emergency system for smart enhanced living environments." *IEEE Cloud Computing*. 2016; 3(6): 54–62.

[8] Batalla, J. M., Krawiec, P., Mavromoustakis, C. X., *et al.* "Efficient media streaming with collaborative terminals for the smart city environment." *IEEE Communications Magazine*. 2017; 55(1): 98–104.

[9] Hadjioannou, V., Mavromoustakis, C. X., Mastorakis, G., *et al.* "Security in smart grids and smart spaces for smooth IoT deployment in 5G." *Modeling and Optimization in Science and Technologies*. 2016; 8: 371–397.

[10] Gajewski, M., Batalla, J. M., Mastorakis, G., *et al.* "A distributed IDS architecture model for smart home systems." *Cluster Computing*. 2019; 22: 1739–1749.

[11] Xu, X., Li, D., Sun, M., *et al.* "Research on key technologies of smart campus teaching Platform based on 5G network." *IEEE Access*. 2019; 7: 20664–20675.

[12] Marin, R. C., Ciobanu, R. I., Dobre, C., Mavromoustakis, C. X., and Mastorakis, G. "A context-aware collaborative model for smart phone energy efficiency over 5G wireless networks." *Computer Networks*. 2017; 129(6): 352–362.

[13] Nikoloudakis, Y., Markakis, E., Mastorakis, G., Pallis, E., and Pallis, E. "An NF V-powered emergency system for smart enhanced living environments." *IEEE Conference on Network Function Virtualization and Software Defined Networks (NFV-SDN)*. Berlin, Germany; 2011. pp. 1–6.

[14] Hanzo, L., Haas, H., Imre, S., O'Brien, D., Rupp, M., and Gyongyosi, L. "Wireless myths, realities, and futures: From 3G/4G to optical and quantum wireless." *Proceeding of the IEEE*. 2012; 100: 1853–1888.

[15] Hemmati, H. "Deep space optical communications." *Deep Space Communications and Navigation Series*. 2006; 85–113. Available from https://www.wiley.com/en-us/Deep+Space+Optical+Communications-p-9780470042403.

[16] Willebrand, H. and Ghuman, B. *Free-space optics: enabling optical connectivity in today's networks*. Indianapolis, IN: Sams Publishing; 2002.

[17] Carbonneau, T. H. and Wisely, D. R. "Opportunities and challenges for optical wireless: The competitive advantage of free space telecommunications links in today's crowded marketplace." *Proceedings of the SPIE, Wireless Technologies and Systems: Millimeter-Wave and Optical*. 1997. pp. 119–128.

[18] Baldemair, R., Dahlman, E., Fodor, G., *et al.* "Evolving wireless communications: Addressing the challenges and expectations of the future." *IEEE Vehicular Technology Magazine*. 2013; 8: 24–30.

[19] Saari, J. I., Krause, M. M., Walsh, B. R., and Kambhampati, P. "Terahertz bandwidth all-optical modulation and logic using multiexcitons in semiconductor nanocrystals." *Nano Letters*. 2013; 13(2): 722–727.

[20] Ghassemlooy, Z., Rajbhandari, S., and Popoola, W. *Optical wireless communications: System and channel modelling with MATLAB*. Boca Raton, FL: Taylor & Francis; 2013.

[21] Batalla, J. M., Mastorakis, G., Mavromoustakis, C. X., and Zurek, J. "On cohabitating networking technologies with common wireless access for home automation system purposes." *IEEE Wireless Communications*. 2016; 23(5): 76–83.

[22] Dimitriou, C. D., Mavromoustakis, C. X., Mastorakis, G., and Pallis, E. "On the performance response of delay-bounded energy-aware bandwidth allocation scheme in wireless networks." *IEEE International Conference on Communications Workshops*. ICC 2013. pp. 631–636.

[23] Mavromoustakis, C., Mastorakis, G., and Batalla, J. M. *Internet of things (IoT) in 5G mobile technologies*. 1st ed. Springer International Publishing; 2016.

[24] Batalla, J. M., Mavromoustakis, C. X., Mastorakis, G., and Sienkiewicz, K. "On the track of 5G radio access network for IoT wireless spectrum sharing in device positioning applications." *Modeling and Optimization in Science and Technologies*. 2016; 8: 25–35.

[25] Mahdy, A. and Deogun, J. S. "Wireless optical communications: a survey." *Proceedings of IEEE Wireless Communications and Networking Conference (WCNC)*. 2004. pp. 2399–2404.

[26] Andrews, L. C. and Philips, R. L. *Laser beam propagation through random media*. Bellingham, WA: SPIE Optical Engineering Press; 2005.

[27] Nadeem, F., Leitgeb, E., Awan, M. S., and Kandus, G. "FSO/RF hybrid network availability analysis under different weather condition." *Third International Conference on Next Generation Mobile Applications, Services and Technologies (NGMAST '09)*. 2009. pp. 239–244.

[28] Nadeem, F., Leitgeb, E., Awan, M. S., and Kandus, G. "Forward prediction of FSO received signal for FSO/RF hybrid network." *International Workshop on Satellite and Space Communications (IWSSC)*. 2009. pp. 196–200

[29] Khalighi, M. A. and Uysal, M. "Survey on free space optical communication: A communication theory perspective." *IEEE Communications Surveys and Tutorials*. 2014; 16: 2231–2258.

[30] Kim, I. I., McArthur, B., and Korevaar, E. "Comparison of laser beam propagation at 785 nm and 1550 nm in fog and haze for optical wireless communications." *Proceedings of SPIE, Optical Wireless Communications III*. 2001. pp. 26–37.

[31] Kaushal, H., Jain, V. K., and Kar, S. *Free space optical communication*. 1st edn. Optical Networks. Springer; 2017.

[32] Miglani, R. and Singh, M. L. "Performance evaluation of free space optical link using mid and far infrared wavelengths in turbulent atmospheric conditions". *4th International Workshop on Fiber Optics in Access Network (FOAN)*. Almaty. 2013. pp. 31–5.

[33] Jurado-Navas, A., Garrido-Balsells, J. M., Francisco, J., Castillo-Vázquez, M., and Puerta-Notario, A. "Impact of pointing errors on the performance of generalized atmospheric optical channels." *Optics Express*. 2012; 20(11): 12550–12562.

[34] Ortiz, G. G., Lee, S., Monacos, S. P., Wright, M. W., and Biswas, A. "Design and development of a robust ATP subsystem for the altair UAV-to-ground lasercomm 2.5-Gbps demonstration." *Proceedings of SPIE Free-Space Laser Communications Technology. XV 4975*. 2003. pp. 1–12.

[35] Yuksel, H., Milner, S., and Davis, C. C. "Aperture averaging for optimizing receiver design and system performance on free-space optical communication links." *Journal of Optical Communications and Networking*. 2005; 4(8): 462–475.

[36] Andrews, L. C., Phillips, R. L., and Hopen, C. Y. "Aperture averaging of optical scintillations: Power fluctuations and the temporal spectrum." *Waves Random Media*. 2000; 10(1): 53–70.

[37] Khalighi, M. A., Schwartz, N., Aitamer, N., and Bourennane, S. "Fading reduction by aperture averaging and spatial diversity in optical wireless systems." *IEEE/OSA Journal of Optical Communication Networking*. 2009; 1(6): 580–593.

[38] Martin, J. M. and Flatté, S. M. "Intensity images and statistics from numerical simulation of wave propagation in 3-D random media." *Applied Optics*. 1988; 27(11): 2111–2126.

[39] Miglani, R. and Malhotra, J. S. "Statistical analysis of FSO links employing multiple transmitter/receiver strategy over double-generalized and Gamma–Gamma fading channel using different modulation techniques." *Journal of Optical Communications*. 2019; 40(3): 295–305.

[40] Simon, M. K. and Vilnrotter, V. A. "Alamouti-type space-time coding for free-space optical communication with direct detection." *IEEE Transactions on Wireless Communications*. 2005; 4(1): 35–359.

[41] Tarokh, V., Seshadri, N., and Calderbank, A. R. "Space-time codes for high data rate wireless communication: Performance analysis and code construction." *IEEE Transactions on Information Theory*. 1998; 44(2): 744–765.

[42] Tarokh, V., Jafarkhani, H. J., and Calderbank, A. R. "Space-time block codes from orthogonal designs." *IEEE Transactions on Information Theory*. 1999; 45(5): 1456–1467.

[43] García-Zambrana, A., Castillo-Vázquez, C., Castillo-Vázquez, B., and Hiniesta-Gómez, A. "Selection transmit diversity for FSO links over strong atmospheric turbulence channels." *IEEE Photonics Technology Letters*. 2009; 21(14): 1017–1019.

[44] Castillo-Vázquez, B. and García-Zambrana, A. "Closed-form BER expression for FSO links with transmit laser selection over exponential atmospheric turbulence channels." *Electronics Letters*. 2009; 45(23): 1185–1187.

[45] Abou-Rjeily, C. "On the optimality of the selection transmit diversity for MIMO-FSO links with feedback." *IEEE Communications Letters*. 2011; 15(6): 641–643.

[46] Abou-Rjeily, C. and Haddad, S. "Cooperative FSO systems: Performance analysis and optimal power allocation." *Journal of Lightwave Technology*. 2011; 299(4): 1058–1065.

[47] Bhatnagar, M. "Performance analysis of decode-and-forward relaying in gamma-gamma fading channels." *IEEE Photonics Technology Letters*. 2012; 24(7): 545–547.

[48] Peppas, K. P., Stassinakis, A. N., Nistazakis, H. E., and Tombras, G. S. "Capacity analysis of dual amplify-and-forward relayed free-space optical communication systems over turbulence channels with pointing errors." *IEEE/OSA Journal of Optical Communications Networkings*. 2013; 5(9): 1032–1042.

[49] Chatzidiamantis, N. D., Michalopoulos, D. S., Kriezis, E. E., Karagiannidis, G. K., and Schober, R. "Relay selection protocols for relay-assisted free-space optical systems." *IEEE/OSA Journal of Optical Communications Networkings*. 2013; 5(1): 92–103.

[50] Kazemlou, S., Hranilovic, S., and Kumar, S. "All-optical multihop free-space optical communication systems." *Journal of Lightwave Technology*. 2011; 29(18): 2663–2669.

[51] Bayaki, E., Michalopoulos, D., and Schober, R. "EDFA-based all-optical relaying in free-space optical systems." *IEEE Transactions on Communications*. 2012; 60(12): 3797–3807.

[52] Kashani, M. A., Rad, M. M., Safari, M., and Uysal, M. "All-optical amplify-and-forward relaying system for atmospheric channels." *IEEE Communications on Letters*. 2012; 16(10): 1684–1687.

[53] Miglani, R. and Malhotra, J. S. "An innovative approach for performance enhancement of 320 Gbps free space optical communication system over turbulent channel." *Optical and Quantum Electronics*. 2019; 51(9): 289.

[54] Miglani, R. and Malhotra, J. S. "Performance enhancement of high-capacity coherent DWDM free-space optical communication link using digital signal processing." *Photonic Network Communications*. 2019; 38(3): 326–342.

[55] Xu, F., Khalighi, M. A., and Bourennane, S. "Coded PPM and multipulse PPM and iterative detection for free-space optical links." *IEEE/OSA Journal of Optical Communications Networkings*. 2009; 1(5): 404–415.
[56] Muhammad, S. S., Javornik, T., Jelovcan, I., Leitgeb, E., and Ghassemlooy, Z. "Comparison of hard-decision and soft-decision channel coded M-ary PPM performance over free space optical links." *European Transactions on Telecommunications*. 2008; 20(8): 746–757.
[57] Hamkins, J. and Moision, B. "Selection of modulation and codes for deep space optical communications." *Proceedings of SPIE Free-Space Laser Communications Technology XVI*. 2004; 5338: 123–130.
[58] Moision, B. and Hamkins, J. *Multipulse PPM on discrete memoryless channels*. IPN Progress Report 42-161. The Jet Propulsion Laboratory, California Institute of Technology. 2005.
[59] Sugiyama, H. and Nosu, K. "MPPM: A method for improving the band-utilization efficiency in optical PPM." *Journal of Lightwave Technology*. 1989; 7(3): 465–471.
[60] Carruthers, J. B. and Kahn, J. M. "Multiple-subcarrier modulation for non-directed wireless infrared communication." *IEEE Journal on Selected Areas Communications*. 1996; 14(3): 538–546.
[61] Ohtsuki, T. "Multiple-subcarrier modulation in optical wireless communications." *IEEE Communications Magazine*. 2003; 41(3): 74–79.
[62] Armstrong, J. "OFDM for optical communications." *Journal on Lightwave Technology*. 2009; 27(3): 189–204.
[63] Shieh, W. and Djordjevic, I. *OFDM for optical communications*. New York, NY: Academic; 2009.
[64] Miglani, R. and Malhotra, J. S. "Investigation on R–S coded coherent OFDM free space optical (CO-OFDM-FSO) communication link over gamma–gamma channel." *Wireless Personal Communications*. 2019; 109(1): 415–435.
[65] Betti, S., Marchis, G. D., and Iannone, E. "Polarisation modulated direct detection optical transmission systems." *Journal on Lightwave Technology*. 1992; 10(12): 1985–1997.
[66] Karbassian, M. M. and Ghafouri-Shiraz, H. "Transceiver architecture for incoherent optical CDMA network based on polarization modulation." *Journal on Lightwave Technology*. 2008; 26 (12): 3820–3828.
[67] Papadakis, S. E., Stykas, V. A., Mastorakis, G., and Mavromoustakis, C. X. "A hyper-box approach using relational databases for large scale machine learning." *International Conference on Telecommunications and Multimedia*. Heraklion, Greece: TEMU; 2014. pp. 69–73.
[68] Mukherjee, B. *Optical WDM networks*. New York, NY: Springer-Verlag; 2006.
[69] Simmons, J. M. *Optical network design and planning*. 2nd edn. Springer International Publishing; 2014.
[70] Monoyios, D. and Vlachos, K. "Multiobjective genetic algorithms for solving the impairment-aware routing and wavelength assignment problem."

IEEE/OSA Journal of Optical Communications and Networking. 2011; 3(1): 40–47.

[71] Wang, X., Brandt-Pearce, M., and Subramaniam, S. "Distributed grooming, routing, and wavelength assignment for dynamic optical networks using ant colony optimization." *IEEE/OSA Journal of Optical Communications and Networking*. 2014; 6(6): 578–589.

[72] Awwad, O., Al-Fuqaha, A., Khan, B., Benhaddou, D., Guizani, M., and Rayes, A. "Bayesian-based game theoretic model to guarantee cooperativeness in hybrid RF/FSO mesh networks." *IEEE Global Telecommunications Conference, GLOBECOM*. 2009. pp. 1–7.

[73] Gosselin, S., Courant, J. L., Tembo, S. R., and Vaton, S. "Application of probabilistic modeling and machine learning to the diagnosis of FTTH GPON networks." *International Conference on Optical Network Design and Modeling (ONDM)*. 2017. pp. 1–3.

[74] Tachibana, T. and Koyanagi, I. "MDP-based lightpath establishment for service differentiation in all-optical WDM networks with wavelength conversion capability." *Photonic Network Communications*. 2010; 20(2): 183–192.

[75] Sue, C., Hsu, Y., and Ho, P. "Dynamic preemption call admission control scheme based on Markov decision process in traffic groomed optical networks." *IEEE/OSA Journal of Optical Communications and Networking*. 2011; 3(4): 300–311.

[76] Reyes, R. R. and Bauschert, T. "Adaptive and state-dependent online resource allocation in dynamic optical networks." *IEEE/OSA Journal of Optical Communications and Networkings*. 2017; 9(3): B64–B77.

[77] Rottenberg, F., Nguyen, T. H., Gorza, S. P., Horlin, F., and Louveaux, J. "ML and MAP phase noise estimators for optical fiber FBMC-OQAM systems." *IEEE International Conference on Communications (ICC)*. 2017. pp. 1–6.

[78] Wu, X., Jargon, J. A., Skoog, R. A., Paraschis, L., and Willner, A. E. "Applications of artificial neural networks in optical performance monitoring." *Journal on Lightwave Technology*. 2009; 27(16): 3580–3589.

[79] Jiménez, T., Aguado, J. C., Miguel, I. D., *et al.* "A cognitive system for fast quality of transmission estimation in core optical networks." *Optical Fiber Communication Conference and Exposition (OFC/NFOEC) and the National Fiber Optic Engineers Conference*. IEEE. 2012. pp. 1–3.

[80] Jiménez, T., Aguado, J. C., Miguel, I. D., *et al.* "A cognitive quality of transmission estimator for core optical networks." *Journal on Lightwave Technology*. 2013; 31(6): 942–951.

[81] Caballero, A., Aguado, J. C., Borkowski, R., *et al.* "Experimental demonstration of a cognitive quality of transmission estimator for optical communication systems." *Optic Express*. 2012; 20(26): B64–B70.

[82] Barletta, L., Giusti, A., Rottondi, C., and Tornatore, M. "QoT estimation for unestablished lighpaths using machine learning." *Optical Fiber Communications Conference and Exhibition (OFC)*. 2017. pp. 1–3.

[83] Gonzalez, N. G., Zibar, D., and Monroy, I. T. "Cognitive digital receiver for burst mode phase modulated radio over fiber links optical communication (ECOC)." *36th European Conference and Exhibition*. IEEE. 2010. pp. 1–3.

[84] Tan, M. C., Khan, F. N., Al-Arashi, W. H., Zhou, Y., and Lau, A. P. T. "Simultaneous optical performance monitoring and modulation format/bit-rate identification using principal component analysis." *Journal of Optical Communications and Networking*. 2014; 6(5): 441–448.

[85] Torres, J. J. G., Chiuchiarelli, A., Thomas, V., Ralph, S. E., Soto, A. M. C., and González N. G. "Adaptive non symmetrical demodulation based on machine learning to mitigate time-varying impairments." *Avionics and Vehicle Fiber-Optics and Photonics Conference (AVFOP)*. IEEE. 2016. pp. 289–290.

[86] Kiran, Y. V., Venkatesh, T., and Murthy, C. S. R. "A reinforcement learning framework for path selection and wavelength selection in optical burst switched networks." *IEEE Journal on Selected Areas in Communications*. 2007; 25(9): 18–26.

[87] Duthel, T., Clarici, G., Fludger, C. R., Geyer, J. C., Schulien, C., and Wiese, S. "Laser linewidth estimation by means of coherent detection." *IEEE Photonics Technology Letters*. 2009; 21(20): 1568–1570.

[88] Kikuchi, K. "Characterization of semiconductor-laser phase noise and estimation of bit-error rate performance with low-speed offline digital coherent receivers." *Optic Express*. 2012; 20(5): 5291–5302.

[89] Zibar, D., Carvalho, L. H. H. D., Piels, M., *et al.* "Application of machine learning techniques for amplitude and phase noise characterization." *Journal of Lightwave Technology*. 2015; 33(7): 1333–1343.

[90] Brunton, S. L., Fu, X., and Kutz, J. N. "Self-tuning fiber lasers." *IEEE Journal of Selected Topics in Quantum Electronics*. 2014; 20(5): 464–471.

[91] Hraghi, A., Chaibi, M. E., Menif, M., and Erasme, D. "Demonstration of 16QAM-OFDM UDWDM transmission using a tunable optical flat comb source." *Journal of Lightwave Technology*. 2017; 35(2): 238–245.

[92] Huang, Y., Samoud, W., Gutterman, C. L., *et al.* "A machine learning approach for dynamic optical channel add/drop strategies that minimize EDFA power excursions." *Proceedings of 42nd European Conference on Optical Communication*. VDE. 2016. pp. 1–3.

[93] Barboza, E. D. A., Bastos-Filho, C. J., Martins-Filho, J. F., Moura, U. C. D., and Oliveira, J. R. D. "Self-adaptive erbium-doped fiber amplifiers using machine learning." *Microwave & Optoelectronics Conference (IMOC)*. SBMO/IEEE MTT-s International IEEE. 2013. pp. 1–5.

[94] Mavromoustakis, C. X., Kormentzas, G., Mastorakis, G., Bourdena, A., Pallis, E., and Rodrigues, J. "Context-oriented opportunistic cloud offload processing for energy conservation in wireless devices." *IEEE Globecom Workshops, GC Workshops*. 2014. pp. 24–30.

[95] Mavromoustakis, C. X., Dimitriou, C. D., and Mastorakis, G. "Using real-time backward traffic difference estimation for energy conservation in wireless devices." *4th International Conference on Advances in P2P Systems*. 2012. pp. 18–23.

[96] Bourdena, A., Mavromoustakis, C. X., Mastorakis, G., Rodrigue, J. J. P. C., and Dobre, C. "Using socio-spatial context in mobile cloud process off-loading for energy conservation in wireless devices." *IEEE Transactions on Cloud Computing*. 2019; 7(2): 392–402.

[97] Mavromoustakis, C. X., Dimitriou, C. D., Mastorakis, G., and Pallis, E. "Real-time performance evaluation of F-BTD scheme for optimized QoS energy conservation in wireless devices." *IEEE Globecom Workshops, GC Workshops*. 2013. pp. 1151–1156.

[98] Mavromoustakis, C. X., Dimitriou, C. D., and Mastorakis, G. "On the real-time evaluation of two-Level BTD scheme for energy conservation in the presence of delay sensitive transmissions and intermittent connectivity in wireless devices." *International Journal on Advances in Networks and Services*. 2013; 6(3): 148–162.

[99] Mavromoustakis, C. X., Mastorakis, G., Papadakis, S., Andreou, A., Bourdena, A., and Stratakis, D. "Energy consumption optimization through pre-scheduled opportunistic offloading in wireless devices." *EMERGING 2014: The Sixth International Conference on Emerging Network Intelligence*. 2014. pp. 22–28.

[100] Mavromoustakis, C. X., Andreou, A., Mastorakis, G., Bourdena, A., Batalla, J. M., and Dobre, C. On the performance evaluation of a novel offloading-based energy conservation mechanism for wireless devices." *6th International Conference on Mobile Networks and Management*. MONAMI 2014. Revised Selected Papers, vol. 141, 2015. pp. 179–191.

[101] Mavromoustakis, C. X., Dimitriou, C., Mastorakis, G., Bourdena, A., and Pallis, E. "Using traffic diversities for scheduling wireless interfaces for energy harvesting in wireless devices." *Modeling and Optimization in Science and Technologies*. 2014; 3: 481–496.

[102] Szafraniec, B., Marshall, T. S., and Nebendahl, B. "Performance monitoring and measurement techniques for coherent optical systems." *Journal of Lightwave Technology*. 2013; 31(4): 648–663.

[103] Thrane, J., Wass, J., Piels, M., Diniz, J. C., Jones, R., and Zibar, D. Machine learning techniques for optical performance monitoring from directly detected PDM-QAM signals. *Journal of Lightwave Technology*. 2017; 35(4): 868–875.

[104] Zibar, D., Piels, M., Jones, R., and Schäeffer, C. G. "Machine learning techniques in optical communication." *Journal of Lightwave Technology*. 2016; 34(6): 1442–1452.

[105] Giacoumidis, E., Wei, J., Mhatli, S., *et al.* "Nonlinear inter-subcarrier intermixing reduction in coherent optical OFDM using fast machine learning equalization." *Optical Fiber Communications Conference and Exhibition (OFC) IEEE*. 2017. pp. 1–3.

[106] Rejeb, R., Pavlosoglou, I., Leeson, M. S., and Green, R. J. "Securing all-optical networks." *Proceedings of 5th International Conference on Transparent Optical Networks*, vol. 1. 2003. pp. 87–90.

[107] Medard, M., Chinn, S. R., and Saengudomlert, P. "Attack detection in all-optical networks." *Optical Fiber Communication Conference and Exhibit (OFC)*. 1998. pp. 272–273.

Chapter 11

Intelligent underwater wireless communications

Elmustafa Sayed Ali[1,2] *and Rashid A. Saeed*[2,3]

Underwater wireless communications play a vital role in marine applications such as oceans' monitoring, exploration, and scientific data collection. Underwater wireless network (UWN) is an important area for the research due to the unique and harsh underwater environment and application which are quite challenging and problematics issues still remain to be addressed. Hence, intelligent solutions are needed for various levels of network architecture. Recently, many academic and industrial researchers have paid attention to the development of state-of-the-art intelligent solutions for future underwater wireless communications and networks. This chapter enlightens and guides the potential research communities about the recent progress in the area of intelligent UWC. This chapter also reviews underwater wireless networks based on acoustics, even electromagnetic, and optical signals. UWN systems also are reviewed for data gathering and recharge docking systems. Intelligent methodologies and modelings are the key drivers for future intelligent underwater communication.

11.1 Introduction

Nowadays, underwater wireless communications are attracting more attention because of many important applications in the military or other underwater environment aspects. The underwater communication networks may consist of different kinds of underwater sensors, autonomous vehicles, and drones which interact with each other for self-coordination and shared information of underwater phenomena, motions, and monitoring purposes [1]. Applications such as coastal surveillance, underwater vehicle operations, marine communications, or underwater monitoring are becoming the most wide research areas for investigation and study. For such applications, a fully wireless underwater communications network can be deployed through several configurations including floating devices, autonomous underwater vehicles (AUVs), seabed-attached

[1]Department of Electrical and Electronics Engineering, Red Sea University (RSU), Portsudan, Sudan
[2]Department of Electronics Engineering, Sudan University of Science and Technology (SUST), Khartoum, Sudan
[3]Department of Computer Engineering, Taif University (TU), Al-Taif, Saudi Arabia

sensors, or processing of towers, submarines, or ships. The communications in underwater networks can be carried by using even electromagnetic (EM), optical, or acoustic waves depending on the bandwidth requirements and distance range. Because of limitations in EM and acoustic waves bandwidth, optical waves can be used to provide higher data rates but some problems underwater optical communication might encounter such as channel absorption, scattering, and turbulence, which may significantly reduce the attainable communication distance ranges [2]. The use of acoustic waves has been considered one of the most stable underwater communication technologies that can communicate very long communication distances.

There are many underwater phenomena that can contribute to affecting underwater wireless networks such as salt concentration, pressure, temperature, wind, and their effects on sea or ocean waves in addition to the degree of water purity which may be considered as distinct challenges in implementing underwater networks applications. These impacts will lead to reduce the performance of underwater communications through path loss, absorption loss, or multipath delay due to shallow water according to the communication technology used [3]. The use of intelligent technology such as a neural network in channel estimation will enable to send image/data without any loss. Moreover, the use of Q-learning method can provide to learn about the environment and react to changes in the network topology, channel fading statistics, or transmission geometry.

11.2 Underwater EM wireless communication

The radiofrequency (RF) waves can be used for transmitting information in an underwater environment which can provide fast communication and can perform best in shallow water when compared with acoustic or optical methods because of its ability to overcome turbidity and pressure problems. But even so, it is affected by the signal attenuation due to water conductivity along long communication distances that require underwater nodes to communicate in short distances not exceeding 10 m in the ocean. In low frequencies, the EM can provide reliable communication over several kilometers distances but with low data rates [4]. Some naval and environment applications can use these low frequencies in ELF and VLF but at the expense of high-power requirements and high equipment costs. The RF signal in underwater communication can travel through multipath and across water/air boundary propagating through seabed which can improve the communication in shallow water by increasing the propagation distance due to multipaths. But it may suffer from the Doppler effect according to salinity and temperature conditions which can attenuate the RF channels. Because of heavy affection by propagation loss, an accurate estimation for RF channels must take the consideration of sensitive operations related to retrieving data, storage, and power transfer to underwater nodes. Moreover, channel and noise estimates must be reliable so that the channel amplitude can be reached [5]. The use of multiwave transmitters and receivers with an intelligent estimate of channel status and channel loads could be a possible solution.

11.2.1 Underwater optical communication

To meet the underwater RF signal attenuation problems by conductivity, optical communication technology can be used in an underwater environment to obtain a high data rate over distances not exceeding tens of meters. Because of the high speed of optical signals in the water, the Doppler spreading effect can be negligible, but nevertheless, the optical underwater communications require that both communication sides must be in line-of-sight which is actually needed for direction tracking to keep the communication path. In addition, the water environment can affect light propagation by two optical properties known as apparent and inherent properties, where they depend on light source characteristics and water medium, respectively, and contribute to spectral absorption and scattering of optical signals. For underwater optic communication, inherent properties are more relevant [6]. The phenomena of optical backscattering can be used to estimate the water quality by knowing the degree of water turbidity for the design of intelligent transmitters, which can change transmission power and data rate accordingly.

11.2.2 Underwater acoustic communication

Underwater communications using radio waves and optical signals are considered a short-range communication application. To meet the requirements of underwater communication for long distances, the acoustic wave is an alternative technology to reach higher distances. The propagation speed is much lower than RF and is affected by water temperature, conductivity, and salinity in addition to water depth which must be estimated carefully in acoustic underwater applications. Moreover, acoustic waves in underwater may face problems of attenuation and delays which depend on acoustic frequency. At higher frequency range, acoustic waves will suffer from high attenuation and less in low frequencies but this falls on account of low propagation speeds and bandwidth, adding to these, channels might have minimum phases which require an intelligent error-based decision equalization [7]. For long-range communication, the acoustic waves will be impacted through spreading loss and absorption loss especially in high frequencies and will incur more losses with the presence of any obstacles or any kinds of acoustic noises.

11.2.3 Comparison of underwater communications technologies

The technologies used for underwater wireless communications, such as EM, optical, and acoustic waves have different features and parameters which are distinguished from each other [8]. Table 11.1 shows the comparison between three different kinds of underwater wireless communication carrier waves.

11.3 Underwater communication networks

Many underwater applications are needed to develop communications networks which are dependent on sensors-based system to effectively handling data

Table 11.1. Underwater communication technologies comparison

Parameters	**EM waves**	**Optical waves**	**Acoustic waves**
Communication distance	100 m	10 to 100 m	Up to 20 km
Data rate	Mbps	Gbps	Kbps
Transmission power	mW to 100 W	Few watts	10 to 100 W
Cost	High	Low	High
Bandwidth	Medium	High	Low
	Range-dispersion limited	Not dispersion limited	Transduction + multipath limited
Propagation speed latency	Medium latency	Low latency	High latency
Frequency range	100 Hz (up to 100 m) 100 kHz (few m)	Blue-green visible light range	500 Hz (long range) 50 kHz (medium range)
Noise sources	• Motor/lighting/pump/solenoid EM noise • Electronic preamp noise	• Sunlight detector “SHOT” and preamp noise	• Rain • Marine life • Thrusters electronic preamp noise
Absorption loss	High-frequency dependent	Medium-high turbidity dependent	Low-frequency dependent
Environmental sensitivity	• Conductivity • Complex permittivity	• Absorption • Scattering/turbidity • Phytoplankton • Organic matter • Bubbles	• Temperature • Salinity • Pressure (at depth) • Bubbles
Applications	• Non-line of sight (NLOS) applications • High energy turbid environments • Media transition (seabed-sea, air-sea)	• Blue water high bandwidth • Low latency LoS	• Long-range command-control • Medium range telemetry and navigation • Medium range integrated

exchanges. Such applications are related to sea or oceans monitoring or management in order to prevent underwater hazardous, pollutions, or for surveillance and military. Underwater networks must be constructed in small size and easy to operate underwater sensors or unmanned autonomous vehicles (UAV) and ensure to self-adapting and optimizing the position during motion by an intelligent route, in addition to having the ability to manage the distributed sensors and optimizing spatial distribution for such underwater networks [9]. Another consideration is related to underwater wireless network security which is very important in applications such as underwater military communications or monitoring because the underwater environment may be highly vulnerable to some malicious attacks due to high error rates, propagation delay, and low channel bandwidth depending on the type of communication technology used, whether it is acoustic, optical, or RF waves [10]. The design of underwater networks must take into calculation the impact against the underwater wireless channels due to the differences in the underwater sensor networks localization, routing, and time synchronization which require intelligent security mechanisms.

Underwater communication networks are varying, depending on the application required. Sensors are often used to perform some operations, such as monitoring for finding underwater information or disaster prevention [11]. There are other operations in which submarines, autonomous underwater vehicles, or robots can be used and remotely controlled for some applications such as ocean sampling, mine reconnaissance, and distributed tactical surveillance (see Figure 11.1).

Usually, the underwater networks are fixed as individual nodes that are connected to anchoring, moored buoys, or seabed. Semimobile underwater networks can also be suspended from buoys that are deployed by ship and used temporarily and can be left for hours or days so that the vertical sensors operate on short-term deployments. In underwater communication networks, the connection may change due to small-scale movement or due to water dynamics, which can directly affect the operation, especially in the energy consumption of the batteries used with the sensors [12]. Sensors connected to AUV devices can provide a navigation feature to maximize sensor coverage through limited devices, but they present localization and network connectivity challenges as well as energy problems.

11.3.1 Underwater wireless sensor network

Underwater monitoring system designs are based on the communication between sensor nodes and the ways of processing, storing, and collaborative capabilities to those sensors according to the communication model used. Acoustic and RF wireless communication technology models can enable the sensor nodes to perform real-time underwater monitoring by taking into considerations of underwater network reconfiguration, failure detection, and report of operational devices [13]. These considerations require efficient underwater networking services and protocols. Underwater sensor network topology can be changed during the underwater environment which will need to provide an intelligent methodology to power control, management of wireless interface, and mobility [13,14]. According to the possibility

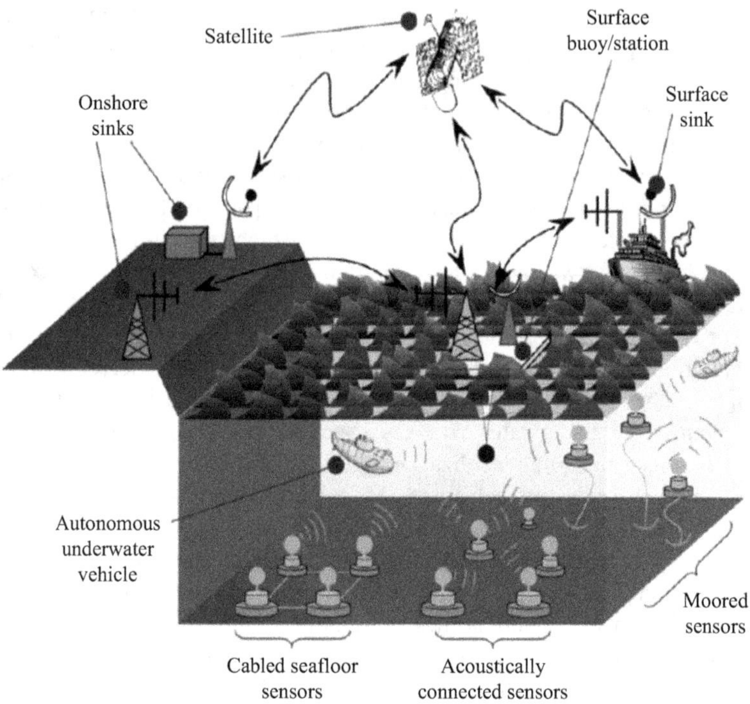

Figure 11.1 Underwater wireless network environment. Adapted from [12]

of some node's movement in the underwater sensor networks due to the surrounding currents, it found that spatial-temporal data can be collected because of the change of the network topology. Since the sensed underwater nodes are limited by energy and the tasks of replacing the underwater contract batteries are very expensive, so the change in network topology requires an algorithm to control energy and conserve it in underwater wireless battery-powered systems. Moreover, these power control algorithms can improve the underwater sensor network throughput with help of increasing bandwidth. The management of underwater sensor communication interfaces is a very important method for topology control and reduces the energy consumption [14]. This control management is depending on the underwater sensors' network operations such as transmitting, receiving, in addition to sensors operation modes such as idle, sleep, and active or power off modes during network lifetime.

Signal processing and medium access control are important factors and those should be carefully calculated for underwater sensor networks design. Both factors will determine the efficiency of channel allocation, bandwidth multiplexing, modulation, and channel coding [15]. Most operations related to energy management and network topology control, as well as signal processing methods, and channel multiple access techniques, all of which were previously mentioned, must be defined and designed depending on the type of underwater sensor network model applied. There are two UWSN models: 2D and 3D models. The 2D model can be deployed for marine observations focusing on coastal regions which consist of underwater sensors with the same structures and functions used for data detection and acquisition from the underwater environment [15]. In this kind of UWSN, the surface station can be used to collect the data from underwater sensors that typically dedicated to only be applied to 2D detection at the seabed as in Figure 11.2. The related sensors are fixed at the seabed. The connection between the surface station and underwater sink can be done by the cabling link for the convenience and reliability of underwater communication. The collected data from these sensors by the surface station will be transmitted to the onshore station. The applications of 2D are more suitable for shallow water environments because of limited communication distances [16].

In a 3D model, sensor nodes can be set at different depths to provide large-scale coverage in an underwater environment. The nodes are distributed in different

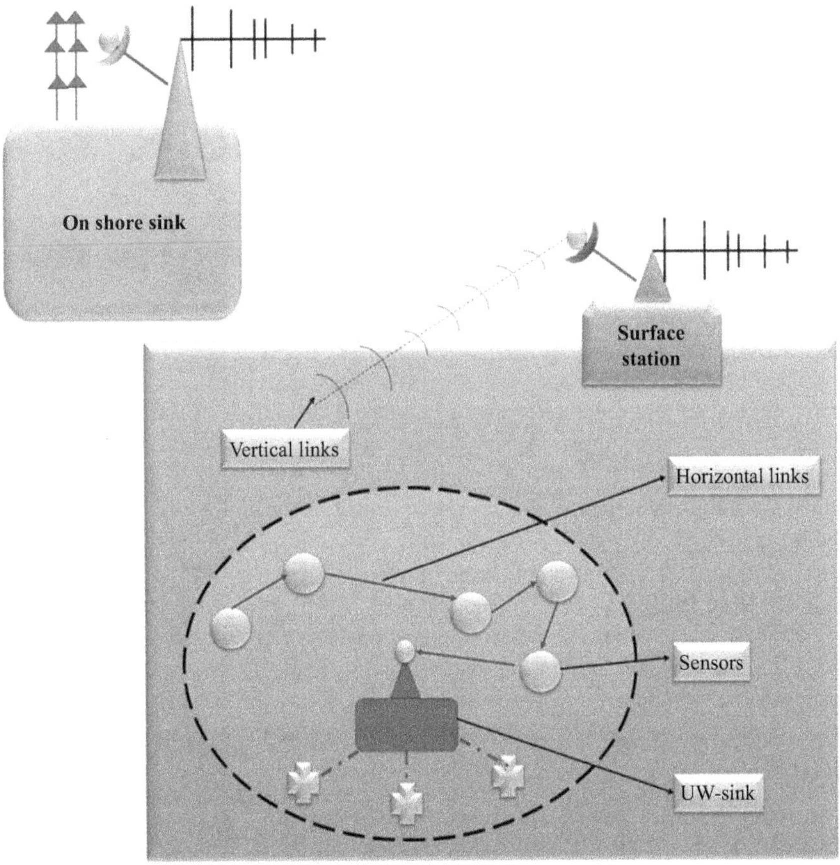

Figure 11.2 Underwater wireless sensor 2D network architecture [16]

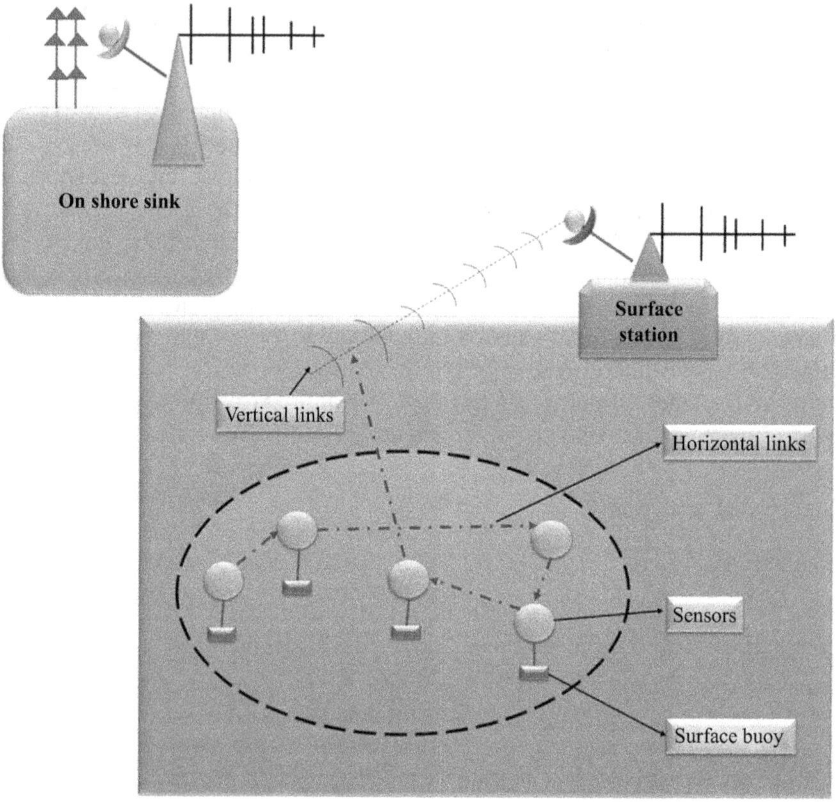

Figure 11.3 Underwater wireless sensor 3D network architecture [16]

depths by using buoy or anchor chain connected to nodes via the cable with the adjustable length (see Figure 11.3). To avoid the appearance of a buoy that may be subject to target detection and affect navigation, nodes can be anchored to the seabed via adjustable anchor chains. Other sensors may be floating on the surface which are needed to be controlled via some connection like attaching each underwater sensor node to a surface buoy by means of wires whose length can be regulated to adjust the depth of the sensor node. Or by anchoring sensor devices to the bottom of the ocean in which each sensor is anchored to the ocean and equipped with a floating buoy that can be inflated by a pump [17]. By these connection types, the depth can be controlled by adjusting the length of the wire that connects the sensor to the anchor through an electronically controlled engine in sensors.

11.3.2 Autonomous underwater vehicles and robots network

The use of AUVs will provide a means of cooperative relay to perform many navigational tasks and exchange information between each other. Due to variant in AUVs motion behavior, the AUVs networks must calculate the position of vehicles to avoid the degradation in the performance of navigation because of the absence of positioning information (GPS) which requires to build an external reference from acoustic bacons at positions, and if not, the vehicle has to rely on compass information, a Doppler velocity logger (DVL), or an inertial navigation system (INS) and perform an estimation process to extract the error in the position [18]. Underwater robotics networks are used for persistent monitoring of an area for many applications such as sea or ocean survey or for ocean sampling. Sensors integrated with a fleet of AUVs can be used to collect data over survey tracks beyond the range of the fixed stations, and for deepwater geophysical surveys, underwater robotics network can be deployed temporarily according to the survey requirements [19]. Robotics also will provide strong support when integrated together with different marine, air, and space systems with access to the command and control center (C2) capable of monitoring the system in real-time, and to implementation of orders and updates for the network itself (see Figure 11.4).

In an underwater robotics network for surveillance purposes, the interface between the robot system such as sensors and actuators will be performed by an executive layer according to the network layering design, which enables to collect data from sensors and pass it to the signal processing units for classification and tracking. Signal processing unit works to discover the echoes of the targets and the process of extracting a group of contacts that need appropriate algorithms according to the underwater scenario, which is characterized by the possibility of a lower

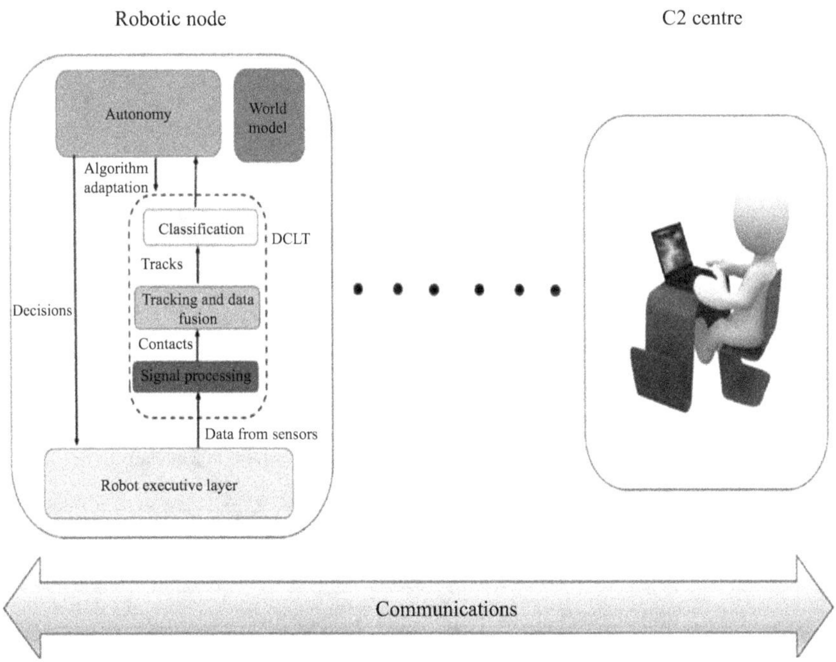

Figure 11.4 Robotic surveillance network concept. Adapted from [19]

target for the detection (PD) [19]. In most underwater robotics applications, there is a need to respond to events that occur in an underwater environment without relying on a continuous contact link with the C2 decision-making center. So, it can depend on the AUV data-driven approach which enables robots to adjust their paths over the Internet in addition to making decisions at a higher level such as switching the mission stage and improving performance (see Figure 11.5) [21]. An intelligent adaptive data-driven robot can provide a means of quality in data collection and will turn to make an efficient decision-making process [20].

11.4 Internet of underwater things

Internet of things and its applications have become the most amazing technologies that can provide a connection of every things to the Internet. For underwater applications, the novel concept of intelligent interconnected underwater objects is related to the Internet of underwater things (IoUT) approaches. Underwater wireless sensor networks can be connected to the Internet and promise to provide many practical applications such as environmental monitoring, surveying, military purpose, and others (see Figure 11.6) [22]. IoUT network may suffer from different impairments such as low bandwidth and reliability in addition to long propagation delay.

In IoUT systems, underwater sensors, vehicles, or monitor stations can collect data for underwater objects and interact with the Internet through the network layer, which contains a coverage network that enables an Internet connection with network administration in addition to cloud computing

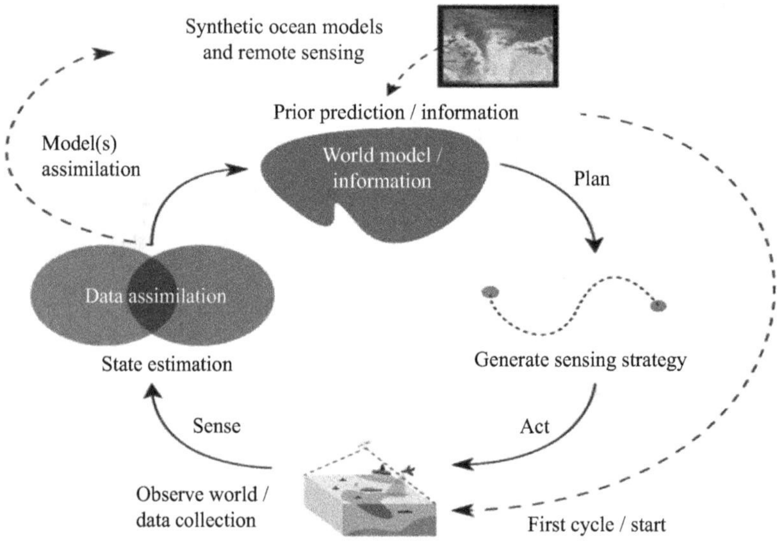

Figure 11.5 AUV-based data-driven and information sampling approached. Adapted from [20]

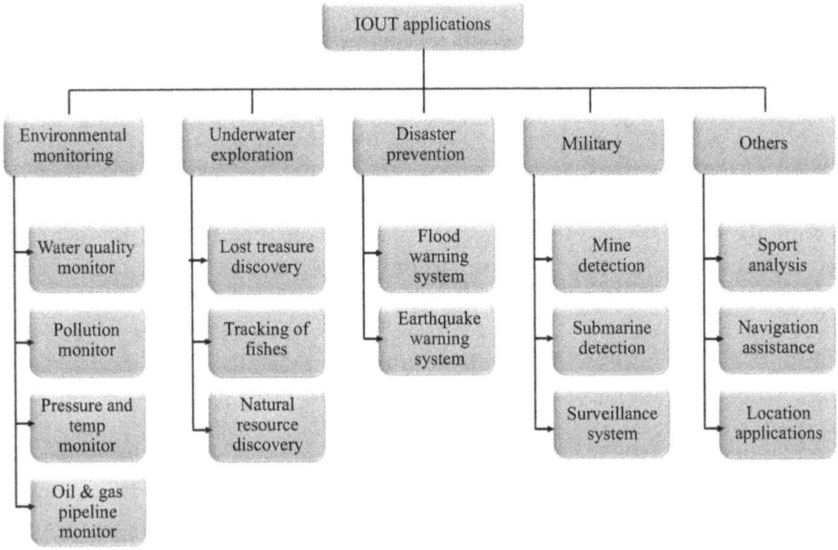

Figure 11.6 IoUT applications. Adapted from [22]

platforms [23]. Information received by the network layer is directly provided to the users by an intelligent application solution. In general, it can be considered for IoUT as a combination between the underwater wireless sensor networks and the Internet of things and for both technologies, many challenges must be overcome to have a highly IoUT operation performance. Such challenges are energy efficiency, underwater vehicle mobility, propagation speed, and security [24–26]. The sensor nodes in IoUT require more power to be communicated along with large transmission distances which need an intelligent algorithm for power adaptation management. Mobility of underwater vehicles will change the underwater network topology which will increase the power consumption during reroute of the communication link. An intelligent algorithm for self-organization and management can ensure to save the IoUT network stability adding to keep better power efficiency. Security is considered more important since the IoUT will be connected to the Internet [23,27]. Due to malicious attacks in an underwater environment, a high bit error rate and propagation delay will occur which will impact the IoUT network connectivity and make sensor nodes and underwater vehicles consume more energy leads to reduce the network lifetime. An intelligent dynamic IoUT network coding and attack recognition will overcome such challenge.

11.5 Underwater intelligent data gathering system

Data collection and gathering in multihop underwear sensor network may suffer from unbalancing power consumption which is required to use a mobile edge system such as an autonomous underwater vehicle (AUV) that can able to consider an efficient edge computing in mobility model for underwater applications [28,29]. The adaption of AUV mobility and velocity can be used to calculate the mobility path for data gathering in order to reduce the power consumption and keep the network utilized for a long lifetime. For the IoUT, massive data are generated and because of the long transmission distance, these data may lose easily during collection. For such applications, data collection is usually taken by the cloud which is far from the source nodes. In addition, for the massive distributed data source, the use of fog-based computing processes will increase the power consumption by the sensing system [28,30]. The operation of underwater data collection is related to AUVs and those need to collect the information from all underwater nodes which depend on the data collection algorithms based on probabilistic neighborhoods and need to construct a minimum probability neighborhood coverage set for aiming to access several selected locations to reduce data latency. Clustering concept may be used to reduce the AUV mobility by using cluster heads or gateway nodes for gathering the information from the other nodes and let AUV visit only the gateways from data collection. With the edge network model, AUV can provide data collection in addition to storing and computing services [31,32].

In most underwater communication networks, the traveling of mobile AUVs and path quality are both more important factors that may impact the information gathering. The improvement in data collection performance is depending on the AUVs' communication with multiple nodes path. The AUV path requires an intelligent algorithm to keep the path always optimized as long as data collection is performed. Moreover, an intelligent mobile traces algorithm is based on a geographical method to determine which locations must be visited by the mobile agent in a minimal amount of time [31]. The geographical trace must consider the amount of information regarding the sensor measurements and that information is gathered according to the locations based on a probabilistic communication model to describe contributions of the neighborhoods. Robotic vessels as services (RoboVaaS) provide a novel technology for shipping and nearshore operations by offering on-demand robotic aided services. Data gathering in such an underwater system can be served by a service request by underwater robotics as a dispatch vessel (see Figure 11.7). The use of the distributed data gathering method by underwater robotics can enable a mobile gathering from random distributed underwater sensor networks grouped in clusters [34]. This scheme will help to reduce the transmission power of sensor nodes by selecting a dedicated node based on path quality.

In a dense underwater sensor network, gateways can provide a route service for the data collected by the sensors enabling to pass the information to the shore centers. Gateways must take decision-making strategy to perform a data uploading scheme depending on the similarity of the previously uploaded data, the compression ratio, the power consumed in transmission and reception, data rates, and locations of sensor nodes [33]. Underwater sensor network nodes provide the gateway with the information about the data compression, sensors module status, in addition to communication status all of which should be

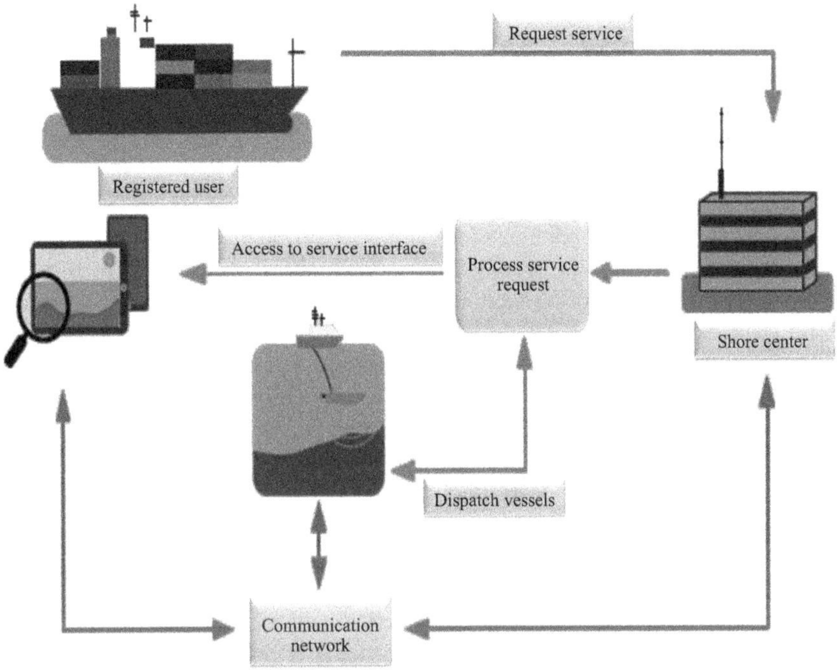

Figure 11.7 Underwater robotic vessels as a services network. Adapted from [33]

considered in decision-making side by side with a power control algorithm to enable gateways to efficiently select the appropriate transmission power according to the channel conditions due to underwater environment [35]. The goal of gateway decision-making is to save energy consumption as minimum as possible for the integrated network [36]. For intelligent decision making, a neural network can be used to decide data uploading after taking calculations for total energy consumption and transmission rate required.

11.6 The underwater intelligent recharge docking system

In most modern marine applications, docking system technologies are required to establish a docking station, underwater communications, transport, and navigation that require a recharge and recovery system for AUVs without the assistance of the mother ship to reinforce the docking system [37]. Underwater docking systems can be classified as docking systems as physical contact or noncontact according to the power transmission methods and the data used. Physical communication can be used as wet underwater conductors accomplish an underwater signal and transmit energy, while the noncontact method uses an electromagnetic coupling and wireless communications [38]. Docking system consists of a docking station that has many devices such as underwater camera, lighting, navigation, observation and depth sensors, and data and power transmission systems.

In an underwater docking system, power can be transferred from the docking station to the AUV through coupling for rectifying and charging and then energy will be stored in batteries. The docking system can route the data between the AUVs and dock through gateways which enable to connect between the AUV and dock terminals such as host or embedded computers [39]. There are a number of control and navigation components that interact with each other in a docking system as shown in Figure 11.8. In addition, the system has a guidance system that has the ability to create a set of possible paths, they are taken perceptively, depending on the circumstantial environment of the operation, depending on an intelligent navigation system, and based on AUV dynamic and operation objectives [40].

The use of an intelligent docking system can provide an underwater recharging and data transfer to the AUVs. The detection operation for the location information on underwater docking stations can be performed by the detection module which enables to recover the position of docking stations in 2D or 3D underwater scenarios. The use of a neural network for docking can enable to establish robust detection of docking stations with an accurate positioning estimation between docking stations and AUVs [41]. A deep neural network in the docking system can guarantee detection performance in the various nonstationary underwater environment while a credible one can improve the docking efficiency.

11.7 Intelligent UWC methodologies

Underwater wireless communication (UWC) implementation may encounter various impairments due to the conditions that impact the channel selectivity, such as multipath channels delay that depends on the water purity if it was clear or shallow. The existence of many boundaries and obstacles may affect the acoustic propagation signals in an underwater environment that will lead to inter-symbol interference because of multireflections which require an efficient channel estimation [42]. The use of a neural network can provide a means of better estimation for selective channel according to the channel's bandwidths [43]. For AUVs applications, locations of multiple AUVs can impact the performance of underwater AUVs network, because the unprecise positioning will affect the communication paths.

An intelligence learning trajectory planning mechanism can enable to extract the sampling locations of multiple AUVs which will help to reduce the instability of underwater communications due to weakness in dynamic AUVs' position detection. Moreover, reinforcement learning (RL) can be used to improve the mechanism for selecting communication pathways between AUVs, in addition to making the transmitters able to recognize channel behavior and adapt the transmission strategy accordingly [44]. The use of Q-learning can contribute to adapt the AUVs' transmission parameters to the temporal variations of the channel that occur according to the underwater communication environment. It can provide self-configuration and adaptation capabilities to enable operations in dynamic

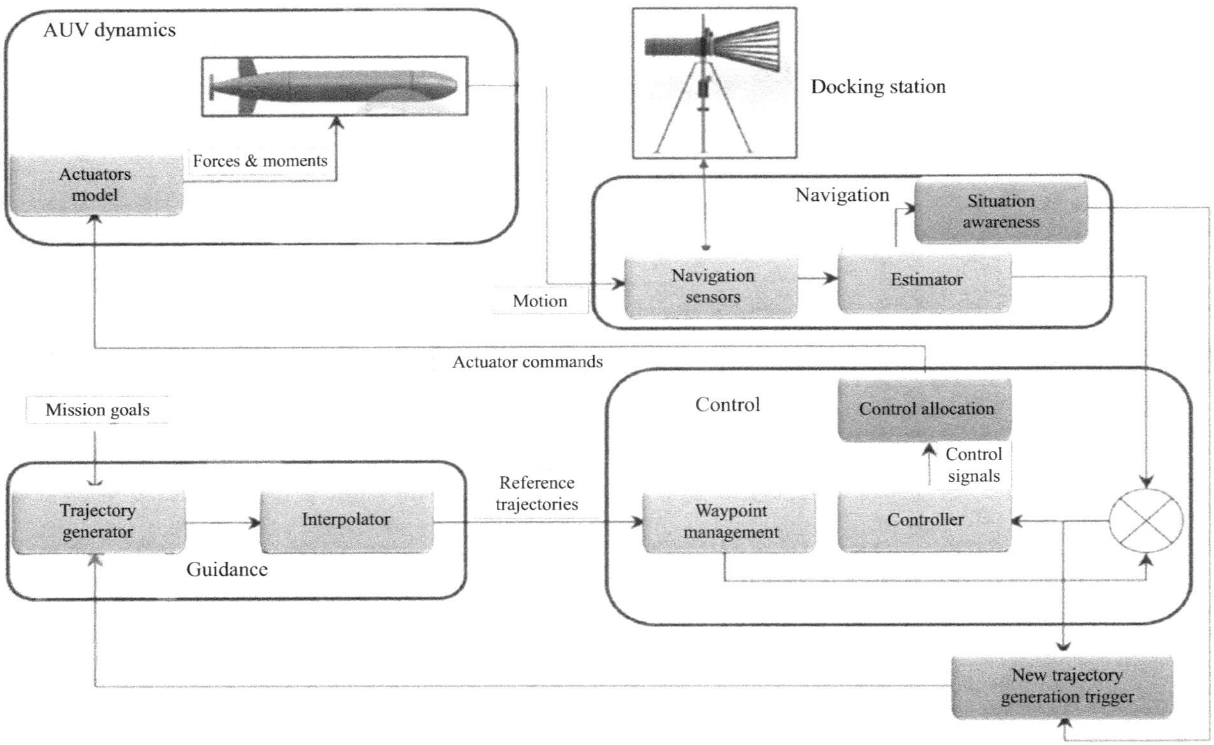

Figure 11.8 AUV docking system architecture. Adapted from [40]

underwater communications network environments by allowing underwater sensor nodes, vehicles, and robotics to learn about the environment and interact with changes in network topology, fading, or transmission statistics [45,46]. In this section, we will talk about some artificial intelligence techniques that can be used to upgrade the performance of underwater communications related to navigation, smart control, distribution mechanisms, and smart adaptation of some UWC applications.

11.7.1 Intelligent navigation and control algorithms

Unmanned underwater vehicles are increasingly used for different underwater services to civil and military applications. The use of autonomous marine vehicles (AMV) to serve in underwater environment requires a means of artificial intelligent platform developed for the execution of marine operations in an efficient and accurate manner. Various AMVs can be deployed which actually should cooperatively interact with each other to exchange the information related to navigation and control. For surface and underwater commination networks, the cooperative navigation and control method will provide an efficient way to improve the performance of marine operations [47]. In underwater environment operations, the use of inertial navigation system (INS) integrated with AMV can enable to calculate the drift position with time for navigation accuracy, and by taking the information of navigation and communication on the surface of the water by using global navigation satellite system (GNSS) will slightly improve the navigational and control operations. But for deep underwater, this method becomes more difficult unless a cooperative method between multiple AMVs is used which will help to gather navigational control information from different navigation and control systems in surface-underwater mobile communication networks [47,48]. Due to different gained navigation and control information, AMVs will be able to adapt and control themselves.

Intelligent sensing is becoming more important in the underwater monitoring field. Underwater robotics and AUVs are expected to sense the surrounding environment and be able to avoid any impediments during its movement. Path planning is a procedure that should be taken by the AUVs to plan the shortest path by taking decisions according to the underwater environment information which will make AUV perform many operations depending on its ability to interact with the environment. An artificial intelligent path planning algorithm can be used to enable AUV to perceive the surrounding environment to implement a number of applications. Such intelligent algorithms are generic, living things colonies, fuzzy logic and neural network-based algorithms. Living things colonies and generic algorithms are considered as a global path planning for underwater robotics [48]. AUV underwater networks can use genetic algorithms to perform random search and monitoring operations of the marine environment by enabling them to adapt to an environment and integrate with their organisms by simulating biological genetics and adapting to their natural development, making them able to deeply understand the underwater environment and the ability to plan a path very efficiently.

Algorithms based on colonies simulation of living organisms enable to perform AUVs networks path control and intelligent planning to effectively perform a number of tasks. Such an algorithm is an artificial bee colony (ABC) algorithm that enables the arrangement of data exchange and confidentiality by controlling the levels of information exchange and authority to perform tasks as it actually happens in bee colonies. ABC algorithm can save information against any intrusion and secure the privacy of AUVs system observing the whole activities of the system mainly for malicious activities and then creates reports for the services under management (see Figure 11.9) [48]. The use of ABC algorithm enables to choose the best path for AUVs and underwater robotics, depends on location and intensity of vehicles and robots, in addition to its presence in a specific region (see Figure 11.10). The ABC algorithm for path searching is like the strategy used by ant colony to select an optimal path that most used according to positive path information [49]. This algorithm enables the use of computing distributed according to marine environment information in a three-dimensional space to map paths that can be used by AUVs.

Algorithms based on the field of objects' location can enable to find optimal paths by simulating the movement of objects depending on the field function in which the object is located. In AUVs networks, artificial capabilities of field algorithms can be used with rapid synthesis to avoid marine currents on underwater navigation by realizing marine influences on the field where the AUVs are located,

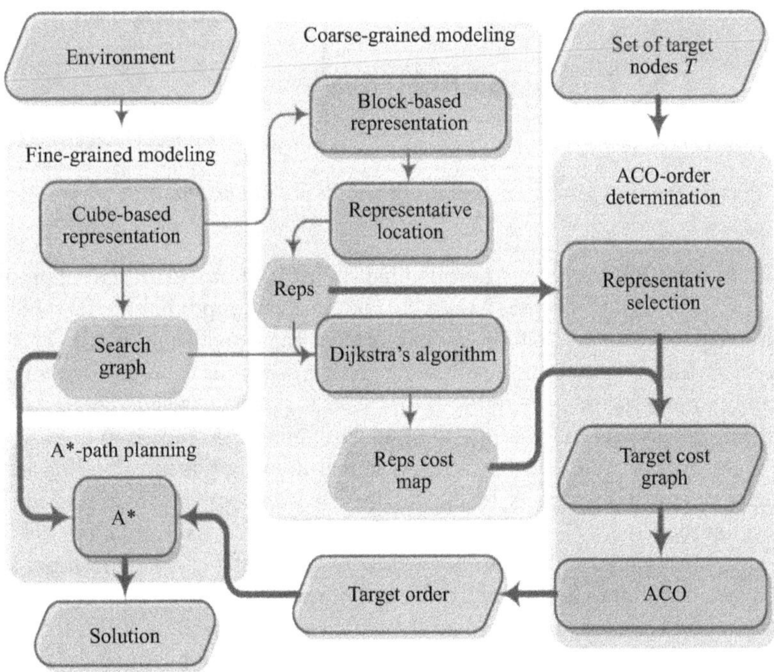

Figure 11.9 Ant colony optimization algorithm modeling. Adapted from [48]

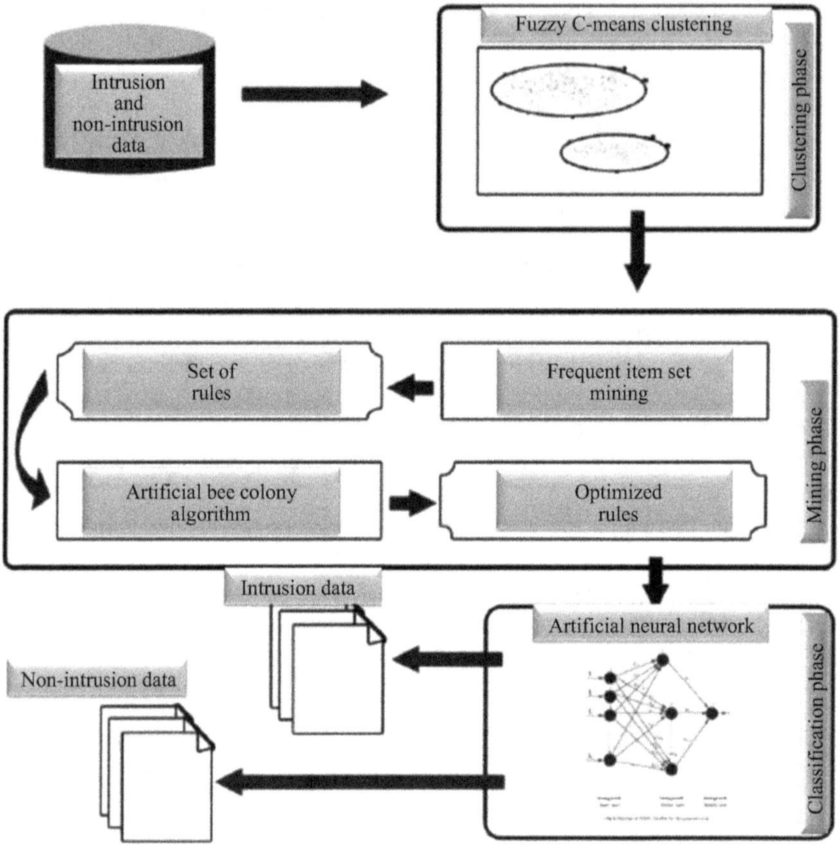

Figure 11.10 ABC-based intrusion detection system. Adapted from [49]

and the information perceived by this algorithm can be used to plan appropriate paths for movements for the AUVs in underwater by promoting learning and decision making with the help of Q learning algorithms and neural networks so that it can make the AUV able to adapt to the new environment through learning and constantly improving its own ability [50]. Fuzzy logic can be used to help create an intelligent underwater navigation mechanism through direct programming that relies on information collected by sensors in real-time such as location information and the movement of objects in the marine environment in 3Ds, in addition to information related to environmental obstacles. Through this information, the fuzzy logic algorithm works by faithfully extracting information about the robot's and AUV's speed and drawing its paths of motion by making fuzzy decisions capable of avoiding underwater obstacles and dynamic obstacles such as strikes of marine currents in the depths [51]. To reach the highest degree of intelligence in underwater navigation control, algorithms of neural and deep networks can provide excellent solutions and a high ability to learn and make decisions.

11.7.2 Q-learning optimization

In underwater wireless sensor networks, an acoustic signal will be transmitted at relatively low and variable speeds because of different underwater environment impacts. The choice of selecting an efficient routing path with low delay is depending on the performance of the underwater routing algorithm that enables to reduce the path delay extending the underwater sensors lifetime. Q-learning routing algorithms can provide an intelligent technique to find an optimal nest hop and best of routing path choice that can able to play adeptly in a dynamic environment. An action utility Q function can be performed for residual energy and propagation delay due to route performance as a routing decision to extract a shortest path that can be used by underwater sensors. Moreover, these routing based on Q learning algorithms (see Figure 11.11) must take into the considerations of delay due to minimum bandwidth cost and latency, and must make nodes to keep its process in an adaptive and distributed way using the reward-penalty mechanism of the Q-learning algorithm [52].

Sensor nodes in underwater communication networks can be equipped with a powerful function of processing data which can update the routing information in the underwater sensor network. The structure of such network is based on clustering sensor nodes arrangement depending on the cluster heads. The use of reinforcement learning can perform to plan an efficient routing of data aggregation. The

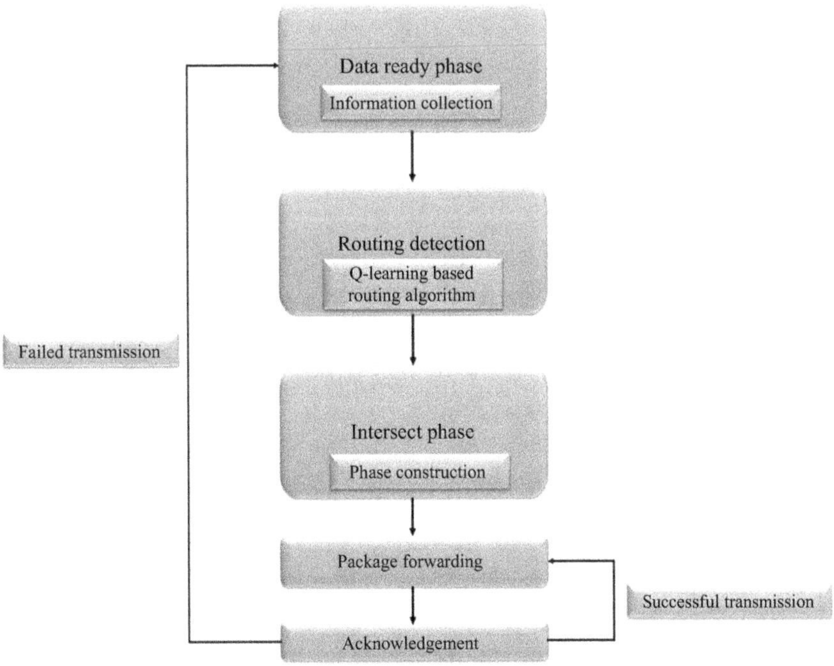

Figure 11.11 Q-learning routing algorithm. Adapted from [52]

routing path structure is based on the data collection and processing (see Figure 11.12) in which several cluster heads may be chosen to gather the monitoring information from sensor nodes and convert these information to stations via multihop path that is determined efficiently by reinforcement learning according to information of data processing and analysis [53].

Data collection process may impact the network lifetime, especially in case of huge amount of data which will increase the energy consumption. The methods of energy use reduction can be obtained by keeping sensor nodes to work in low power states. The operation of data collection must be intelligently related with the senor nodes sleeping and waking up states and self-adapted by listening to the channel state to take decisions of data transmission or reception. Synchronization process must be ensured between adjacent sensor nodes for sleep and listening operations, and this can be performed by making cluster heads broadcasting the scheduling information to keep time synchronization among nodes and allocating different time slots to different nodes for data aggregation.

11.7.3 Intelligent distributed virtual system

In large-scale underwater networks, sensors, autonomous vehicles and robots are deployed in an underwater environment, and are able to exchange information in an integrated heterogeneous network architecture. One of the intelligent underwater communication systems is based on use the software define network which can contribute to implement highly virtualizable underwater network having an advantage of high flexibility to improve the network management and reduce the

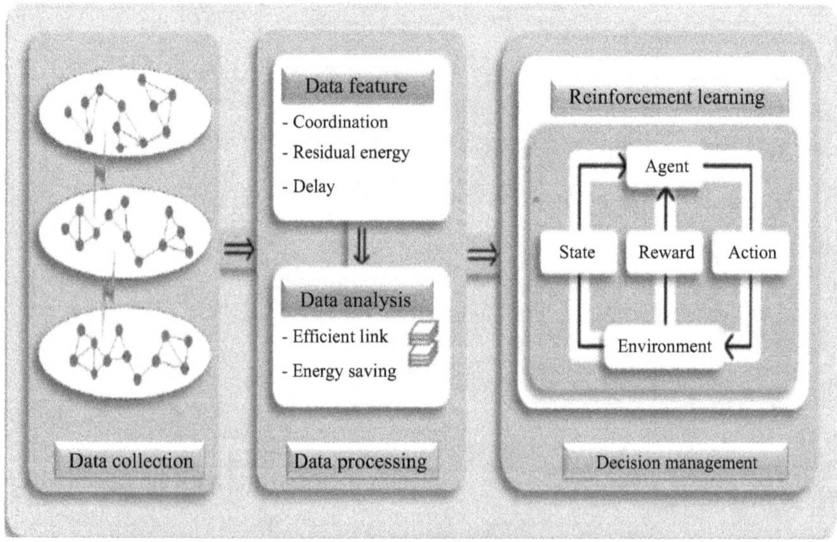

Figure 11.12 Data processing-based decision management learning in UWSN. Adapted from [53]

operation energy and costs. The concept of network virtualization in the underwater network systems can contribute to maximize the network capacity in addition to adding robustness and energy efficiency to provide scalable underwater networking services which will enable to support different kinds of underwater applications based on many management and control tools (see Figure 11.13) [54]. The concept of the distributed virtual underwater system can dynamically operate different system configurations to optimize the underwater network resources for efficient spectrum management and energy computations. Moreover, it will enable to design underwater network virtualized system to allow multiple applications to be operated on the same physical underwater communication system [54,55]. The use of multiple virtual wireless underwater networks can manage many applications under separated protocols based on demand which will simplify the network monitoring and management for heterogeneous underwater network.

In IoUT, distributed virtual networks based on sensor-cloud integrated resource will enhance the real-time processing and storing. Virtual sensors and software-based resources in underwater communication systems can be divided into several virtual networks to enable cooperation between different underwater

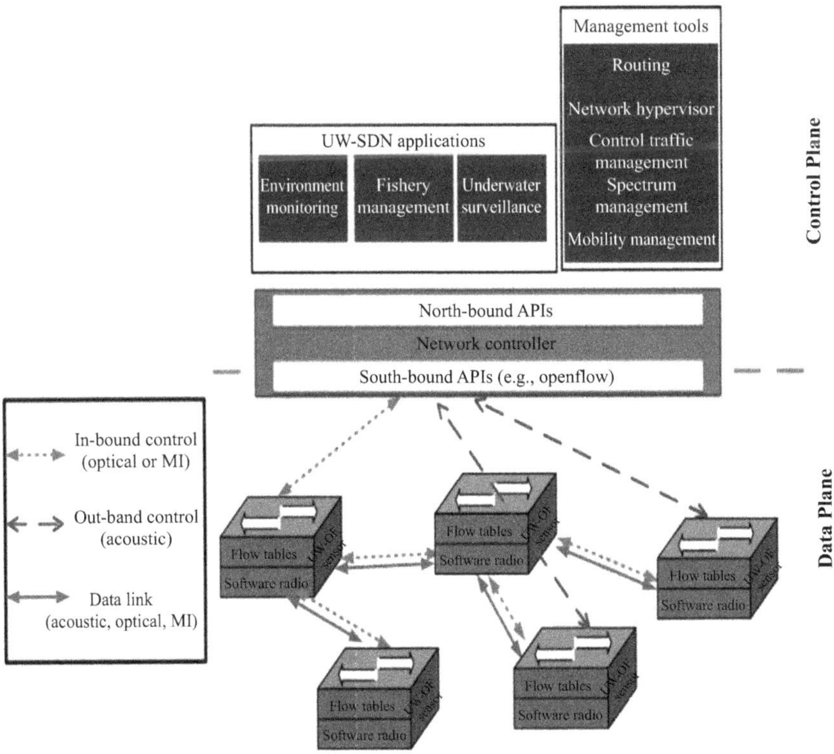

Figure 11.13 Soft water model based on distributed network function visualization. Adapted from [54]

network services as a concept of sensing-as-a-service which will enhance the sharing of underwater network infrastructure in order to reduce the cost of underwater sensor networks design [55]. The use of a virtualized underwater system can be also deployed in a cluster-based underwater sensor network to detect the distributed sparse events by providing event measurements for accurate detection with low sensing overhead.

11.7.4 Adaptive networks architecture

Underwater networks may consist of various massive nodes positioned in three-dimensional space along with different depth levels and operated in many configurations according to the underwear topography environment. Signals in such a network may face many reflections and refraction variations in addition to dynamic underwear environment and relative motion of mobile vehicles. Channel conditions in underwater networks must be able to adapt its communication modem parameters to overcome the effect of signal channel variations. In addition, energy-efficient commination must take into underwater network architecture considerations in order to design the adaptive physical network layer [56]. Depending on channel conditions, data rate and modulation must ensure maximum channel capacity and efficiency and keep saving energy. Channels must be selected in an adaptive manner and alternatively due to the decision of using low or high data rates which depends on the effect of the underwear channel impacts such as interference from multipath.

In underwater communications monitoring, the use of mobile data collectors improves the communication energy. Moreover, the strategy of repositioning the underwater sensors improves the network's collective coverage. These methods give the underwater network the ability to adapt to communication channels based on the sensor node positions [56]. In underwater sensor networks based on cluster heads, sensor nodes must able to adapt its operations according to the cluster head (CH) functions to ensure low energy consumption. Adaptive underwater networks must be self-organized for any underwater dynamics change. The concepts of biological adaptive networks such as swarm optimization algorithms can be used in design the underwater adaptive networks architecture. Algorithm such as Cuckoo birds' search can be implemented to find out the best and stronger signal to the CH. The sensor nodes can choose among worse cases to find the best and decide CH communication cost taking a comparison if the cost is less in case of directly communicating to the base station or through CH [57]. Ant colony optimization algorithms also can contribute to select the best route between underwater sensor nodes depending on different underwater conditions and information collected from the underwater environment.

11.7.5 Intelligent fish-like underwater network

Many underwater communication applications have got more attention in marine environment researches such as fish motioning. IoUT plays an important role and opened new opportunities in the advanced big data analytic for the underwater environment [58,59]. By using underwater sensors, autonomous platforms can help to collect data of marine lives for scientists, and with an artificial intelligence and machine learning, it will be easy to solve most complex operations related to massive data collected for analysis and resource management. In an underwater network, the communications-based acoustic (fish sounds) equipment can be used to transmit several frequencies and receive different signal responses which will provide many information about fish species, sizes, and orientations [60]. Underwater nodes with camera can detect the fish colors by the reflection of optics waves. For fish underwater imaging, marine environment will affect the quality of the images taken under the water, making it difficult to determine the types of fish, which reduces the performance of the classification method. Neural network methods, with the use of a copresence matrix at the gray level and contrast-adaptive color correction technology, can provide intelligent solutions in precisely identifying the fish types. Classification methods can be affected by the data related to the characteristics of color, shape, and texture, and are related to the data sources [61]. Therefore, it is necessary to overcome the improvement of the quality of data sources while determining the type of fish. Also, analysis and correction of new colors compatible with the contrast in the image to be improved and can be done through the use of neural networks.

In underwater images and video processing, the estimation of fish biomass and assemblage need to develop an intelligent fish sampling technique that will enable to overcome the variations due to orientation of fish, seabed structures, movement of aquatic plants in the background, and image diversity in addition to weak luminosity. For fish movement detection applications, the use of neural networks and machine learning can solve the problem of generic object detection and localization [62]. The use of a neural network can help to capture color and texture information for localizing and detecting fish instances in the images [63]. The abundance estimation of fish from various underwater images recorded can be done by deep learning algorithms.

11.7.6 Intelligent recognition of noise target

Generally, underwater communications are more reliable when using acoustic sensor networks instead of optical or RF communications due to their great impact on ocean monitoring and exploration. Underwater target detection and localization nowadays have become one of the most important marine applications. For underwater localization, the use of GPS cannot meet the requirement of the efficient underwater application because of its high attenuation by water and poor ability to propagate underwater especially in sea or ocean waters [64]. The use of sound in water can enable to implement the underwater localization due to its nature propagation in water. The use of accurate distance and angle-based algorithms can perform sensor localization operations. Most applications are requiring that underwater systems navigations should be able to make self-localization depending on direction and speed measurements. Localization algorithms can perform this task with help of more than two reference nodes as an anchor at a known position to easy self-localization estimation. Deep learning neural network can help to measure the acoustic signals to locate the communications channels for the purpose of estimating the received sound pressure levels [65].

Underwater sonar systems are very important for target recognition in civil and military applications related to a marine biological survey or marine exploration. Underwater recognition can help to identify unknown targets such as submarines, mines, or other underwater objects. Gaussian mixture model is one of the widely used mechanisms for target recognition and with a combination to deep Boltzmann machine, they will improve and overcome the problem of distribution mismatch. Gaussian mixture model can modify the structure of the deep neural network to improve the target recognition accuracy when it is applied to underwater recorded dataset that might contain ships, mammals, and background noise with weak targets [66]. Intelligent noise target recognition is related on how to recognize a small size target based on ship noise. The recognition scheme requires an unsupervised pertain to use unlabelled data with small number of labelled data. By using an artificial neural network for auto coding mechanism, efficient learning of data coding can be achieved in order to pre-train a large number of unlabeled data to fully learn the underlying laws from ship-radiated noise to extract features from ship-radiated noise as the unlabeled data always contain the general features of the interested underwater acoustic targets to identify a specific class of ships. For further identification and features separation, fine-tuning learn help to extract the classifications by training the separated features to complete the target recognition [67].

11.8 Intelligent UWC modeling

Underwater environment is characterized by several features that will make the communications unique when compared to the terrestrials and atmosphere environment. Self-configuration and adaptation must be considered in underwater sensor network design to enable automatically drive the wireless link with the underwater conditions in addition to sensor node location management. Moreover, the distance of transmission and underwater obstacles may impact the performance of underwater commutations because of many channel impairments such as multiple paths and signal attenuations due to the propagation model if it was RF, optical, or acoustic signals. Channel modeling contributes to design an efficient underwater communications network by making mathematical calculations in order to show up the performance of underwater signal propagation with the effect such as signal scattering, absorption, and spreading in the underwater environment.

11.8.1 Mathematical modeling

For underwater communication applications, the underwater acoustic channel is known as a common model used, but it faces challenges related to acoustic signal propagation. Signal attenuation is the most important problem which occurs due to absorption and attenuation losses and requires to be carefully calculated in underwater communication network design. By taking oceanic parameters such as temperature, pressure, acidity, and salinity with a frequency used (f) in kHz, absorption can be calculated as follows [68]:

$$\alpha = \frac{A_1 P_1 f_1 f^2}{f^2 + f_1^2} + \frac{A_2 P_2 f_2 f^2}{f^2 + f_2^2} + A_3 P_3 f^2 \tag{11.1}$$

where f_1 and f_2 in kHz represent the relaxation frequency of the boric acid and the magnesium sulfate, respectively. A_1, A_2, and A_3 represent the component in seawater of boric acid, magnesium sulfate, and pure water, respectively. P_1, P_2, and P_3 are respectively the depth pressure in seawater of boric acid, magnesium sulfate, and pure water. More calculation can be done to extract the absorption coefficient according to the frequency used. Equations (11.2) and (11.3) show the absorption coefficient for a hundred Hz and lower frequencies, respectively, as follows:

$$\alpha = \frac{0.1f^2}{f^2 + 1} + \frac{40f^2}{f^2 + 4100} + 2.75 * 10^{-4} f^2 + 0.003 \tag{11.2}$$

$$\alpha = 0.002 + 0.11 \frac{f^2}{f^2 + 1} + 0.011 f^2 \tag{11.3}$$

Underwater acoustic communication transmission loss can be calculated depending on the amount of sound intensity, spreading factor, and water depth in addition to sound source signal level (SL). The transmission loss (TL) caused by the spreading loss and absorption is given in the following equation [68]:

$$\mathrm{TL} = k * \log(r) + \alpha * 10^{-3} \tag{11.4}$$

where k is the spreading factor which depends on the water depth and describes the propagation geometry. Path loss that occurs due to attenuation occurring in a single path over a distance d using signal frequency f can be calculated as follows:

$$A(d,f) = \left(\frac{d}{d_r}\right)^k \alpha(f)^{(d-d_r)} \tag{11.5}$$

where d_r is the distance reference. For long-distance propagation through multipath, acoustic signals may face refraction resulting from the underwater spatial variability of sound speed in deep water. For shallow water, as it is considered a general case, temperature, salinity, and pressure are the most important parameters that should be considered in calculations. The velocity medium can be calculated as follows [68]:

$$C = 1412 + 3.21T + 1.19S + 0.0167z \tag{11.6}$$

where S represents the salinity (in parts/1000), T is the temperature by (C), and z represents the depth in meter. For optical signals, propagation in underwater is affected by the absorption and scattering. The extinction coefficient $c(\lambda)$ is represented by the summation of absorption coefficient $a(\lambda)$ and scattering coefficient $b(\lambda)$. The scattering coefficient can be calculated as follows [69]:

$$b = 2\pi \int \beta(\theta) \sin(\theta) d\theta \tag{11.7}$$

where $\beta(\theta)$ represents the volume scattering function (VSF) defined as the fraction of light intensity deviated from bean due to scattering within the range of solid angle centered on scattering angle θ. For radio frequency signals (RF), the mathematical model is depending on wavelength, dielectric constant, and path loss. Measurements of underwear channel for electromagnetic signals can be calculated only for short-range communications because of high attenuation due to RF signal propagation in underwater [70].

11.8.2 Monte Carlo simulations

In the underwater communications network with respect to the communication link design, special attention should be paid to the properties of the underwater channel. Underwater communication network is important to accurately characterize the underwater channel by taking into account different underwater phenomena impacts. Based on a precise channel model, one can adjust the system parameters appropriately in order to create a high-quality link between the network nodes. The Monte Carlo method can be used to model the impulse response of the underwater wireless communication link [71]. It also can help to understand the impact of underwater environment and uncertainty in prediction and forecasting the channel models which will able to construct an estimated probability density function for a surface or underwater source.

For underwater acoustic communication, a Monte Carlo simulation modeling can be considered as a vertical line array (VLA) by using the underwater acoustic propagation model. To make a Monte Carlo simulation of the underwater communication scenario, it can consider a normal model of an underwater point source located at depth Z_s (see Figure 11.14) to represent the acoustic pressure field in the cylindrical oceanic waveguide [71]. The effect of an underwater ocean or sea pressure against acoustic waves can be calculated by the following equation:

$$P(r,z) = \frac{\sqrt{2\pi}e^{j\pi/4}}{p(Z_s)} \sum_{m=1}^{M} \Phi_m(Z)\Phi_m(Z_s)\frac{e^{jk_m r}}{\sqrt{k_m r}} \tag{11.8}$$

where k_m and Φ_m represent the horizontal wavenumber (normal mode) and the mode function from mth mode, respectively. P is the water density and M is the number of propagation modes. This equation is calculated depending on the passively receiving a target signal using the VLA in the ocean environment.

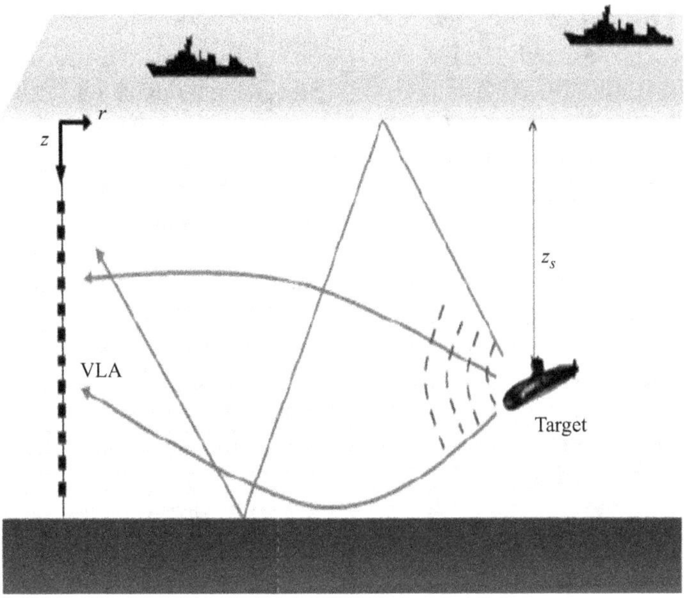

Figure.11.14 Vertical line array (VLA) for passive target signal receiving. Adapted from [71]

In underwater optical communications, Monte Carlo method can help to simulate and track the absorption and scattering of each emitted photon interacted with the medium. By using radiation transfer equation with Monte Carlo methods, it can be easy to characterize and evaluate the optical channel by estimating the impulse response depending on a large number of source photons according to the emitted light intensity distribution. By considering photon position in Cartesian coordinates (x, y, z) and describing the transmission direction by zenith angle θ and azimuth angle φ, in addition to propagation time t and weight by w, the photon's propagation distance d and the photon weight can be calculated by the following equations [72]:

$$d = \frac{-\ln\xi}{k_a + k_s} \tag{11.9}$$

$$w' = w\left(1 - \frac{k_s}{k_a + k_s}\right) \tag{11.10}$$

where ξ is a uniform random variable between 0 and 1. k_s and k_a represent the spectra absorption and scattering coefficient, respectively. Due to the interaction of the optical wave with water particles, the photon will lose a fraction of its initial weight by scattering phenomena.

11.8.3 Hidden Markov model

For underwater spectral features and channel classifiers, the use of a hidden Markov model (HMM) can help in modeling the random fluctuations in underwater wireless signals. It contributes to derive the relationship with the state of past and present of underwater channels which provide an effective estimation of underwater communication signals [73]. For underwater communication channel modeling, it is important to calculate the statistical distribution of channel errors by using a statistical or probabilistic model. Such a model is known as the Fritchman model which can provide a better overall model for channel error distributions and patterns. It helps to identify the good states that represent error-free transmissions and the bad states of transmission where channel error always occurs [74].

In applications for classification of underwater targets, the aspects that are concerned with the development of classification systems should be taken into account in order to increase the performance of communication systems by carefully identifying the features of the sources. Discrete sine transforms (DST) can be used to extract the basic properties of the target from the underwater signal. With the help of a hidden Markov model, sources of noise that affect underwater signals can be classified [74]. For underwater signal classification, the DST based feature set for underwater signals can be modified by Hermite polynomials with the main goal of solving the mystery function. HMM-based classifier system will help to take decisions in order to identify the unknown signal based on previously learned information. Decisions can be done according to HMM classifier modeling by using the training set comprising the extracted features from the underwater signals [75]. Then matching process enables to measure the correspondence between the features extracted from the unknown signal and the system trained estimated data (see Figure 11.15).

The hidden Markov model classifier utilizing the extracted features has to be trained using a set of underwater signals and good success rates have been obtained for the assumed underwater source signals [75]. The process of underwater target classification using this approach involves second degree Hermite polynomial modified DST-based feature extraction, HMM modeling and training, as well as validating the classifier performance in addition to system performance under Gaussian ambient noise conditions.

11.9 Summary

Underwater communications find many applications related to oceanographic data collection, navigation, disaster prevention, surveillance, and tactical and military fields. Technology such as acoustic wireless networks will enable these applications but it may encounter difficulties due to the underwater environmental changes which will affect in both signal quality and communication distance. These difficulties directly affect the applications of underwater communications networks, and according to the type of devices used, the fixed underwater sensor nodes networks can be more stable in communication compared to the autonomous

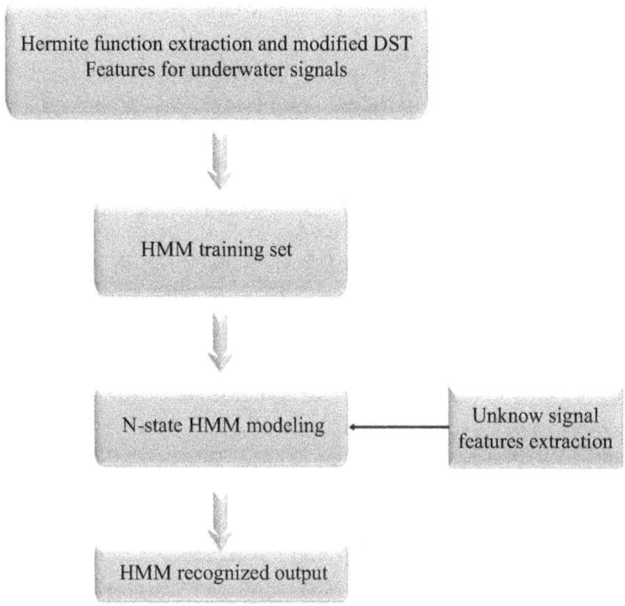

Figure.11.15 Hidden Markov modeling for underwater target classifier. Adapted from [75]

underwater vehicles and robots' applications. Methodologies such as intelligent navigation and control, distributed virtual system, and adaptive network architecture, in addition to Q-learning optimization and deep learning methods can achieve a degree of intelligence in selecting routing paths and provide an intelligent way of data collection and processing. Through these intelligent methods, it is necessary to model underwater communication networks by mathematical methods and simulation methods in addition to statistical or probabilistic model to extract all impacts to the underwater communication channels from different kinds of underwater environments and the possibility of an error occurring during the implementation of the underwater wireless network. With the great advancement in the IoT, the technology of IoUT can provide a world-wide network of smart interconnected underwater objects that enable to monitor vast unexplored water areas. It will show great potential for the future applications such as climate regulation, marine animals' study, monitoring, detection, and unmanned operations.

References

[1] Ranjan, A., and Ranjan, A. "Underwater wireless communication network." *Advance in Electronic and Electric Engineering*. 2013; 3(1): 41–46.

[2] Saeed, N., Celik, A., Al-Naffouri, T. Y., and Alouini, M.-S. "Underwater optical wireless communications, networking, and localization: A survey."

Ad Hoc Networks. 2019; 94: 101935. Available from https://doi.org/10.1016/j.adhoc.2019.101935.

[3] Gussen, C. M. G., Diniz, P., Campos, M., Martins, W., Costa, F., and Gois, J. "A survey of underwater wireless communication technologies." *Journal of Communication and Information Systems*. 2016; 31(1). Available from https://doi.org/10.14209/jcis.2016.22.

[4] Sendra, S., Lloret, J., Rodrigues, J. J. P. C., and Aguiar, J. M. "Underwater wireless communications in freshwater at 2.4 *GHz*." *IEEE Communications Letters*. 2013; 17(9): 1794–1797. doi: 10.1109/LCOMM.2013.072313.131214.

[5] Qureshi, U. M., Shaikh, F. K., Aziz, Z., *et al.* "RF path and absorption loss estimation for underwater wireless sensor networks in different water environments." *Sensors (Basel)*. 2016; 16(6): 890. doi: 10.3390/s16060890.

[6] Kaushal, H. and Kaddoum, G. "Underwater optical wireless communication." *IEEE Access*. 2016; 4: 1518–1547. doi: 10.1109/ACCESS.2016.2552538.

[7] Climent, S., Sanchez, A., Capella, J. V., Meratnia, N., and Serrano, J. J. "Underwater acoustic wireless sensor networks: Advances and future trends in physical, MAC and routing layers." *Sensors*. 2014; 14: 795–833. Available from https://doi.org/10.3390/s140100795.

[8] Saeed, N., Celik, A., Al-Naffouri, T. Y., and Alouini, M.-S. "Energy harvesting hybrid acoustic-optical underwater wireless sensor networks localization." *Sensors*. 2018; 18: 51. Available from https://doi.org/10.3390/s18010051.

[9] Kebkal, K. G., Mashoshin, A. I., and Morozs, N. V. "Solutions for underwater communication and positioning network development." *Gyroscopy and Navigation*. 2019; 10: 161–179. Available from https://doi.org/10.1134/S2075108719030040.

[10] Domingo, M. C. "Overview of channel models for underwater wireless communication networks." *Physical Communication*. 2008; 1(3): 163–182.

[11] Kiran, J. S., Sunitha, L., Rao, D. K., and Sooram, A. "Review on underwater sensor networks: Applications, research challenges and time synchronization." *International Journal of Engineering Research and Technology (IJERT)*. 2015; 4(5): 1329–1332.

[12] Hidemann, J., Milica, S., and Michele, Z. "Underwater sensor networks: Applications, advances and challenges." *Philosophical Transactions of the Royal Society A*. 2012; 370: 158–175. doi:10.1098/rsta.2011.0214.

[13] Lloret, J. "Underwater sensor nodes and networks." *Sensors*. 2013; 13: 11782–11796. doi:10.3390/s130911782

[14] Coutinho, R. W. L., Boukerche, A., Vieira, L. F. M., and Loureiro, A. A. F. "Underwater wireless sensor networks: A new challenge for topology control–based systems." *ACM Computing Surveys*. 2018; 51(1): 36.

[15] Sharif-Yazd, M., Khosravi, M., and Moghimi, M. "A survey on underwater acoustic sensor networks: Perspectives on protocol design for signaling, MAC and routing." *Journal of Computer and Communications*. 2017; 5: 12–23.

[16] Bhambri, H. and Swaroop, A. "Underwater sensor network: Architectures, challenges and applications." *International Conference on "Computing for Sustainable Global Development."* 5th–7th March 2014.

[17] He, M., Chen, Q., Dai, F., and Zheng, X. "Topological configuration and optimization in underwater acoustic sensor networks: A survey." *International Journal of Distributed Sensor Networks*. 2018. doi:10.1177/1550147718795359.

[18] Bahr, A., Leonard, J. J., and Fallon, M. F. "Cooperative localization for autonomous underwater vehicles." *The International Journal of Robotics Research*. 2009; 28(6): 714–728.

[19] Ferri, G., Munafò, A., Tesei, A., *et al*. "Cooperative robotic networks for underwater surveillance: An overview. *IET Radar, Sonar and Navigation*. 2017; 11(12): 1740–1761. doi: 10.1049/iet-rsn.2017.0074.

[20] Fossum, T. O. "Intelligent autonomous underwater vehicles: A review of AUV autonomy and data-driven sample strategies." *Applied Underwater Robotic Laboratory Norwegian University of Science & Technology (NTNU.)*, Trondheim, Norway. 2016.

[21] Munafò, A., Simetti, E., Turetta, A., *et al*. "Autonomous underwater vehicle teams for adaptive ocean sampling: A data-driven approach." *Ocean Dynamics, Springer-Verlag*. 2011; 61: 1981–1994. Available from https://doi.org/10.1007/s10236-011-0464-x.

[22] Kao, C.-C., Lin, Y.-S., Wu, G.-D., and Huang, C.-J. "A comprehensive study on the internet of underwater things: Applications, challenges, and channel models." *Sensors*. 2017; 17: 1477. doi:10.3390/s17071477

[23] Domingo, M. C. "An overview of the internet of underwater things." *Journal of Network and Computer Applications*. 2012; 35: 1879–1890.

[24] Nayyar, A., Ba, C. H., Cong Duc, N. P., and Binh, H. D. *Smart-IoUT 1.0: A smart aquatic monitoring network based on internet of underwater things (IoUT)*. In T. Duong, and N. S. Vo (eds.), *Industrial Networks and Intelligent Systems. INISCOM* 2018. Lecture Notes of the Institute for Computer Sciences, Social Informatics and Telecommunications Engineering, vol. 257. Springer, Cham.

[25] Qu, J., Zhang, Z., Cui, Y., Wang, J., and Mastorakis, G. "Research and application of multi-node communication and energy consumption prediction control in underwater acoustic network." *IEEE Access*. 2019; 7. DOI: 10.1109/ACCESS.2019.2907376.

[26] Mavromoustakis, C. X., Batalla, J. M., Mastorakis, G., Markakis, E., and Pallis, E. "Socially oriented edge computing for energy awareness in IoT architectures." *IEEE Communications Magazine*. 2018; 56(7): 139–145. doi: 10.1109/MCOM.2018.1700600.

[27] Goşman, C., Cornea, T., Dobre, C., Mavromoustakis, C. X., and Mastorakis, G. "Secure model to share data in intelligent transportation systems." *18th Mediterranean Electrotechnical Conference (MELECON)*; Lemesos, 2016, pp. 1–8. doi: 10.1109/MELCON.2016.7495426.

[28] Hassan, M. B., Ali, E. S., Mokhtar, R. A., Saeed, R. A., and Chaudhari, B. S. "NB-IoT: Concepts, applications, and deployment challenges". In B. S. Chaudhari, and M. Zennaro (eds.), *LPWAN technologies for IoT and M2M applications*. Elsevier; 2020. Chapter 6.

[29] Papadokostaki, K., Mastorakis, G., Panagiotakis, S., Mavromoustaki, C. X., Dobre, C., and Batalla, J. M. "Handling big data in the era of Internet of things (IoT)." In C. Mavromoustakis, G. Mastorakis, and C. Dobre (eds.), *Advances in Mobile Cloud Computing and Big Data in the 5G Era. Studies in Big Data*, vol. 22. Springer, Cham; 2017.

[30] Hadjioannou, V., Mavromoustakis, C. X., Mastorakis, G., *et al.* "Security in smart grids and smart spaces for smooth IoT deployment in 5G." In C. Mavromoustakis, G. Mastorakis, and J. Batalla (eds.), *Internet of Things (IoT) in 5G Mobile Technologies. Modeling and Optimization in Science and Technologies*, vol. 8. Springer, Cham; 2016.

[31] Hollinger, G. A., Choudhary, S., and Qarabaqi, P., *et al.* "Underwater data collection using robotic sensor networks." *IEEE Journal on Selected Areas in Communications*. 2012; 30(5): 899–911. doi: 10.1109/JSAC.2012.120606.

[32] Markakis, E., Mastorakis, G., Negru, D., Pallis, E., Mavromoustakis, C. X., and Bourdena, A. "A context-aware system for efficient peer-to-peer content provision." In C. Dobre, and F. Xhafa (eds.), *Intelligent Data-Centric Systems, Pervasive Computing*, Academic Press; 2016, Pages 131–151.

[33] Signori, A., Campagnaro, F., Steinmetz, F., Renner, B.-C., and Zorzi, M. "Data gathering from a multimodal dense underwater acoustic sensor network deployed in shallow fresh water scenarios." *Journal of Sensor and Actuator Networks*. 2019; 8: 55.

[34] Khan, J. U. and Cho, H.-S. "A distributed data-gathering protocol using AUV in underwater sensor networks." *Sensors (Basel)*. 2015; 15(8): 19331–19350. doi:10.3390/s150819331.

[35] Huang, X., Sun, S., and Yang, Q. "Data uploading strategy for underwater wireless sensor networks." *Sensors (Basel)*. 2019; 19(23): 5265. doi:10.3390/s19235265.

[36] Mavromoustakis, C. X., Mastorakis, G., Papadakis, S., Andreou, A., Bourdena, A., and Stratakis, D. "Energy consumption optimization through pre-scheduled opportunistic offloading in wireless devices." *Sixth International Conference on Emerging Network Intelligence, EMERGING*. 2014; 22–28.

[37] Liu, S., Xu, H., Lin, Y., and Gao, L. "Visual navigation for recovering an AUV by another AUV in shallow water." *Sensors (Basel)*. 2019; 19: 1889. doi:10.3390/s19081889.

[38] Li, B., Xu, Yx., Fan, Ss., *et al.* "Underwater docking of an under-actuated autonomous underwater vehicle: system design and control implementation." *Frontiers of Information Technology and Electronic Engineering*. 2018; 19: 1024–1041. Available from https://doi.org/10.1631/FITEE.1700382.

[39] Lin, R., Li, D., Zhang, T., *et al.* "A non-contact docking system for charging and recovering autonomous underwater vehicle." *Journal of Marine Science and Technology*. 2019; 24: 902–916. Available from https://doi.org/10.1007/s00773-018-0595-6.

[40] Yazdani, A. M., Sammut, K., Yakimenko, O., and Lammas, A. "A survey of underwater docking guidance systems." *Robotics and Autonomous Systems*. 2020; 124: 103382. Available from https://doi.org/10.1016/j.robot.2019.103382.

[41] Liu, S., Ozay, M., Okatani T., Xu, H., Sun, K., and Lin, Y. "A vision based system for underwater docking." *arXiv:1712. 04138*; 2017.
[42] Xi, Z., Kan, X., Cao, L., Liu, H., Manogaran, G., and Mastorakis, G. "Research on underwater wireless sensor network and MAC protocol and location algorithm." *IEEE Access*. 2019; 7: 56606–56616. doi: 10.1109/ACCESS.2019.2901375.
[43] Zhao, J. "Neural network predictive control for autonomous underwater vehicle with input delay." *Journal of Control Science and Engineering*. 2018; 2018: 8. Available from https://doi.org/10.1155/2018/2316957.
[44] Wang, C., Wei, L., Wang, Z., Song, M., and Mahmoudian, N. "Reinforcement learning-based multi-AUV adaptive trajectory planning for under-ice field estimation." *Sensors (Basel)*. 2018; 18: 3859. Available from https://doi.org/10.3390/s18113859.
[45] Centelles, D., Soriano-Asensi, A., Martí, J. V., Marín, R., and Sanz, P. J. "Underwater wireless communications for cooperative robotics with UWSim-NET." *Applied Science*. 2019; 9: 3526. Available from https://doi.org/10.3390/app9173526.
[46] Gajewski, M., Latoszek, W., Batalla, J. M., Mastorakis, G., Mavromoustakis, C. X., and Pallis, E. "ID-based communication for access to sensor nodes." In J. Batalla, G. Mastorakis, C. Mavromoustakis, and E. Pallis (eds.), *Beyond the internet of things*. Springer; 2017.
[47] Hu, C., Fu, L., and Yang, Y. "Cooperative navigation and control for surface-underwater autonomous marine vehicle." *IEEE 2nd Information Technology, Networking, Electronic and Automation Control Conference (ITNEC)*, Chengdu; 2017, pp. 589–592. doi: 0.1109/ITNEC.2017.8284801.
[48] Yu, X., Chen, W.-N., Gu, T., Yuan, H., Zhang, H., and Zhang, J. "ACO-A*: Ant colony optimization plus A* for 3D traveling in environments with dense obstacles." *IEEE Transactions on Evolutionary Computation*. 2019; 23(4).
[49] Khan, Q., Asger, M., and Muheet Ahmed, B. "Improving information integrity using artificial bee colony based intrusion detection system." *International Journal of Computer Applications*. 2015; 130(6): 12–23.
[50] Patle, B. K., Ganesh Babu, L., Anish, P., Parhi, D. R. K., and Jagadeesh, A. "A review: On path planning strategies for navigation of mobile robot". *Defence Technology*. 2019; 15(4): 582–606.
[51] Yao, T. T., He, T., Zhao, W. L., and M. Sani, A. Y. "Review of path planning for autonomous underwater vehicles." *Proceedings of the 2019 International Conference on Robotics, Intelligent Control and Artificial Intelligence (RICAI 2019)*. Association for Computing Machinery, New York, USA, 482–487.
[52] Jin, Z., Ma, Y., Su, Y., Li, S., and Fu, X. "A Q-learning-based delay-aware routing algorithm to extend the lifetime of underwater sensor networks." *Sensors (Basel)*. 2017; 17(7): 1660. doi:10.3390/s17071660.
[53] Wang, S. and Shin, Y. "Efficient routing protocol based on reinforcement learning for magnetic induction underwater sensor networks." *IEEE Access*. 2019; 7: 82027–82037. doi: 10.1109/ACCESS.2019.2923425.

[54] Akyildiz, I. F., Wang, P., and Lin, S.-C. "Soft water: Software-defined networking for next generation underwater communication systems." *Ad Hoc Networks*. 2016; 46: 1–11. Available from https://doi.org/10.1016/j.adhoc.2016.02.016.

[55] Luo, H., Wu, K., Ruby, R., Liang, Y., Guo, Z., and Ni, L. M. "Software-defined architectures and technologies for underwater wireless sensor networks: A survey." *IEEE Communications Surveys and Tutorials*. 2018; 20(4): 2855–2888.

[56] Wu, L., Trezzo, J., Mirza, D., *et al*. "Designing an adaptive acoustic modem for underwater sensor networks." *IEEE Embedded Systems Letters*. 2012; 4(1): 1–4. doi: 10.1109/LES.2011.2180013.

[57] Sofi, S. A. and Mir, R. N. "Natural algorithm based adaptive architecture for underwater wireless sensor networks." *International Conference on Wireless Communications, Signal Processing and Networking (WiSPNET)*; Chennai, 2017, pp. 2343–2346. doi:10.1109/WiSPNET.2017.8300179.

[58] Saeed, R. A. and Ahmed, E. S. A. "Big data distributed systems management." In S. Yu, X. Lin, J. Misic, and X. S. Shen (eds.), *Networking for big data*; Boca Raton, FL: Chapman and Hall/CRC. 2016, Chapter 4, pp. 57–71.

[59] Mavromoustakis, C. X., Mastorakis, G., and Batalla, J. M. *Internet of things (IoT) in 5G mobile technologies*. Springer; 2016.

[60] Malde, K., Handegard, N. O., Eikvil, L., and Salberg, A.-B. "Machine intelligence and the data-driven future of marine science." *ICES Journal of Marine Science*. 2020; 77(4): 1274–1285.

[61] Pramunendar, R. A., Wibirama, S., Santosa, P. I., Andono, P. N., and Soeleman, M. "A robust image enhancement techniques for underwater fish classification in marine environment." *International Journal of Intelligent Engineering and Systems*. 2019; 12: 116–129.

[62] Papadakis, S. E., Stykas, V. A., Mastorakis, G., and Mavromoustakis, C. X. "A hyper-box approach using relational databases for large scale machine learning." *International Conference on Telecommunications and Multimedia (TEMU)*; Heraklion, 2014, pp. 69–73. doi:10.1109/TEMU.2014.6917738.

[63] Salman, A., Siddiqui, S. A., Shafait, F., *et al*. "Automatic fish detection in underwater videos by a deep neural network-based hybrid motion learning system." *ICES Journal of Marine Science*. 2019; 77(4): 1295–1307. doi:10.1093/icesjms/fsz025.

[64] Ullah, I., Chen, J., Su, X., Esposito, C., and Choi, C. "Localization and detection of targets in underwater wireless sensor using distance and angle based algorithms." *IEEE Access*. 2019; 7: 45693–45704. doi: 10.1109/ACCESS.2019.2909133.

[65] Parras, J., Zazo, S., Pérez-Álvarez, I. A., and Sanz González, J. L. "Model free localization with deep neural architectures by means of an underwater WSN." *Sensors*. 2019; 19: 3530. Available from https://doi.org/10.3390/s19163530.

[66] Wang, X., Liu, A., Zhang, Y., and Xue, F. "Underwater acoustic target recognition: A combination of multi-dimensional fusion features and modified

deep neural network." *Remote Sensing*. 2019; 11: 1888. Available from https://doi.org/10.3390/rs11161888.

[67] Ke, X., Yuan, F., and Cheng, E. "Underwater acoustic target recognition based on supervised feature-separation algorithm." *Sensors*. 2018; 18: 4318. doi:10.3390/s18124318.

[68] Jouhari, M., Ibrahimi, K., Tembine, H., and Ben-Othman, J. "Underwater wireless sensor networks: A survey on enabling technologies, localization protocols, and internet of underwater things." *IEEE Access*. 2019. doi: 10.1109/ACCESS.2019.2928876

[69] Dong, Y., Zhang, H., and Zhang, X. "On impulse response modeling for underwater wireless optical MIMO links." *IEEE/CIC ICCC 2014 Symposium on Optical Communication Systems and Networks.*

[70] Singh, G. and Kumar, M. *Channel modeling of underwater wireless sensor network using EM waves*. JETIR; 2019.

[71] Choi, J., Choo, Y., and Lee, K. "Article acoustic classification of surface and underwater vessels in the ocean using supervised machine learning." *Sensors*. 2019; 19: 3492. doi:10.3390/s19163492

[72] Dong, F., Xu, L., Jiang, D., and Zhang, T. "Monte Carlo-based impulse response modeling for underwater wireless optical communication." *Progress in Electromagnetics Research M*. 2017; 54: 137–144.

[73] Mohammed, S. K., Supriya, M. H., and Suraj, K. "A GTCC-based underwater HMM target classifier with fading channel compensation." *Hindawi Journal of Sensors*. 2018; 14. Article ID6593037.

[74] Holmes, D. G., Cheng, L., Shimaponda-Nawa, M., Familua, A. D., and Abu-Mahfouz, A. M. "Modelling noise and pulse width modulation interference in indoor visible light communication channels." *AEUE – International Journal of Electronics and Communications*. 2019; 106: 40–47.

[75] Binesh, T., Mohankumar, K., Supriya, M. H., and Saseendran Pillai, P.R. "Underwater target classifier using a modified transform based feature set." *International Journal of Electronics and Communication Engineering*. 2013; 2(3): 21–32.

[76] Nikoloudakis, Y., Panagiotakis, S., Markakis, E., Pallis, E., and Mastorakis, G. "A fog-based emergency system for smart enhanced living environments." *IEEE Cloud Computing*. 2016; 3(6): 54–62. doi: 10.1109/MCC.2016.118.

Chapter 12

Machine learning algorithms for smart data analysis in the Internet of things: an overview

Mohammed H. Alsharif[1], Anabi Hilary Kelechi[2], Imran Khan[3], Mahmoud A. Albreem[4], Abu Jahid[5], Constandinos X. Mavromoustakis[6], George Mastorakis[7], Evangelos Pallis[8] and Evangelos K. Markakis[8]

Machine learning (ML) techniques will benefit immensely from the avalanche of data readily available from various (IoT) applications considered as the major contributor of new data for future intelligent network. Based on this new concept, network systems will further magnify their capacity to exploit variety of experimental data across a plethora of network devices, study the data information, obtain knowledge and make informed decisions based on the dataset at their disposal. Smart IoT data analysis are performed utilizing supervised learning, unsupervised learning and reinforced learning. This study is limited to supervised and unsupervised ML techniques. In other to achieve the set objectives, reviews and discussions of substantial issues related to supervised or unsupervised machine learning techniques were executed, highlighting the advantages and limitations of each algorithm as well presenting the recent research trends and recommendations for future study.

12.1 Introduction

It is obvious that the paradigm shift towards smart technology will snowballed into the generation of large quantities of various data network system leading to big data era.

[1]Department of Electrical Engineering, Sejong University, Seoul, South Korea
[2]Department of Aerospace Engineering, Air Force Institute of Technology (AFIT), Kaduna, Nigeria
[3]Department of Electrical Engineering, University of Engineering and Technology Peshawar, Khyber Pakhtunkhwa, Pakistan
[4]Department of Electronics and Communications Engineering, A'Sharqiyah University, Ibra, Oman
[5]Department of Electrical and Computer Engineering, University of Ottawa, Ottawa, ON, Canada
[6]Department of Computer Science, University of Nicosia, Nicosia, Cyprus
[7]Department of Management Science and Technology, Hellenic Mediterranean University, Heraklion, Greece
[8]Department of Electrical and Computer Engineering, Hellenic Mediterranean University, Heraklion, Greece

Big data is defined and characterized by the five Vs: large-volume, high-velocity, large-variety, high-veracity, and large-value. To make meaningful insight from these avalanche of raw data, there must be a well-defined mechanism to extract these data, process them via process automation and utilized these processed data for problem-solving tasks [1]. The Internet-of-things (IoT) is a veritable platform for the creation and generation of novel datasets from various human sensory attributes of hearing, vision, and thinking. With the upcoming tactile Internet which can extract and process human sensory characteristics in real time, the concept of machine-to-machine (M2M) interface will be developed for exchange of mandatory data and execution of intelligent decisions with little human intervention [2]. The system obtains useful information from the raw data which are generated from different locations in the network. Machine learning (ML) is not a domain specific technology and hence, can be applied to diverse domains such as data models, data types, and probabilistic datasets [3]. The non- domain specific attribute of ML has endeared it to lots of applications and hence, a valuable tool for the actualization of smart IoT concept [4].

ML cannot be executed without artificial intelligence (AI) which is applied to machines to enable them to learn the ongoing process [5]. This learning is an important part of the ML functionality without which the task cannot be completed because it plays the role of input function mapping to determine the features of the raw data in the model. The learning scheme is also called the training model or training set and it is categorized into three groups and depicted in Figure 12.1 [6]:

1. Supervised learning: In supervised learning, algorithms work with "labeled" data to determine a function which assigns the data as input and assigns them the appropriate output label. It predicts the output value to a new variable by the "data history" lookup mechanism. For example, a spam detector analyzes the message history, seeing what function it can represent, according to the input parameters that are defined (the sender, if the recipient is an individual or part of a list if the subject contains certain terms, etc.), assigning the label "spam" or "not spam." Once this function is defined, when introducing a new

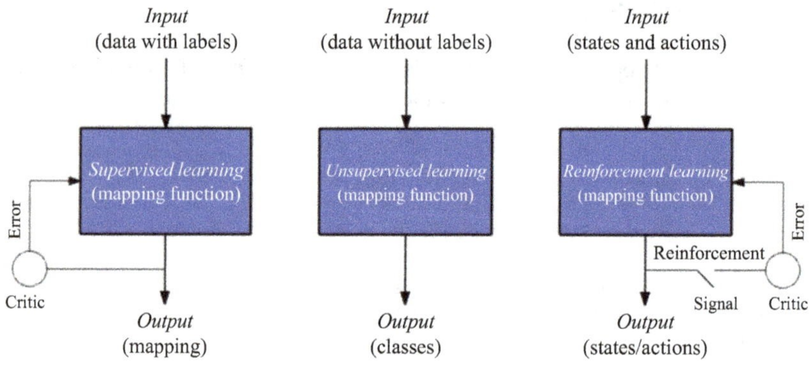

Figure 12.1 Classification of learning models

untagged message, the algorithm can assign it the correct label. The major applications are in regression problems and classification problems and these two are distinguished by the type of target variable [5].

2. Unsupervised learning: It is deployed when "labeled" data are not available for training and only the input data are known, but there is no output data that correspond to certain input and only can determine the data structure. Therefore, they are exploratory in nature. For example, clustering tasks look for groupings based on similarities, but nothing guarantees that these have any meaning or utility. Sometimes when exploring the data without a definite goal, you can find curious but impractical spurious correlations. Unsupervised learning is often used in clustering problems, cooccurrence groupings, or profiling. However, problems that involve finding similarity, predicting links or reducing data, may or may not be monitored [7].
3. Reinforcement learning: Not all ML algorithms can be classified as supervised or unsupervised learning algorithms. There is a "no man's land" which is where reinforcement learning techniques fit in. This type of learning is based on improving the response of the model using a feedback process. The algorithm learns from the surrounding environment by trial and error. It is not a type of supervised learning, because it is not based strictly on a set of labeled data, but on the monitoring of the response to the actions taken. It is not unsupervised learning, since when we model our "apprentice," we know in advance what the expected reward is [8].

In this work, we deploy supervised and unsupervised learning algorithms which are effective for the data analysis of smart IoT applications [5]. The key contributions of this work are to provide a detailed related literature analysis on supervised and unsupervised schemes. The analysis is performed to obtain an accurate, concise, and concrete conclusion based on the subdomains as well as advantages and limitations. The proposed study also highlighted the ongoing research issues in smart IoT data.

The rest of this chapter is organized as follows. Section 12.2 discusses the taxonomies of supervised and unsupervised machine learning techniques based on their respective subdomains with their advantages and limitations. Section 12.3 reviews the research trends and open issues. Section 12.4 elaborates the conclusions and recommendations.

12.2 Taxonomies of supervised and unsupervised ML algorithms

In practice, the ML deploys supervised learning as the datasets are called "true" datasets or "correct" datasets in supervised learning and such input datasets are used to "train" the learning algorithm. The learning process is performed until an acceptable accuracy is achieved [9]. A cost function is generated, adjusted, and satisfied by ML scheme which quantifies the error between the calculations and the "ground truth." ML aims to reduce the cost function value which directly impacts the accuracy improvement and the minimization is performed through a gradient descent method [10]. Various gradient descent methods are deployed to train the

ML schemes such as momentum-based gradient descent, Nesterov accelerated gradient descent, and stochastic gradient descent [11]. To elaborate it further, consider an example in which 'm' denotes the number of training and each training can be represented in a pair (x, y), where x represents the data input and y denotes the label for the class identifier. The dimension of data input x is n, while the individual dimension links to a specific variable or an explicit feature. In this example, the ML scheme is aligned with a specific sensor system embedded in the program to operate the IoT application [12]. The supervised learning problems can be further categorized into regression and classification [10].

The summary of supervised and unsupervised ML schemes is given in Figure 12.2. In Sections 12.2.1 and 12.2.2, a detailed discussion is presented.

12.2.1 Supervised ML algorithm

12.2.1.1 Classification tasks

Classification tasks are primarily concerned with the issue of assigning data classes to their distinct and unique labels [13]. In order to execute the classification task, one out of the numerous classification algorithms are utilized and hence are discussed below:

K-nearest neighbors

K-nearest neighbors (KNN) algorithm is one of the extensively deployed supervised ML algorithm capable of handling both classification and regression

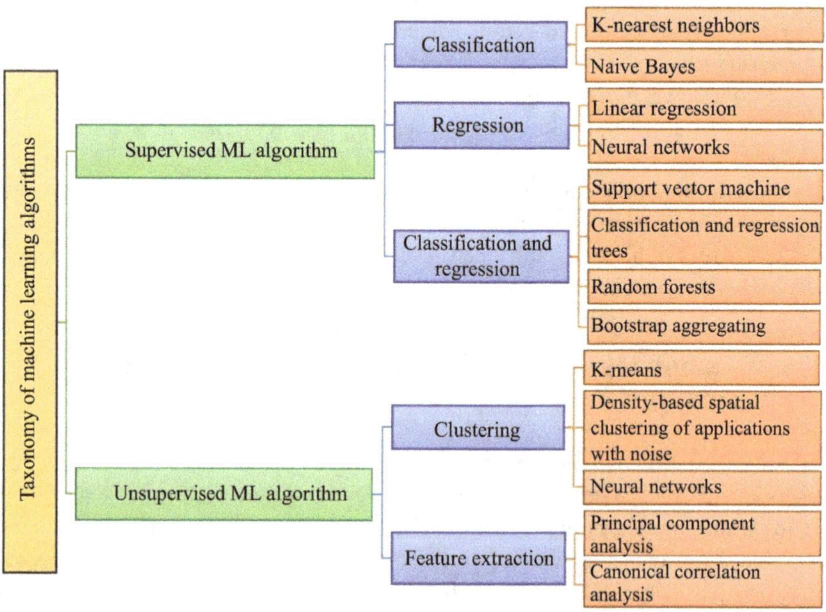

Figure 12.2 Summarized taxonomy of supervised and unsupervised ML algorithms

problems. The algorithm is designed to address both classification and regressions problems but studies have indicated that deployability has tilted toward solving mostly classification problems [14]. The key drivers of the algorithm are as follows: (1) output results are easy to analyze and interpret, (2) reduced computation time, and (3) predictive capability. The algorithm has reduced computational complexity, and the results can be easily inferred and interpreted with relative computation time. In jobs requiring classification and regression outcomes, it is desirable to have the input dataset to be k closest to the training datasets as obtained in the featured set. The output is dependent on whether KNN is deployed to function as a classification or regression algorithm: (1) when it is deployed for the classification task, the ensuing outcome reduces to mere class membership tasks [5]. The voting right is reserved for the members who choose their neighbors and subsequently classify them accordingly as their k nearest neighbor. Evidently, k must not be a negative number. Interestingly, there will be occasion that the cardinality of k will be $k = 1$. It is tractable that the object is projected to a single available neighborhood function. Then, when the KNN algorithm is deployed to solve regression problems, the objective function reduces to finding the average number of k's nearest neighbors. The underlying principle behind the algorithm is based on distance metrics. Many distance metrics are often deployed in locating the k of a data point. The notable distance metrics include: Euclidean distance, $L\infty$ norm, angle, Mahalanobis distance, or Hamming distance [15,16]. Consider a typical KNN model as depicted in Figure 12.3. For $k = 3$, in this instance, the test point (star) falls into the class B domain whereas, for $k = 6$, the point falls into to class A domain. In the scenario being considered, KNN is a nonprobabilistic and nonparametric model [17]. As a rule of thumb, KNN is considered an optimal classification algorithm to use when there is sparse information on the data distribution. In the case being discussed, KNN provides all the labeled input points. When there is no label attached to a dataset, how can this dataset be manipulated? It

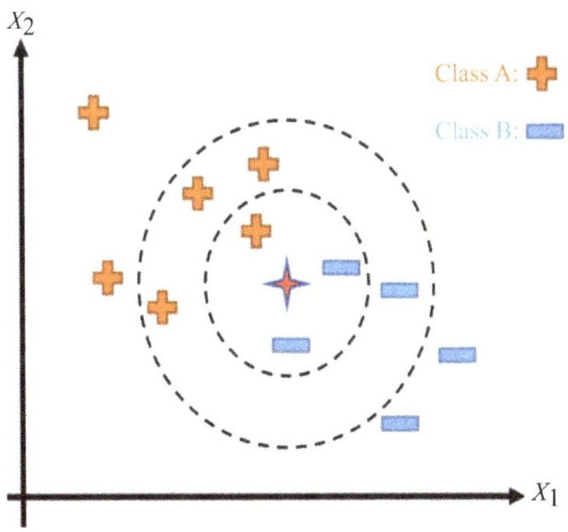

Figure 12.3 A simple KNN model for different values of k

is expected that solving this problem will result in heavy computational penalties. The best approach to tackle this type of classification issue is through the deployment of a distance metric otherwise known as a similarity measure. Hence, any unknown sample is subjected to k nearest neighbors' votes who classify the data sample by simply majority vote. It is expected as the complexity increases, the dimension cardinality also increases. Hence, the major problem of KNN reduces to find a novel scheme to reduce the surging dimensionality issue [18]. Inadvertently, this must be considered before using the KNN algorithm. Ideally, strategies must be implemented to reduce occasions that will decrease data dimensionality. It is on record that KNN has been deployed on stress detection in human physiological signals [19] as well as in detecting seizure activity in epilepsy patience [20].

Based on Figure 12.3, the problem can be reformulated as the input dataset be denoted as let x, with the K nearest neighbors be described as $Nk(x)$. Then, the estimated class label identifier for x be denoted as y, with the class variable using unassigned variable t. In addition, 1(.) can indicate the attribute: 1(s) = 1 if s is true and 1(s) = 0 conversely [5]:

$$p(t = c|x, K) = \frac{1}{K} \sum_{i \in N_k(x)} 1(t_i = c)$$

$$y = \max\nolimits_c p(t = c|x, K) \tag{12.1}$$

Irrespective of the numerous advantages ensuing from the KNN algorithm, notably the absence of training time permits unseen data to be introduced effortlessly without adverse effects on the algorithm accuracy. The major limitation of the KNN algorithm is the need for large storage memory to warehouse the training set data. Expectedly, this has been a clock in the wheel for the widespread adoption of KNN more specifically in the face of high dataset dimensionality. Hence, it is difficult for the algorithm to determine the distance between the dimensions. Moreover, the KNN is highly responsive to an aberration in the dataset [21]. Missing data and outliers are also another impediment that must be taken care of. To overcome the shortcoming of a large dataset, a tree-based search consisting of one-off computation has been proposed [22]. Similarly, the authors in [23] proposed a scheme that can acquire numerous parameters based on robust and specific KNN classifier. Other authors, [24] conversely, proposed a novel classification scheme driven by rough-set-algorithm to understand travel pattern regularities has become a necessity. Based on the flip side of the present KNN algorithm, there have been efforts to enhance the existing algorithm with the emergency of the wavelet-based KNN partial distance search (WKPDS) algorithm [25], equal-average equal-norm nearest neighbor code-word search (EENNS) algorithm, and the equal-average equal-variance equal-norm nearest neighbor search (EEENNS) algorithm [26].

Naive Bayes

Exploiting the rich mathematical attributes of probability theorem, naive Bayes classifier was formulated. The algorithm is based on Bayes' theorem and naive assumption that data attributes are statistically independent and uncorrelated. Hence, the term "naïve" because there is no solid knowledge about the intentions of the other attributes in the dataset. Therefore, given a virgin experimented data point (input vector) $x = (x_1, \ldots, x_M)$, the goal is to choose an algorithm that predicts the projected outcome with minimal error. Consequently, naive Bayes classifiers adopt a postulation of choice when independence among the predictors are desired. This implies that given a set of features in a class, the outcome is totally independent of the other features. Essentially, the likelihood of inferring more information about other variables in the presence of additional variables is very minimal. The algorithm easily translates a word problem into a straightforward statistical problem and solves it using established probability principles. Using Bayes' theorem, the form can be expressed as follows [27]:

$$p(t = c|x_1, \ldots, x_M) = \frac{p(x_1, \ldots, x_M|t = c)p(t = c)}{p(x_1, \ldots, x_M)} \tag{12.2}$$

Invoking the naive independence model concept and after some simplifications, the result is:

$$p(t = c|x_1, \ldots, x_M) \propto p(t = c)\prod_{j=1}^{M} p(x_j|t = c) \tag{12.3}$$

The form of the classification task can be expressed as follows [28]:

$$y = \max_c p(t = c)\prod_{j=1}^{M} p(x_j|t = c) \tag{12.4}$$

where y connotes the estimated class identifier for x. Various naive Bayes classifiers adopt various schemes and distributions to forecast $p(t = c)$ and $p(xj|t = c)$.

Unlike the other discussed classifier algorithm, naive Bayes classifier is not a data-hungry algorithm and hence, it can address the issue of data reduction dimensionality issue adequately, thus, making the algorithm highly robust and dynamic [29]. Motivated by these advantages over other classifiers, the algorithm has been deployed to solve numerous real-world problems such as spam filtering [28], text categorization, and automatic medical diagnosis [30]. The algorithm has also been deployed in addressing trust value and calculating the last numerical trust value of the farm produce [31]. Despite the benefits that can be achieved by this classifier, the major flip side of the algorithm is independent predictors' assumption and zero frequency issue and is generally regarded as a bad estimator. Practically, it is almost infeasible to have a set of predictors which are completely independent [32]. The issue of zero frequency arises when some of the data are not detected by the training dataset. Hence, it will be assigned a zero frequency. This flip side has been addressed via the Laplacian estimation [33].

12.2.1.2 Regression tasks

Classification is deployed when the required outcome must take a discrete value or mode of "yes or no" whereas regression models are needed when the prediction result must take a continuous value. In the following, we present detailed discussions on the various regression categories and fundamental concepts.

Linear regression

Linear regression is another class of ML algorithm suitable for performing supervised learning in which the goal is to understand the intrinsic relationship between a dependent variable and a single or numerous variables. Regression models are designed to generate prediction values by making use of the independent variables. The model is based on fitting a linear equation to a set of experimental observed data [34]. From a modeling and compact representation perspective, linear regression can be denoted as a function $f(x,w)$. Thus, leading to plotting the following: f: $\phi(x) \rightarrow y$. This is the linear amalgamation of a set of fixed linear or nonlinear functions from the input variable. This can be symbolized as the basic function $\phi_i(x)$ [27]:

$$f(x, w) = \phi(x)^T w \tag{12.5}$$

With w signifying the weight vector (i.e., matrix), the equation would be conveyed as $w = (w_1, \ldots, w_D)^T$, and $\phi = (\phi_1, \ldots, \phi_D)^T$. A broad range of basic functions exist to assist in creating this application. For example, polynomial, Gaussian, radial, or sigmoidal basic functions could be used in this application [35].

The major issue with the algorithm is selecting the ideal training algorithm as there exist many variants. Several approaches are available: ordinary least square, regularized least squares, least-mean-squares (LMS), and Bayesian linear regression. The LMS scheme is the most desirable because it has fast execution time and performs optimally in the presence of big data set. The algorithm has the capability to acquire the parameter attributes via the Internet by using stochastic gradient descent (sequential gradient descent) [36]. Using the appropriate basic function, random nonlinearities in the mapping from input variable to output variable can be identified. However, the use of fixed basis functions can lead to significant shortcomings (e.g., an upsurge in input space dimensionality leads to a precipitous increase in the cardinality of the fundamental functions) [37]. Linear regression algorithms have a high execution rate [38]. For example, this algorithm is adept at analyzing and predicting buildings' energy usage.

In contrast, neural networks are effective in addressing certain fundamental functional issues as well as permitting the model to acquire the system parameters of the fundamental functionality. In addition, neural networks have high computational capability in the face of novel data, due to their compact nature. Additionally, they are properly tuned to solve regression and classification tasks. The flip side of the algorithm is the stringent requirement for a huge experimental training dataset to train the model [39]. Neural network has gained traction in recent times because of the wide acceptance and high accuracy. Hence, there are various neural network architectures desirably for unique and specific use case scenarios.

12.2.1.3 Combining classification and regression tasks

Support vector machine

Classical support vector machine (SVM) is a support-vector network that can be utilized with supervised learning models. This model is a nonprobabilistic, binary classifier that can be used to identify the hyperplane that divides classes of the training set. This provides a maximized margin. The predicted label of a previously unobserved data point can be determined by the side of the hyperplane on which it falls [40]. The major attraction of SWM is with few training points, a high degree of accuracy is ensured. These training points are support vectors that can categorize any novel data point in the network. SVMs not only perform binary classification, they are also able to do multiclass classification. Four such models are all-vs-all (AVA) SVM, one-vs-all (OVA) SVM, structured SVM [41], and the Weston and Watkins version [42]. Besides linear classification, SVMs can perform nonlinear classification. This can be useful for finding the hyperplane of a nonlinear functioning input variable. For example, an input variable can be mapped into a high-dimensional feature space. This process is referred to as a kernel trick [43]. To design this task, identify the typical vector of the hyperplane as w and the parameter for controlling the offset of the hyperplane as b. To safeguard that, SVM will be able to control for outliers in the data, a variable ε_i can be introduced for every training point x_i. This is a slack variable that determines the distance that the training point encroached upon the margin in units of $|w|$. In this example, a binary linear classification task can be designated as a constrained optimization problem in the following manner [44]:

$$\min_{w,b,\varepsilon} f(w,b,\varepsilon) = \frac{1}{2}w^T w + C\sum_{i=1}^{n} \varepsilon_i \tag{12.6}$$

subject to $y_i(w^T x_i + b) - 1 + \varepsilon_i \geq 0$ $i = 1, \ldots n$; $\varepsilon_i \geq 0$.

Where parameter $C > 0$ determines how heavily a violation is punished. Furthermore, the parameter C is a hyperparameter whose choices are implemented either from cross-validation or Bayesian optimization. There exist various strategies to address the constrained optimization problem in (12.6). P-pack SVM [45], quadratic programming optimization [46], and sequential minimal optimization [47] are techniques that can be applied to this problem. SVM is an excellent supervised learning model that can efficiently address high dimensional data sets. It is particularly effective in addressing memory usage because it utilizes support vectors to facilitate prediction. However, this model has a significant drawback in that it lacks the technique to directly stipulate probability estimates. SVM is very suitable in numerous practical applications including hand-written identification problem [48], image recognition [49], and protein arrangement [50]. Finally, it is possible to train SVMs in an online fashion as discussed in [51]. The authors in [52] suggested a technique using Intel Lab Dataset, data set composed of four basic environmental attributes of (temperature, voltage, humidity, and light) obtained via S4 Mica2Dot sensors. The authors in [53] applied SVM to classify traffic data.

Classification and regression trees (CART)

Classification and regression trees (CART) is a fast training algorithm that has found applicability in classifying smart citizen behaviors [54]. Though some of the already discussed algorithms are deployed in modeling machine learning, decision trees are unique. In the family of the classical decision tree and its variant algorithms, random forest is among the popular approaches in use. In CART algorithm, the input domain is divided into bloc-aligned cuboid sections R_k, and then a distinct classification or regression scheme is applied to each section to estimate the character of the data points located in that section [55]. Presented with a novel untested input experimental vector (data point) x, the goal of estimating the appropriate target attribute could be described as a binary tree mechanism which corresponds to a successive decision-making approach. The primary purpose of a classification algorithm is to predict a specific attribute for a given bloc, while regression algorithm focuses on predicting a constant for each bloc. Mathematically, the classification task is formulated to recognize an attribute variable using a noncontinuous random variable t and the estimated attribute identifier for x by y. The classification task is denoted as follows [27]:

$$p(t = c|k) = \frac{1}{|R_k|}\sum_{i\in R_k} 1(t_i = c)$$

$$y = \max_c p\,(t = c|x) = \max_c p(t = c|k) \tag{12.7}$$

Equation (12.7) connotes that it will be tagged by the most significant mode in its appropriate bloc [27].

Similarly, to model the regression task, let us represent the output vector by a number t and the estimated output vector for x by y. Then, the regression task is stated as

$$y = \frac{1}{|R_k|}\sum_{i\in R_k} t_i \tag{12.8}$$

The output vector for x is the average of the output vectors of the data set in a specific region.

For CART training, the tree topology should be derived using the training set. This implies obtaining the fragmented property at an individual point, together with the limiting parameter figure. Locating an ideal tree topology is an NP-complete problem. In literature, this is known as a greedy heuristic, which fashions the tree in a top-down approach and chooses the optimal fragmented point by point to train CART. The problem of overfilling and attaining better generalization requires that some stopping criteria are necessary for the tree design. Some of the potential terminating benchmarks are the highest depth attained, if the branch distribution is unadulterated, if the gains of separation are lower than a certain benchmark, and if the cardinality of samples in every single branch is lower than the criteria benchmark. Additionally, the pruning technique is effective in dealing with the problem of overfitting [54,56]. The main advantage of CART is that it is rapid and adjustable to big data sets. Unfortunately, it is responsive to the training set selected [57]. A significant drawback of this technique limits to unsmoothly identifying the input domain since each bloc of the input domain is related to a unique identifier [58].

Random forests

Random forests otherwise known as random decision forests are collaborative learning techniques suitable for classification and regression. Similarly, other algorithms that function by designing a horde of decision trees at training instance and presenting the class that is the mode of the classes (classification) or average estimation (regression) of a singular tree. In random forests, rather than training a singular tree, an array of trees are trained simultaneously. Individual tree is trained on a subcategory of the training a larger family, randomly picked with substitution, deploying an arbitrarily selected subcategory of M input variables (features) [59]. There are two situations for estimating the attribute of a novel, unexplored data point: (1) in classification tasks, it is tuned as the highest occurring identifiers predicted by each tree; and (2) in regression tasks, it is tuned as the average of the estimated identifier by an individual tree. A balance exists between various figures of M. A figure of M that is insignificant results in arbitrary trees with low estimation capability, while a bogus figure of M might result in very familiar arbitrary trees.

Random forests are noted to be extremely accurate, nevertheless, at the expense of meaningful human interpretability [60]. However, they are agile and dynamic to big data sets and have many practical usages, typically body pose recognition [61] and body part classification.

Bootstrap aggregating

Bootstrap aggregating, often referred to as bagging, is a collaborative whose goal is to enhance and improve the precision as well as the robustness of ML algorithms resulting in a decrease in overfitting issues. Using this approach, K novel M sized training datasets are arbitrarily selected from the raw dataset with substitutions. Consequently, the newly selected data training set is trained using an ML model [62]. The estimated identifier of a novel untrained data point is denoted to be the mode of the identifiers estimated by an individual scheme in classification assignment and is denoted to be the average of regression assignments. By using different ML schemes such as CART and neural networks, their bagging schemes enhance results. Although bagging deteriorates the operation of robust models, for example, KNN, typical real-life usage scenarios include customer attrition prediction and preimage learning [27].

12.2.2 Unsupervised ML algorithm

Unsupervised ML algorithms decode data morphology from a dataset without reference to already identified results. Different from supervised machine learning, unsupervised machine learning methods are not ideal for regression and classification tasks based on the fact that there is an opaqueness of the expected outcome. Hence, it is difficult to train the model. Unsupervised learning finds applicability in decoding the fundamental structure of the data. Of all the different types of unsupervised algorithms, clustering is widely used. A detailed discussion about the types of unsupervised machine learning algorithms is given in Sections 12.2.2.1 and 12.2.2.2.

12.2.2.1 Clustering

K-means

The modus operandi of K-means algorithm lies toward grouping unidentified data set into K clusters or constellations simply by arranging datasets with the same property into one cluster and otherwise. Using the traditional K-means model, the norm between datasets denotes the degree of resemblance. Hence, K-means set out to discover K cluster centers, represented as $[s_1, \ldots, s_k]$, with the norm between datasets and their closest center being reduced [54,56]. Grouping data points into clusters centers can be implemented via a set of binary indicator variables $\pi_{nk} \in [0,1]$. On the occasion that the data point x_n is designated to the cluster center s_k, then $\pi_{nk}=1$. This can be modeled as thus

$$\min_{s,\pi} \sum_{n=1}^{N} \sum_{k=1}^{k} \pi_{nk} ||x_n - s_k||^2 \tag{12.9}$$

subject to $\sum_{k=1}^{K} \pi_{nk} = 1, n=1, \ldots, N.$

K-means is a highly efficient and flexible algorithm with a fast convergency rate. Note that an online stochastic version of K-means exists along with the offline version. Studies [50,52] analyzed a strategy to deploy the K-means algorithm toward the smart city and smart home data management with impressive outcomes. Unfortunately, this technique is confronted with numerous setbacks due to the deployment of the Euclidean norm as the measure of comparison. Typically, the limitations arose because of the categories of data variables being used, and cluster centers are unstable against data point located outside the main region. Furthermore, the K-means model designates individual data point uniquely to a cluster, which has the tendency of resulting in wrong clusters [12]. Reference [45] used Map Reduce to study the several minute data sets suggesting a cluster approach for a big capacity of minute data driven by the K-means algorithm. Reference [47] deployed K-means in categorizing and grouping traveling arrangement uniformities. Reference [60] utilized a real-time event processing and clustering model to sensor data via OpenIoT middleware which serves as an interface for state-of-the-art analytical IoT applications.

Density-based spatial clustering of applications with noise (DBSCAN)

Density-based approach to spatial clustering of applications with noise (DBSCAN) is another clustering algorithm that dwells in the structuring of data from unlabeled data labels and has found functionality in clustering citizen intelligent conduct [39]. In a DBSCAN, the primary goal lies grouping a set of unidentified data groups using density data points as metrics. Using this approach, clusters of complex data points (data points with many close neighbors) are regarded as groups and blocs of data points with low-density and are not in the main region of concentration [17]. Reference [63] fashioned an algorithm to train a DBSCAN model.

From the viewpoint of efficiency in large dataset and robustness against outliers, DBSCAN performs optimally. Furthermore, it can detect clusters of random shape (i.e., spherical, elongated, and linear). Additionally, in contrast to K-means which required a specified number of clusters, DBSCAN determines the number of clusters based on the density of the data points [52]. However, DBSCAN faces with some challenges which include unstableness when presented with data with large disparities in densities, thus leading to poor results. Furthermore, the algorithm fluctuates rapidly in the face of distance metric which is a criterion to infer the density of a bloc [53]. Notwithstanding the discussed limitations, DBSCAN, one of the most popular clustering algorithms is deployed in practical use-cases of anomaly detection in temperature data [55] and X-ray crystallography [57]. The authors of [31] are of the opinion that expertise in data streams discovery streams is crucial for research and business. They deployed DBSCAN to a data stream exposing the cardinality of current classes and thereafter identifying the data. Similarly, [52] deployed this model to infer the group random shape. The DBSCAN model yields domains of random shape and outlying objects.

12.2.2.2 Feature extraction

Principal component analysis

Principle component analysis (PCA) is one of the most important preprocessing techniques in machine learning. PCA is driven by the theorem of orthogonal projection in which data points are projected onto L dimensional linear subspace, called the principal subspace, possessing utmost projected discrepancies. Similarly, the aim can be construed as locating a comprehensive orthonormal group of L linear M-dimensional basis vectors $\{w_j\}$ and the equivalent linear projections of data points $\{z_{nj}\}$ are designed in such a manner that, there is a reduction in the mean reconstruction error, where x is the mean of all data points [55]:

$$J = \frac{1}{N} \sum_{n} ||\check{x}_n - x_n||^2 \tag{12.10}$$

$$\check{x}_n = \sum_{j=1}^{L} z_{nj} w_j + \overline{x} \tag{12.11}$$

PCA application consists of data compression, whitening, and data visualization. Some of the real-world applications of PCA are face recognition, interest rate derivative portfolios, and neuroscience. Be informed that a kernelized version of PCA, called KPCA, is available in the open domain specifically designed for locating nonlinear principal components [44,46]. The pros of PCA are reduction in the size of data, allowing estimation of probabilities in high-dimensional data, and rendering a set of uncorrelated components. A high computational cost is considered the main disadvantage of this algorithm.

Canonical correlation analysis (CCA)

Canonical correlation analysis (CCA) is a linear dimensionality reduction technique that is closely related to PCA. Reference [51] compared PCA with CCA for discovering sporadic faults and identifying masking malfunctions of enclosed settings. CCA is considered superior in functionality to PCA from the perspective of being capable to handle two or more variables simultaneously which contrasts PCA that handles a single variable at a time. The main purpose is to locate an equivalent pair of extremely cross-correlated linear subspaces. In the corresponding perfect pair subspaces, there exists a correlation between an individual element and an individual element from another subspace. An ideal result is derived from resolving a generalized eigenvector problem [55].

Given two column vectors $X = (x_1, \ldots, x_n)'$ and $Y = (y_1, \ldots, y_m)$ of random variables with finite second moments, one may define the cross-covariance $\sum XY = \text{cov}(X,Y)$ to be the $n\times m$ matrix whose (i, j) entry is the covariance $\text{cov}(x_i, y_j)$. In practice, it can estimate the covariance matrix based on sampled data from X and Y. Canonical-correlation analysis seeks vectors a ($a \in \mathbb{R}^n$) and b ($b \in \mathbb{R}^m$) such that the random variables a^TX and b^TY maximize the correlation $\rho = \text{corr}(a^TX, b^TY)$. The random variables $U = a^TX$ and $V = b^TY$ are the first pair of canonical variables. The solution set returns to finding that vector which maximizes the equivalent correlation subject to the constraint that they are uncorrelated with the first pair of canonical variables, this provides the second pair of canonical variables. This procedure may be continued up to min $\{m, n\}$ times.

$$(a', b') = \max_{a,b} \text{corr}(aTX, bTY) \tag{12.12}$$

12.2.3 Neural networks approach

Recent years have been witnessed tremendous efforts of exploiting the neural networks (NN) in several research issues related to image processing, drug detection, etc. There is a plethora of research-based NN in wireless communication systems such as channel estimation, data detection, and precoders design. However, perceptron (MLP) has a significant impact in practical deployment where Figure 12.4 shows the MLP archetype with a simple two-layer system. The variables of the input vector x are units (neurons) in the input layer, $\varphi_i^{(1)}$ are the hidden layer units, and $\varphi_i^{(2)}$ are the output layer units that output y. The nonlinear function of the actions is modified based on the units in each layer. In ML, $\varphi(.)$ is termed as

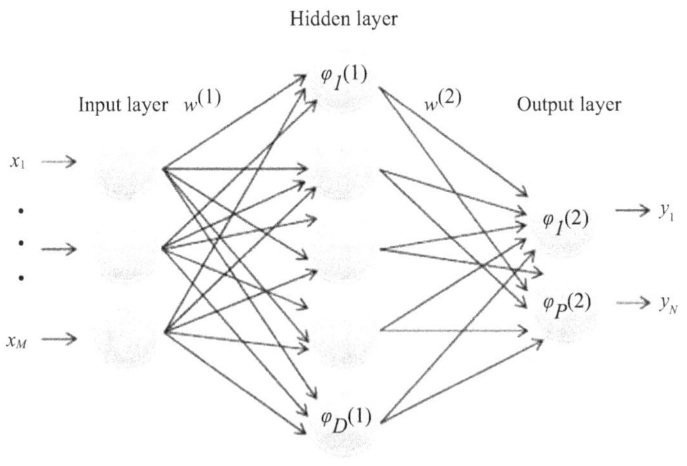

Figure 12.4 A visual of the MLP archetype with a simple two-layer model

an activation function. In NNs, the activation function receives the linear input data and converts them to nonlinear data. For estimation assignment, a linear activation function is exploited and for multiclass classification, a softmax activation function is utilized [55,57]. Equation (12.13) presents the form of classification or regression task:

$$f\left(x, w^{(1)}, w^{(2)}\right) = \varphi^{(2)}\left(\varphi^{(1)}\left(x^T w^{(1)}\right)^T w^{(2)}\right) \tag{12.13}$$

where $w^{(1)} = (w_1{}^{(1)}, \ldots, w_M{}^{(1)})^T, \varphi^{(1)} = (\varphi_1{}^{(1)}, \ldots, \varphi_D{}^{(1)})^T, w^{(2)} = (w_1{}^{(2)}, \ldots, w_D{}^{(2)})^T$, and $\varphi^{(2)} = (\varphi_1{}^{(2)}, \ldots, \varphi_P{}^{(2)})^T$.

The MLP having a minimum of two layers can normalize a random mapping which has been generated from a finite input domain to a finite output domain if sufficient hidden unit exists [21–23]. It is well known that the ideal set of weights w of the MLP can be presented as an NP-complete optimization problem [24]. The literature is rich with algorithms which can be utilized to train neural network models. For instance, stochastic gradient descent, adaptive delta, adaptive gradient, adaptive moment estimation, Nesteroy's accelerated gradience, and RMSprob can be utilized. However, several generalization challenges can be addressed by exploiting the weight decay, weight-sharing, early stopping, Bayesian fitting of neural nets, dropout, and generative pretraining [29]. Therefore, densely denoted functions with one layer admit an exponential size with l-1 layers. Thus, a different method might consider an MLP having more than a single hidden layer, that is, a deep NN (DNN), and the various high-level functionality is accessed by the low-level features [53]. In NNs algorithms, it is assumed that the de facto model in ML models has been buoyed by the overwhelming outcomes [57]. Authors in [26] suggested a scheme of estimating the states of IoT components by exploiting the artificial NN. The analyzed architecture of the NN is an incorporation of MLP and a probabilistic NN. In [21], deployment of the MLP for processing health data is presented. In addition, the MLP has been exploited for future energy consumption where predicting future energy data, data generation, and how the redundancy of these data will be removed are comprehensively presented [26].

12.3 Research trends and open issues

12.3.1 Privacy and security

In IoT, there are hundreds or thousands of interconnected nodes to each other. In addition, a huge amount of data is exchanged among nodes. However, IoT use cases can be classified based on specific characters and attributes. For IoT use-cases to be implemented for data analysis correctly, many challenges must be handled. Firstly, the privacy of the collected data must be guided jealously as the data may contain highly sensitive data such as personal, health, and business-related data. Therefore, the privacy issue has to be addressed. Secondly, as the number of data sources increases and for the simplicity of IoT hardware, it has become essential to overcome the constraints of security, that is, network security and data encryption. Unfortunately, an unsecured network could take place if measures are inadequate to incorporate in the strategy and execution.

12.3.2 Real-time implementation and data analysis

It is well known that the big data is analyzed by analytical algorithms to have a unique attribute and conclusions of smart systems. In other words, models and algorithms to analyze real-time data captured from nodes are essential for the research community. In addition, deep learning can play a crucial role to obtain a high degree of accuracy in real-time implementation. Regrettably, deep learning and NN-based algorithms suffer from an easy corruption by noisy smart data and inaccurate analysis, respectively. Equivalently, semisupervised algorithms can benefit the IoT data analysis.

12.4 Conclusions and recommendations

This study addresses the supervised and unsupervised machine learning techniques that are considered the main pillars of the IoT smart data analysis. Obtaining optimal results in smart data analysis require an in-depth understanding of data structure, discovering abnormal data points, estimating parameters, estimating categories, and extracting salient data features. For sequenced data prediction and classification, the linear regression and SVM methods are the two most frequently applied algorithms. The goals of the deployed models lie in processing and training high-velocity data. Another fast training algorithm is the classification and regression tree. Discovering abnormal data points and irregularities in smart data, these notable algorithms can be executed, namely, the one-class SVM or PCA-based anomaly detection method. Both can train anomalies and noisy data with a high degree of accuracy. SVM is one of the commonly used classification algorithms. The algorithm has the capacity to handle huge datasets classifying them into various categories. Based on this unique attribute of SVM, it is widely deployed in areas where the data have huge volume, data sources coming from various sources, and where smart data processing algorithms are needed. To discover the structure of unlabeled data, clustering algorithms can provide the most appropriate tools. K-means is the most widely used clustering algorithm and it is equipped to work with huge data volume cutting across a broad spectrum of data sources. PCA and CCA are the two most prominent algorithms for feature extraction. Furthermore, CCA has the capacity to depict the correlation between two groups of data. PCA and CCA are both ideal to locate data anomalies. To predict the categories of data, neural networks are suitable learning models for function approximation problems. Moreover, because smart data should be accurate and requires a long training time, a multiclass neural network can provide an appropriate solution. Research on the ML indicates that several challenges remain. This article has highlighted some shortcomings and challenges that exist with respect to some aspects of supervised and unsupervised machine learning techniques and future research may prove beneficial in pursuing this vision as a useful technology.

References

[1] Sharakhina, L. V. and Skvortsova, V. "Big data, smart data in effective communication strategies development." *2019 Communication Strategies in Digital Society Workshop (ComSDS)*. IEEE; 2019.

[2] Al-Fuqaha, A., Guizani, M., Mohammadi, M., Aledhari, M., and Ayyash, M. "Internet of things: A survey on enabling technologies, protocols, and applications." *IEEE Communications Surveys & Tutorials*. 2015; 17(4): 2347–2376.

[3] Alzubi, J., Nayyar, A., and Kumar, A. "Machine learning from theory to algorithms: an overview." *Journal of Physics: Conference Series*. 2018; 1142: 012012.

[4] Jagannath, J., Polosky, N., Jagannath, A., Restuccia, F., and Melodia T. "Machine learning for wireless communications in the Internet of things: A comprehensive survey." *Ad Hoc Networks*. 2019; 93: 101913.

[5] Kashyap, R. "Machine learning for Internet of things." *Next-generation wireless networks meet advanced machine learning applications*. IGI Global; 2019; 57–83.

[6] Martínez, A. M., Ramos-López, D., Cabañas, R., *et al.* "AMIDST: A Java toolbox for scalable probabilistic machine learning." *Knowledge-Based Systems*. 2019; 163: 595–597.

[7] Buskirk, T. D., Kirchner, A., Eck, A., and Signorino, C. S. "An introduction to machine learning methods for survey researchers." *Survey Practice*. 2018; 11(1): 2718.

[8] Luong, N., Hoang, N., Gong, S., *et al.* "Applications of deep reinforcement learning in communications and networking: A survey." *IEEE Communications Surveys & Tutorials*. 2019; 21(4): 3133–3174.

[9] Schrider, D. R. and Kern, A. D. "Supervised machine learning for population genetics: A new paradigm." *Trends in Genetics*. 2018; 34(4): 301–312.

[10] Lee, J. H., Shin, J., and Realff, M. J. "Machine learning: Overview of the recent progresses and implications for the process systems engineering field." *Computers & Chemical Engineering*. 2018; 114: 111–121.

[11] Singh, A., Thakur, N., and A. Sharma. "A review of supervised machine learning algorithms." *2016 3rd International Conference on Computing for Sustainable Global Development (INDIACom)*. IEEE; 2016.

[12] Osisanwo, F. Y., Akinsola, J. E. T., Awodele, O., Hinmikaiye, J. O., Olakanmi, O., and Akinjobi, J. "Supervised machine learning algorithms: Classification and comparison." *International Journal of Computer Trends and Technology (IJCTT)*. 2017; 48(3): 128–138.

[13] Qu, G. and Li, N. "Accelerated distributed nesterov gradient descent for smooth and strongly convex functions." *2016 54th Annual Allerton Conference on Communication, Control, and Computing (Allerton)*. IEEE; 2016.

[14] Lee, J., Stanley, M., Spanias, A., and Tepedelenlioglu, C. "Integrating machine learning in embedded sensor systems for Internet-of-things applications." *2016 IEEE International Symposium on Signal Processing and Information Technology (ISSPIT)*. IEEE; 2016.

[15] Kanj, S., Abdallah, F., Denœux, T., and Tout, K. "Editing training data for multi-label classification with the k-nearest neighbor rule." *Pattern Analysis and Applications*. 2016; 19(1): 145–161.

[16] Maillo, J., Ramírez, S., Triguero, I., and Herrera, F. "kNN-IS: An iterative spark-based design of the k nearest neighbors classifier for big data." *Knowledge-Based Systems*. 2017; 117: 3–15.

[17] Chomboon, K., Chujai, P., Teerarassamee, P., Kerdprasop, K., and Kerdprasop, N. "An empirical study of distance metrics for k nearest neighbor algorithm." *Proceedings of the 3rd International Conference on Industrial Application Engineering*. 2015.

[18] Prasath, V., Abu Alfeilat, H., Hassanat, A., *et al*. "Distance and similarity measures effect on the performance of k-nearest neighbor classifier: A review." 2017. arXiv preprint arXiv:1708.04321.

[19] Berisha, V., Wisler, A., Hero, A. O., and Spanias, A. "Empirically estimable classification bounds based on a nonparametric divergence measure." *IEEE Transactions on Signal Processing*. 2015; 64(3): 580–591.

[20] Azar, A. T. and Hassanien, A. E. "Dimensionality reduction of medical big data using neural-fuzzy classifier." *Soft Computing*. 2015; 19(4): 1115–1127.

[21] Ghaderi, A., Frounchi, J., and Farnam, A. "Machine learning-based signal processing using physiological signals for stress detection." *2015 22nd Iranian Conference on Biomedical Engineering (ICBME)*. IEEE; 2015.

[22] Sharmila, A. and Geethanjali, P. "DWT based detection of epileptic seizure from EEG signals using naive Bayes and k-NN classifiers." *IEEE Access*. 2016; 4: 7716–7727.

[23] Garcia, L. P., de Carvalho, A. C., and Lorena, A. C. "Effect of label noise in the complexity of classification problems." *Neurocomputing*. 2015; 160: 108–119.

[24] Lu, W., Du, X., Hadjieleftheriou, M., and Ooi, B. C. "Efficiently supporting edit distance based string similarity search using B+-trees." *IEEE Transactions on Knowledge and Data Engineering*. 2014; 26(12): 2983–2996.

[25] Do, C.-T., Douzal-Chouakria, A., Marié, S., and Rombaut, M. "Multiple metric learning for large margin kNN Classification of time series." *2015 23rd European Signal Processing Conference (EUSIPCO)*. IEEE; 2015.

[26] Ma, X., Wu, Y.-J., Wang, Y., Chen, F., and Liu, J. "Mining smart card data for transit riders' travel patterns." *Transportation Research Part C: Emerging Technologies*. 2013; 36: 1–12.

[27] Wang, H., Zhang, Y., Waytowich, N. R., *et al.* "Discriminative feature extraction via multivariate linear regression for SSVEP-based BCI." *IEEE Transactions on Neural Systems and Rehabilitation Engineering*. 2016; 24 (5): 532–541.

[28] Jadhav, S. D. and Channe, H. "Comparative study of K-NN, naive Bayes and decision tree classification techniques." *International Journal of Science and Research*. 2016; 5(1): 1842–1845.

[29] Mahdavinejad, M. S., Rezvan, M., Barekatain, M., Adibi, P., Barnaghi, P., and Amit, P. S. "Machine learning for Internet of Things data analysis: A survey." *Digital Communications and Networks*. 2018; 4(3): 161–175.

[30] Xu, S. "Bayesian naïve Bayes classifiers to text classification." *Journal of Information Science*. 2018; 44(1): 48–59.

[31] Singh, G., Kumar, B., Gaur, L., and Tyagi, A. "Comparison between multinomial and Bernoulli naïve Bayes for text classification." *2019 International Conference on Automation, Computational and Technology Management (ICACTM)*. IEEE; 2019.

[32] Hilbe, J. M. *Practical guide to logistic regression*. Chapman and Hall/CRC; 2016.

[33] Pham, B. T., Prakash, I., Jaafari, A., and Bui, D. T. "Spatial prediction of rainfall-induced landslides using aggregating one-dependence estimators classifier." *Journal of the Indian Society of Remote Sensing*. 2018; 46(9): 1457–1470.

[34] Han, W., Gu, Y., Zhang, Y., and Zheng, L. "Data driven quantitative trust model for the internet of agricultural things." *2014 International Conference on the Internet of Things (IOT)*. IEEE; 2014.

[35] Cherian, V. and Bindu, M. "Heart disease prediction using naive Bayes algorithm and Laplace smoothing technique." *International Journal of Computer Science Trends and Technology (IJCST)*. 2017; 5(2).

[36] Weichenthal, S., Ryswyk, K., Goldstein, A., Bagg, S., Shekkarizfard, M., and Hatzopoulou, M. "A land use regression model for ambient ultrafine particles in Montreal, Canada: A comparison of linear regression and a machine learning approach." *Environmental Research*. 2016; 146: 65–72.

[37] Hoffmann, J. P. and Shafer, K. *Linear regression analysis*. Washington, DC: NASW Press; 2015.

[38] Montgomery, D. C., Peck, E. A., and Vining, G. G. *Introduction to linear regression analysis*. Vol. 821. John Wiley & Sons; 2012.

[39] Robert, C. *Machine learning, a probabilistic perspective*. Taylor & Francis; 2014.
[40] Derguech, W., Bruke, E., and Curry, E. "An autonomic approach to real-time predictive analytics using open data and internet of things." *2014 IEEE 11th International Conference on Ubiquitous Intelligence and Computing and 2014 IEEE 11th International Conference on Autonomic and Trusted Computing and 2014 IEEE 14th International Conference on Scalable Computing and Communications and Its Associated Workshops*. IEEE; 2014.
[41] Glorot, X. and Bengio, Y. "Understanding the difficulty of training deep feedforward neural networks." *Proceedings of the Thirteenth International Conference on Artificial Intelligence and Statistics*. 2010.
[42] Shifei, D., Bingjuan, Q., and Hongyan, T. "An overview on theory and algorithm of support vector machines." *Journal of University of Electronic Science and Technology of China*. 2011; 40(1): 2–10.
[43] Nikam, S. S. "A comparative study of classification techniques in data mining algorithms." *Oriental Journal of Computer Science & Technology*. 2015; 8(1): 13–19.
[44] Alber, M., Zimmert, J., Dogan, U., and Kloft, M. "Distributed optimization of multi-class SVMs." *PLoS One*. 2017; 12(6): e0178161.
[45] Ponte, P. and Melko, R. G. "Kernel methods for interpretable machine learning of order parameters." *Physical Review B*. 2017; 96(20): 205146.
[46] Utkin, L. V., Chekh, A. I., and Zhuk, Y. A. "Binary classification SVM-based algorithms with interval-valued training data using triangular and Epanechnikov kernels." *Neural Networks*. 2016; 80: 53–66.
[47] Lee, C. -P. and Roth, D. "Distributed box-constrained quadratic optimization for dual linear SVM." *International Conference on Machine Learning*. 2015.
[48] Huang, X., Shi, L., and Suykens, J. A. "Sequential minimal optimization for SVM with pinball loss." *Neurocomputing*. 2015; 149: 1596–1603.
[49] Díaz-Morales, R. and Navia-Vázquez, Á. "Distributed nonlinear semiparametric support vector machine for big data applications on spark frameworks." *IEEE Transactions on Systems, Man, and Cybernetics: Systems*. 2018; 50(11): 4664–4675.
[50] Azim, R., Rahman, W., and Karim, M. F. "Bangla hand-written character recognition using support vector machine." *International Journal of Engineering Works*. 2016; 3(6): 36–46.
[51] Liu, P., Choo, K-K., Wang, L., and Huang, F. "SVM or deep learning? A comparative study on remote sensing image classification." *Soft Computing*. 2017; 21(23): 7053–7065.
[52] Cang, Z., Mu, L., Wu, K., Opron, K., Xia, K., and Wei, G.-W. "A topological approach for protein classification." *Computational and Mathematical Biophysics*. 2015; 3(1): 140–162.
[53] Wahab, O. A., Mourad, A., Otrok, H., and Bentahar, J. "CEAP: SVM-based intelligent detection model for clustered vehicular ad hoc networks." *Expert Systems with Applications*. 2016; 50: 40–54.

[54] Khan, M. A., Khan, A., Khan M. N., and Anwar, S. "A novel learning method to classify data streams in the internet of things." *2014 National Software Engineering Conference*. IEEE; 2014.
[55] Nikravesh, A. Y., Ajila, S., Lung, C., and Ding, W. "Mobile network traffic prediction using MLP, MLPWD, and SVM." *2016 IEEE International Congress on Big Data (BigData Congress)*. IEEE; 2016.
[56] Breiman, L. *Classification and regression trees*. Routledge; 2017.
[57] Krzywinski, M. and Altman, N. *Points of significance: Classification and regression trees*. Nature Publishing Group; 2017.
[58] Wuest, T., Weimer, D., Irgens, C., and Thoben, K.-D. "Machine learning in manufacturing: advantages, challenges, and applications." *Production & Manufacturing Research*. 2016: 4(1): 23–45.
[59] Hagenauer, J. and Helbich, M. "A comparative study of machine learning classifiers for modeling travel mode choice." *Expert Systems with Applications*. 2017; 78: 273–282.
[60] Belgiu, M. and Drăguţ, L. "Random forest in remote sensing: A review of applications and future directions." *ISPRS Journal of Photogrammetry and Remote Sensing*. 2016; 114: 24–31.
[61] Biau, G. and Scornet, E. "A random forest guided tour." *Test*. 2016; 25(2): 197–227.
[62] Selvi, S. T., Karthikeyan, P., Vincent, A., Abinaya, V., Neeraja, G., and Deepika, R. "Text categorization using Rocchio algorithm and random forest algorithm." *2016 Eighth International Conference on Advanced Computing (ICoAC)*. IEEE; 2017.
[63] Hassan, A. R. and Bhuiyan, M. I. H. "Computer-aided sleep staging using complete ensemble empirical mode decomposition with adaptive noise and bootstrap aggregating." *Biomedical Signal Processing and Control*. 2016; 24: 1–10.

Chapter 13

Artificial intelligence and machine learning aided blockchain systems to address security vulnerabilities and threats in the industrial Internet of things

Karanjeet Choudhary[1], Gurjot Singh Gaba[1], Rajan Miglani[1], Lavish Kansal[1] and Pardeep Kumar[2]

Advent of digital sensors and machines led to a significant acceleration in industrial evolution. The desire to automate industrial processes with minimum human intervention paved the way for the onset of a new era of technological nomenclature called the industrial Internet of things (IIoT). A remarkable feature of IIoT is its underlying architecture which allows the managers/engineers/supervisors to remotely operate and access the performance of their machines. Industries ranging from healthcare, finance, logistics, and power have witnessed a major performance increment and quality stabilization by transforming themselves into an IIoT empowered smart environment. However, this transformation has brought with itself a whole new set of challenges with cybersecurity being the paramount. The vulnerabilities like bugs and broken processes can lead to a serious compromise or even collapse of security mechanisms of IIoT networks. Such a situation will have a devastating impact on the financial health, reputation, and credibility of companies. After an extensive review of existing technologies, we believe that blockchain, artificial intelligence (AI), and machine learning (ML) can complement each other in building a revolutionary deterrent to negate malicious activities that in any form intend to harm the system. While, blockchain offers public/private/consortium relationships, ML and AI, on the other hand, follow the principle of supervised/unsupervised/reinforcement learning and reactive/memory approaches, respectively. Based on the distributed ledger system, blockchain mechanisms can be aided with self-learning algorithms which will update and strengthen the database by learning each time the system suffers new forms of network attacks and intrusions. This process of learning will help build a robust system which can learn to optimize

[1]School of Electronics and Electrical Engineering, Lovely Professional University, Phagwara, India
[2]Department of Computer Science, Swansea University, Swansea, United Kingdom

its deterrence procedures against different forms of attacks. It is due to these overwhelming benefits, blockchain, AI, and ML find applications in smart logistics, predictive maintenance, autonomous vehicles, intelligent manufacturing, and smart grid maintenance.

13.1 Introduction

The industrial Internet of things (IIoT) enables the managers of industries to connect different sensors, instruments, and machines to the Internet for instant access. This allows the user/manager to communicate with any device in the industry and also facilitates data access from a remote location. Industrial Internet of things can also help in increasing the production thus leading to enhanced economic benefits to the industry. The principal usefulness of this technology is that it allows the user to fetch the data directly from the cloud without interrupting any node in the industrial IoT. To suit commercial implementations, IIoT uses an array of different technologies such as cybersecurity, cloud computing, edge computing, mobile technology, machine-to-machine communication, big data, and cognitive computing [1].

More recently, the IIoT is being used in a variety of different industries, which include health care, finance, manufacturing, and retail, and the list is unending. However, amongst all of these, the IIoT finds a major relevance and main application in smart cities as it helps the people to reduce their domestic waste and energy consumption. Industrial Internet of things enables the manager to monitor the industrial process, improves customer satisfaction and experience, enhances employee productivity, and further assists in enhancing revenue generation.

Figure 13.1 illustrates the overall industrial revolution since the past two centuries. The evolution of the industry started in the 18th century with the utilization of steam engines. The advancement in industries, that is, the second revolution took place in the 1870s. This era brought electric-powered machines which revolutionized the concept of mass production. The third revolution made use of programmable logic controllers (PLC) in the 1970s. In this revolution, computers were used for automating the industry. Industry 4.0, the fourth generation of the industrial revolution, brought a massive change wherein the administrators of the industry can manage the operation of the machines remotely.

The entire operation of IIoT sits on a cumulative functionality of four different layers, that is, content layer, service layer, network layer, and device layer. The device layer works on physical components, sensors, and machines. The network layer consists of physical buses that transport the data to the service layer. The service layer consists of an application that converts the data and combines it into information which is displayed further on the driver dashboard of the content layer [2,3].

13.2 Birth of industrial Internet of things

The Internet of things (IoT) is a system of intercorrelated computing devices, mechanical, and digital machines. The devices have the ability to transfer the data

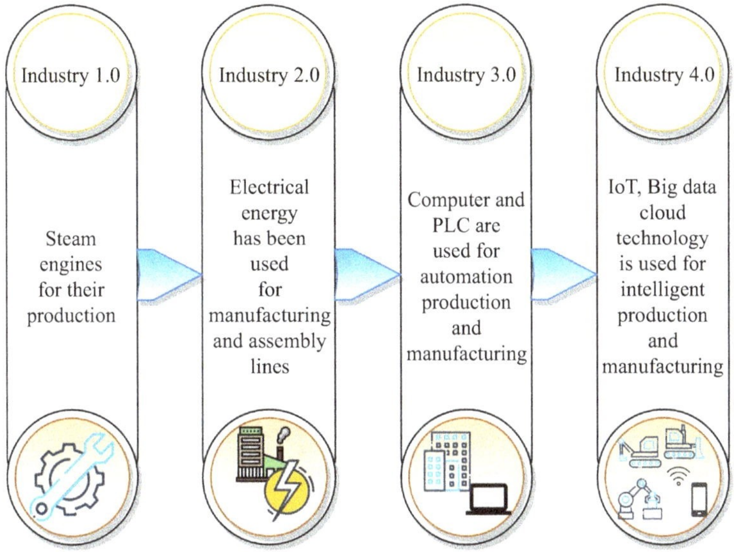

Figure 13.1 Evolution of industrial IoT

over a network without requiring human-to-computer interaction. The main function of IoT is to help people live and work smarter, and have complete control over their objects. IoT offers many services in day-to-day lives to make lives easier than before; IoT provides automotive homes, and real-time businesses to look at how the company's system is really working. IoT reduces the cost of labor because of the fully automated process and also provides transparency to the customers.

IIoT is an integration of cyber and physical technology where machines are enabled with the Internet. Figure 13.2 demonstrates the composition of various elements of IIoT. But as the industry progresses, the challenges to secure the industry also increase. The attackers in the cyber-physical world may harm the industry from anywhere, despite being physically away [4].

Industries 1.0: The first generation of the industrial revolution was in the form of steam engines known as Industry 1.0, introduced by James Watt in the 18th century. In this revolution, all industries were operated with steam engines. During that era, the industries used less human resource and power for the production and manufacturing of the goods, for example, humans used the spinning wheel to make threads, thus consuming a lot of time and manpower.

Industries 2.0: The second generation of the industrial revolution took place at the beginning of the 20th century. In this century, electricity became the primary source of everything, including the industrial revolution. So it became easier to replace the Industries 1.0 with the Industries 2.0, wherein electricity was used rather than water and steam engines for producing industrial goods. During this era, the efficiency and effectiveness of the manufacturing rate in industries escalated due to the development of management programs in industries. With the advent of

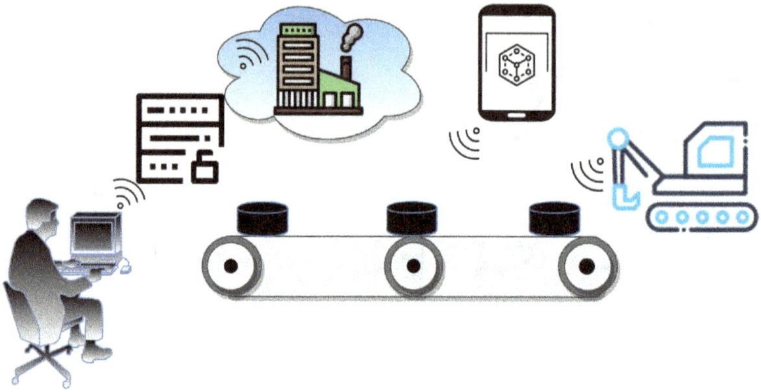

Figure 13.2 Industrial Internet of things

the second generation, mass production became possible in a short period of time and with lesser human resources.

Industries 3.0: The third generation of industrial revolution took place in the last few decades of 20th century. In this revolution, electricity continued to play an important role in the production and manufacturing of goods in industries; but the main transformation in the Industry 3.0 is the automation of the industries which was made possible by programmable logic controllers (PLC) and computers. With this transformation, the automated industries handled the entire production process without any human assistance. The increase in the production rate of industries was also witnessed. The cost of the production was also minimized due to the less involvement of human manpower. Robots, the highlight of this revolution, performed programmed sequences without human interference and produced and manufactured more accurate products with less delay and defect. Despite these advantages, the third revolution raised a wide spectrum of security issues in industries.

Industries 4.0: Industry 4.0 directly connects the Internet of things (IoT) with the manufacturing techniques for enabling the systems to share and analyze the information from the remote location area. In this revolution, the owner of the industry or manager having access to controlling the industrial machines can control the industrial production and manufacturing from the remote location area. It also incorporates cutting-edge technologies, which include additive manufacturing, robotics, artificial intelligence and other cognitive technologies, and advanced materials. The industrial systems have now emerged as "cyber-physical production systems", also known as smart factories or industries, in which the production system of components and the people communicate via a wireless network, thus resulting in industrial production to be nearly autonomous [5,6].

Advantages of industrial Internet of things:

- Improved accuracy
- Optimized processes
- Predictive maintenance and analysis

- Increased efficiency
- Remote monitoring
- Robust system security
- Network stability
- Drastic reduction in manufacturing and processing time
- Efficient power consumption

13.3 Application areas and current examples of its use

Internet of things (IoT) is playing a big role in making the smart world. Industrial Internet of things is an application of IoT system which focuses on the different resource concentrate devices and machines used in the industrial environment. Industrialists and manufacturers of every sector have a great opportunity not only to monitor but also to automate or configure complex processes involved in manufacturing through the Internet. Nowadays, the IIoT is being used in most of the industries for manufacturing, producing, controlling, and analyzing the process of the industry from a remote location or being in the industry without touching any part of a machine or any sensor placed in industries. The application areas of the IIoT are as follows.

13.3.1 Production flow monitoring

The production plays the most important role in industries as the increased and good quality production enhances revenue. Industrial Internet of things-enabled manufacturing is capable of monitoring and controlling an entire production rate, that is, refining the process of manufacturing to final packaging in industries. This complete process of monitoring takes place in real-time. The real-time measurements provide us the scope to recommend any adjustments in operations for the better management of the industry's operational cost. Since the monitoring of production flow rate is done through the computer-aided tools, it can produce a large number of goods with precision in a lesser time than before.

13.3.2 Connected factories

Connecting factories with each other is an effective solution to improve all areas of the industrial environment and its operations in different fields. The main concept of connecting industries with each other is to make a big network for easier management of the production rate and the manufacturing rate of industries, thereby making it easier to control and analyze the industrial network from any part of the world. In addition, it allows the industries to interact with each other securely to share the information which may lead to increased revenue and in turn escalated the economic growth of the country.

13.3.3 Power management

Power management is one of the crucial factors which the industries are currently facing. Industrial Internet of things can deliver a better solution for power management in an industrial environment. It makes switching decisions based on the interrupts generated by the sensors. The sensors are deployed with a wireless communication unit that senses the requirements of lights, machines, motors, etc., and informs the IIoT device. This process allows the efficient use of power in industries, thus saving costs and resources.

13.3.4 Autonomous vehicles

Large-scale industries are producing enormous goods in one day; thus they require massive manpower to transport the goods from one location to another. This freight-in adds a further cost to the goods. Moreover, human efficiency is limited to physical attributes. Humans have time constraints; they need a break for meals. All such issues have been limiting the product flow in the market. Therefore, autonomous vehicles have been brought in to replace the requirement of humans in transporting the goods. These vehicles are directly connected to GPS and wireless technologies for communicating and controlling the whole process of moving goods from one place to another without any difficulty. Smart vehicles in industries can also detect traffic congestions along their path and make a deviation to reach their final destination in the shortest time period. These smart vehicles are equipped with multiple smart sensors to perform their work inside the industries, and they abide by the instructions of the operator.

13.3.5 Predictive maintenance

Smart industries are highly equipped with the smart sensors interfaced with the industrial machines to provide the feedback of system, components, and machinery of the industry. If a critical issue arises in the system, the smart sensor will detect it and provide a warning to the control system before the entire system fails [7–9].

Some of the currently used applications of IIoT are:

1. ***Komatsu: innovation in mining and heavy equipment***
   Komatsu, a Japanese firm, is playing an important role in effective mine management. The firm is focusing not only on increasing productivity and reducing the cost of operation but also is providing a safe work environment for their workers to achieve high benefits with fewer risks. Japan is promoting autonomous mining for the last three decades, and safety is the main priority of the Komatsu organization. One can depend on the most trusted clock reliability of modular mining dispatch fleet management system for the electric drive haul trucks. Autonomous haulage system (AHS) haul trucks are equipped with the high precision global navigation satellite positioning system (HP-GNSS), which is a highly advanced sensor and integrated controller for remotely controlling their trucks from the control room. They are operated by a single operator through a very high-performance wireless network, that is, long-term evolution (LTE). The autonomous vehicles perform all the tasks inside the pits without the human manpower, thereby ensuring the safety of the workers [10]. The working of Komatsu has been depicted in Figure 13.3.

2. ***KUKA: connected robotics***
   KUKA is a cloud-based software platform through which the clients can access and analyze the working of their robots and machines from any remote location

Figure 13.3 Komatsu: a driverless mining vehicle

area. KUKA is set up on an open global standard, connected to the cloud servers with the use of big data analytics to provide their customers maximum visibility and support. Due to this maximum visibility from KUKA, the customers are able to react and respond to the information received from KUKA robots to optimize their operational processes. KUKA also connects to the subscription-based platform that requires zero software installation and also gives freedom to its customers for accessing the new feature and its functionalities through the system updates. This platform also enhances the ecosystem of KUKA and the third-party apps to enhance the overall experience. KUKA Connect uses the features of a user interface that allows the administrative users to construct workflows with their accessibility to the multiple layers of robot data. KUKA Connect provides two different service tiers to their customers with *KUKA connect lite* and *KUKA connect plus* [11].

3. ***Magna Steyr: smart automotive manufacturing***
   Magna Steyr automotive manufacturing industry from Austria is one of the innovative industries nowadays. This industry comes in the list of one of the best smart industries in the world. There are more than 161,000 employees who are internationally working in it for the last many years. This organization can precisely track anything, from their tools to vehicle parts and from any part of the industry when it comes to automatically order a part of car replenishment in urgency. Magna is also enhancing its "smart packaging" system, which will help them to enhance the productivity of their manufacturing material with Bluetooth and to track every component present in their warehouses. This industry also has automated vehicles within their organization to help in carrying heavy components within the plants during the assemblage of components, and these vehicles optimize their paths dynamically with the command from the control room. This organization has a good history for its custom automation manufacturing and production of different types of cars on the same assembly line since 1999. These days, the company is using new

wearable technology to help their employees to guide in the production area of vehicles. Recently, the company has won the Automation Innovations Award from the Centre of Automotive Management and PwC, Germany [7,8].

4. ***Maersk: intelligent logistics***
   Maersk is a company that ships millions of containers to more than 121 countries all over the world. This company with the assistance of the Internet of things keeps a track of their containers, optimizes the fuel usage, and finds the routes for its ships to deliver their containers worldwide at the right time. Their technology of shipping containers has proven to be especially useful for refrigerated containers without any loss and delay. This organization spends nearly $1 billion yearly for transporting empty containers throughout the world. The company is equipped with high sensors and data analytics applications to inform how to store and locate the containers anywhere in the world. The organization is using the blockchain technology for optimizing its supply chain and further operations for shipping these containers [12].

13.4 Implementation challenges of IIoT

The IIoT is also known as Industry 4.0. This technology has brought change to the industrial environment; the inspiration behind Industry 4.0 is the leading industries who have reached their top levels in a short span of time because of automation. Nowadays, industries can escalate the level of productivity, efficiency, and performance through machine-to-machine communications. This big revolution though has created a number of opportunities but also has posed some challenges for industry owners and associates.

The surveys reveal that the smart industries fear losing their market share in the next few years if they do not implement their big data strategy and automation. The nonrealization of few challenges by the smart industries is creating some problems in their marketing, and consequently, some of these problems listed below are resulting in holding back these industries.

13.4.1 Data storage

Data storage is one of the major and big challenges not only for the smart industries but also for the small-scale industries. Nowadays, most of the industries are connected to the Internet for their production and manufacturing process, and they are equipped with a number of smart sensors, machines, and devices. Every device is connected to the Internet to collect the data through the network. The two or more industries communicating with each other share their information but the main problem is where to store that information. These industries not only communicate with each other but also communicate with other millions of sensor nodes, smart devices, and machines too. Every device, sensor, and machine produces its own data and sends it to the user or to some storage device. Some of the large-scale industries have overcome this challenge by securely storing their big data in the cloud.

13.4.2 Security

Security also poses a big challenge that has been faced by Industry 4.0. As we know that every smart industry is connected to the Internet for their communication and manufacturing process both within and outside the company. Industry 4.0 wants to share its data with a few industries and tries to maintain its own secrecy from others. So it is a big challenge to secure the personal information and data from a number of cyberattacks that may turn out to be a big threat to the enterprise. In order to overcome this issue of security, the industries need to implement security procedures such as securing the communication, providing high authentication servers to industries. But this solution may require a high-cost setup of new security measures in the industry and their maintenance.

13.4.3 Delivering values to customers

Delivering satisfaction to customers is another big challenge for the industries. Every organization requires an understanding of its customers' needs and fulfilling these needs by providing good services at an appropriate time. The different IIoT solution plans can definitely impact on efficiency, customer satisfaction, and long-run productivity in any industry. The new technology needs a great understanding to fill the gap of customer's problem statements. Hence, it becomes more important for the industries to figure out the key performance that measures and improves overall customer satisfaction [13].

13.4.4 Technology infrastructure

For making industry a fully smart industry, the owners need to build good infrastructure. This may require high expenses due to the installation of new technology (both hardware and software) and machines in the industry for further growth. Rather this installation will help industries to secure their data and enhance their production as compared to the other industries. But adopting the new technology in the industry is becoming a big challenge because of an increase in expenditure.

13.4.5 Immaturity of IIoT standards

There are some devices and smart sensors in industries that help the industries to communicate with each other or enable one machine to communicate with the other. As industrial machines are connected to the Internet, they thereby can act as an open-source for everyone to get access to such industries. Absentia of security standards may lead to a high loss in the industry as someone intentionally starts tampering with the machines of industries. The immaturity of the lack of standards in IIoT is the main reason why the industry owners doubt investing their money in these areas for the betterment of industries. Although some of the IIoT standards have been developed, still there is a lack of flexibility in IIoT standards which has to be achieved.

13.4.6 Visibility and connectivity

Once the industry is fully equipped with high technologies and ultra-modern machinery, the next piece of business is to check whether the best output from the machines is coming or not and that too at an appropriate time. Keeping an eye on the monthly or annual update of the machine is challenging before the error comes regarding the wellbeing of the machine. Industrial Internet of things might bring more challenges to the industrial operators in case they feel glitches in synchronization of two machines and resulting in some technical errors. In such cases, engineers may have to remove the devices that are connected to the network [14].

13.5 Vulnerabilities, threats, and risks

13.5.1 Vulnerability

Vulnerability is also known as weakness, a weakness of any system which helps the attacker to exploit the system easily. Even a small vulnerability in any system may lead to threats. The system may be exploited by a single vulnerability. There are a number of countless threats that Internet-connected machines may encounter daily and many of these vulnerabilities exist due to the outdated security system. Vulnerabilities may be in the form of some software bugs, broken processes of the system, hardware flaws, and human error. For example, in a man-in-the-middle attack, the attacker puts himself in between the two communicating entities and fetches the most important data because of the dearth of secure medium or encryption to their message being exchanged.

The easy way to fix such types of vulnerabilities in any system, network, or smart device connected to the Internet is that the security team of that enterprise must check the important updates, and at least inspect all the security patches twice a week [15,16].

Tables 13.1 and 13.2 show the security protocols which are able to protect the Internet of energy (IoE) and Internet of sensors (IoS) from the different types of attacks. This analysis is very useful as it may help the administrator of the firm to use the best available security protocol based on requirements.

13.5.2 Threats

The threat is an event that has the power of destroying valuable resources. A threat may be done intentionally or accidentally which may lead to the destruction of the asset. Threats may also be associated with personal life and cyber-security too. It becomes more dangerous when an attacker exploits one or more than one vulnerability of the system for gaining full access to the system, including the operating system. Threats may be internal from an employee of the industry or externally from an attacker.

In other words, threats are the events performed by an intruder by exploiting the vulnerabilities of the system by injecting some harmful malware in the system in order to gain access to the system and to fulfill his intentions of damaging the

Table 13.1 Protection of Internet of energy (IoE) networks from various attacks

Adversary model	[17]	[18]	[19]
Side-channel attack	X	X	*P*
Man-in-the-middle attack	*P*	*P*	✓
Replay attack	*P*	✓	✓
Impersonation attack	*P*	*P*	*P*
Eavesdropping attack	*P*	*P*	*P*

Notes: **X**, vulnerable; ***P***, partial resistance; ✓, complete protection.

Table 13.2 Protection of Internet of sensor (IoS) networks from various attacks

Attacks	[20]	[21]	[22]	[23]
Man-in-the-middle	*P*	*P*	*P*	*P*
Replay	✓	*P*	✓	✓
Impersonation	✓	✓	*P*	✓
Eavesdropping	*P*	*P*	*P*	*P*

Notes: ***P***, partial resistance; ✓, complete protection.

system permanently. For example, in cybersecurity, the phishing attack may help the intruder to install a fake web page or Trojan to help him steal all personal information and use it afterward.

Figure 13.4 defines that the attacker may gain access to the network using the unsecured medium, thus posing threats to network reliability.

The IIoT is connected to the number of machines, sensors, and actuators, for coordinating the manufacturing and production rate of the industry. Smart industries need more connectivity and interoperability to enhance their performance, which may create pathways to the various attacks. Attacks on the industrial environment raise a strong security concern because they may cause a huge loss of property and life. Few attacks explained below can be encountered in the smart industries:

1. ***Denial of service attack (DoS)***: DoS attacks are performed on the network to restrict the server from communicating with the legitimate devices. These attacks may disrupt the communications permanently or temporarily. Also, the server may end utilizing all its resources in the processing of fake requests. For avoiding such attacks, the user needs better authentication and authorization system protocols that can help to avoid such attacks.
2. ***Cloud malware injection attacks:*** Every industry is connected to the cloud for storing its large amount of data and fetching the data from clouds. The intruder tries to inject the malicious service or virtual machine into the cloud system so

Figure 13.4 Threats to industrial IoT

that he can access all the services and data of the valid users. In this type of attack, the intruder creates its own malicious service and tries to add it to the cloud system. Then, the intruder behaves like a valid user so that he can make it a valid service to the cloud system among the valid instances. If the intruder succeeds in this attack, then the cloud automatically accepts the requests as a valid user to the malicious service, and the intruder starts fetching the precious data from the cloud.

3. ***Authentication attacks/impersonation attacks:*** Authentication or impersonation attacks are those attacks in which the intruder tries to behave like a real cum valid user of the industrial environment and to communicate with the industrial nodes. The intruder tries to use the same user id or password of the valid user or he may use some other methods to prove himself as a legitimate user of that industrial environment. Once the intruder gets successful, he may steal useful information or he may perform operations that are authorized to legitimate entities only.
4. ***Man-in-the-middle attacks:*** In the man-in-the-middle attack, the attacker puts himself between the two communicating entities and keeps an eye on both while they are communicating with each other. Whenever the attacker gets a chance, he can modify the important information sent by one user to another user. He can interrupt the communication of both users for obtaining their confidential data in order to get access to their personal information in the future.
5. ***Phishing attacks:*** Phishing attack is the attack in which the intruder makes the same web page as the original one to trick the user to interact with the fake-looking web page or emails for gaining confidential or personal information.

6. ***Brute force attacks:*** In brute force attack, the intruder tries a number of methods to break the code, such as the hit and trial method, to get the real password or information of the valid user to gain access to secret information.
7. ***Remote code execution:*** Remote code execution is the attack in which the intruder tries to exploit the industrial system remotely. He may introduce the malware or inject the virus into the system to control the target remotely.
8. ***Replay attacks:*** In this attack, the intruder tries to capture the information exchanged between the two machines or between the manager and the machine, and tries to send the same message again and again to the end-user/machine in the industrial network, to gain the partial or full access to the industrial environment.
9. ***IP spoofing attacks:*** In this attack, the intruder tries to spoof the logical address (Internet protocol address) of the valid user and later once the legitimate device stops using the device, then the attacker creates the IP packets with a false source IP address with an intention to impersonate. The intruder's intention may be to fetch the important data of the industrial storage devices [24].

The easy way to fix threats is to make a good authentication protocol that helps the user or devices to know that someone is trying to impersonate your device or network.

13.5.3 Risk

Risks are basically the consequences what an industry or an individual has to face if the vulnerabilities are exploited by the attackers. Table 13.3 provides the relationship between vulnerabilities, threats, and risks. The industries are at risk when their system is weak in the context of security measures. In this scenario, the industrial operator or owner is worried about losing important and personal data or even his life too if the machines start behaving abnormally [25].

13.6 Security considerations

Nowadays, the industries are converting to smart industries as they can operate their machines from a remote location area by using a smart device and may control

Table 13.3 Relationship between vulnerabilities, threats, and risks

Vulnerabilities	Threats	Risks
• Software bugs	• Criminals	• Financial loss
• Human error	• Competitors	• Loss of privacy
• Broken process	• Hackers	• Loss of life
• Hardware flaws	• Cyberattacks	• Legal penalties
• Ineffective controls	• Dishonest employs	• Business disruption

all the stuff such as closing the door, lights, and machine of the home and industries from a remote location [26].

For example, if someone tampers with the network of the industry, the owner who is out of town will get the notification. And the owner will also get the notification of that third person if he tries to interrupt the server for a long time. It will help the owner to get alert for further actions. There are several security measures that can be adopted by the industries to prevent them from fraudulent abuses from the attackers.

13.6.1 Authentication

Authentication is one of the best ways of effective management and protection of smart devices, machines, and other important cum personal data of organizations. Most of the big organizations already have strong authentication systems, like active directory certificate service in their network which helps them to allow legitimate users only to access the network. Authentication is used to check whether the parties are legitimate or not before the communication session is established between them. Nowadays, the certificates are used instead of passwords for authenticating the trusted connection between the multiple organizations and the end networks [27]. Table 13.4 provides various approaches to controlling the man-in-the-middle attack with the use of mutual authentication.

13.6.2 Encryption

Encryption is one of the best ways of hiding the information from the plain text to the other form. Encryption based on complex algorithms is known as ciphers; the main purpose of the encryption is to make your private data more secure from other entities by processing the data in random or pseudo-random generators. It is impossible to read the cipher text without decrypting it with special algorithms and secret keys.

In a simplified way, encryption is the way of scrambling the readable data into the noninterpretable format. Encryption technique is one of the best ways of

Table 13.4 Approaches for detecting and avoiding the man-in-the-middle attack

Protocol	**Data attacked**	**Approach**
Lai *et al.* [28]	Communication channel between the mobile management and the home server	Mutual authentication and key agreement provided between the multiple M2M devices and to the core network
Cespedes *et al.* [29]	Impersonate of mobile router or a relay router	Keys are generated at the mobile router and the relay router for authentication and are based on the concept of symmetric polynomials
Farash *et al.* [30]	Data between the sensor node, users, and gateway node	Mutual authentication
Jiang *et al.* [31]	Data between the sensor node, users, and gateway node	Mutual authentication

protecting the data. Without encryption, a piece of data or message is readable to anyone and it is a threat of losing the personal information; the intruder can simply get access to the medium, intercept the information, and interpret it to gain access or steal the important data from your communications [32]. There are different types of encryption algorithms which are being used by the organizations for protecting their data, for example, data encryption standard (DES), triple DES, two fish, and advanced encryption standard.

13.6.3 Digital signatures

Digital signatures are the signatures which are generated through the public key encryption that is attached electronically to the transmitted document to verify its content and contain the full information of the sender identity and make it simple for the receiver to identify that the message has come from the legitimate entity only. The digital signatures are used in the security of any organization because it is used to verify the legitimacy of the entities. Digital signatures are prepared by the private key of the certification authority (CA) and are verified using the public key of the CA. As the CA has never shared the private key with anyone so no one else can prepare the digital signatures. The recipients can verify the certificates by using the public key of the CA and can extract the correct details, that is, a public key of the source from the digital certificates.

13.6.4 Hashing

Hash is a security procedure used to preserve the message integrity. Hash algorithms take the information as input and produce a message digest of fixed length size which is concatenated along with the message and sent along with the information to the recipient. The recipient generates the hash of the information again and compares it with the hash received; a similar hash indicates that the message is unaltered. The secure hash algorithms have been developed by the National Institute of Standards and Technology. Few prominent hashing methods are: SHA-1 which produces a message digest of 160 bits, SHA-256 produces a message digest of 256 bits, SHA-384 produces a message digest of 384 bits, whereas SHA-512 produces a message digest of 512 bits. Hashing can be done without encryption as well. This will reduce the processing burden of the application which does not require confidentiality [33].

13.6.5 Server security

Server security differs from operating system security. The huge number of attacks and hacks only occur when the service is left unpatched or unsecured. Encrypted algorithms are still not completely secure but still serve as a deterrent for most of the hacked networks. For securing the server or service, firewall ports address translation rules that can be used [34].

13.7 Future of IIoT

The future of the IIoT is very bright and limitless. We know that day by day advancement in IIoT is accelerating the production and manufacturing rate of the industries. Automation is making the lives of the managers and the engineers simpler than ever. According to the IIoT analytics in 2016, there are 4.7 billion IIoT devices connected to the Internet. But in today's scenario, there are 11.6 billion IIoT nodes connected to the Internet and it is expected that by 2025 there will be more than 25 billion IIoT nodes connected to the Internet. The future of the industrial IoT is going to be Industry 5.0, where all the machines and the smart devices are going to be connected to the Internet and the owner of the industry will keep an eye on the industry from a remote location area. The industries are going to be fully automated and the machines in the industry will work at the command of its manager or owner. The data will be also stored in the cloud [35–39] and only the legitimate user can fetch the data from the cloud, and the owner will not disturb any IIoT node or smart device for fetching the data as he can directly contact to cloud for the update and the data. Every single detail, such as updates, warnings, production, and manufacturing rate of the industry, will be provided by the sensor to the company owner's smart devices. This will increase privacy but may make this system prone to some different kinds of attacks by the intruder. Bringing the future of IIoT to everyone's life will require very close synergy between the IIoT and the different network platforms [40–46].

13.8 Introduction to blockchain

A blockchain is a technology that is digitally decentralized and uses the public and private distributed ledger to record the transactions made by computers [47–51]. The process is robust enough to prevent intruders. The blockchain is the technology that has been used for the transaction of cryptocurrency over the world and is growing rapidly over the past 2 years, for example, the blockchain acts as a digital ledger which stores all the bitcoin transactions that have been performed. The blockchain is the network that continuously grows by adding new blocks to the chain in the system. The block in a blockchain carries all the information of the previous block along with the current user address, the current message, the previous block hashed value, and the address of the previous block [52,53].

A blockchain is a distributed data structure network system that is disseminated among the members of a network. Whenever the transaction starts between the users, an empty data block is set up to record the content of the first user and others. Every block in the network is assessed by the numerous nodes of the network, and the valid blocks are time stamped. Therefore, the association is ascertained between the blocks for creating a chain of blocks in the network. Each and every block in the network is identified by its cryptographic hash function. There is a valid record of every transaction in the network which is also authenticated by other participants of the network chain. If an intruder tries to tamper the data of the block, he needs to modify more than 51% of the record of transactions in the network, which is almost near to impossible. This means it is very hard to tamper the data of the blocks of the chain when it attains a particular size limit in the network.

In the blockchain network, every user negotiates through his own personal node; these nodes are connected by a peer-to-peer network. Each user device uses the blockchain technology through the use of asymmetric cryptography. Every signed transaction is announced by the user's node to one of its peer connections. All valid transactions are relayed continually until the entire network is covered, while every invalid transaction is discarded from the network. In the mining process of blockchain, every validated transaction is packaged into a time-stamped user block. The new block will be added to the chain of blocks if it contains valid transactions and refers to the correct previous block on the chain network, otherwise, it is discarded from the network [53].

Figure 13.5 illustrates the capabilities of blockchain to provide plenty of security measures to the Internet of things and its applications. The development of the blockchain brought systematic mechanisms and decentralized IoT networks. Due to the decentralization, the communicating nodes became capable of self-management which eliminated the need for centralized databases, thereby dropping the cost of IoT network deployment and management. A peer-to-peer interconnection is provided by the blockchain for the data communication in the IoT networks, without involving the CPU of the system. In this way, the distributed computation and communications happen to process billions of transactions in the blockchain over the whole network. Therefore, the computing power and storage capacity of the blockchain is geographically distributed and the idle devices are utilized to process the transactions worldwide [54].

13.8.1 Types of blockchain

13.8.1.1 Private blockchain

The private blockchain is regulated by an owner or member of the blockchain who has any authority of transaction in the network. In a private blockchain, every new user requires an entity to join the network. Therefore, the private blockchain networks are not centralized because they are administered by the owner or a member of

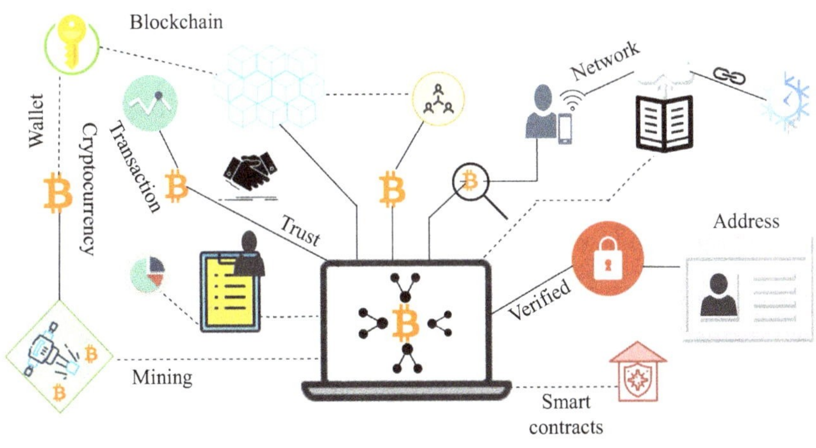

Figure 13.5 Security applications of blockchain

a reserved community. Even transactions are not transparent in the private blockchain because the shared secrets are not available to the outsiders, hence they cannot interpret and validate. The private blockchain is considered as the decentralized operating network where the blockchain participants are well known to each other.

13.8.1.2 Public blockchain

In this type of blockchain network, the user does not need the consent of the entity to join the network. In a public blockchain, everyone can publish and validate their transactions in the network. The miners in the network always receive the reward for their validation work. The public blockchains are mostly used where massive consumer device interaction is required.

13.8.1.3 Federated or consortium blockchain

The consortium blockchain is a semiprivate blockchain where only a selected set of authorized nodes can participate to make or read transactions. The consensus is managed by the set of the people of the enterprise. There are various consensus algorithms being used but the most significant are Tendermint and Proof of Authority (PoA). As only selected nodes are permitted to validate, therefore the consortium blockchain provides high transaction speed per second. There are a few platforms available (Multichain, Hyperledger, Ethermint, etc.) where own consortium can be formed [55].

13.8.2 Role of blockchain in industrial IoT

Industries are operated at large and have involvement of many entities such as machines, IIoT nodes, suppliers, and clients. Various sorts of communication happen between intelligent vehicles, smart machines, clients, etc. over the unsecured medium. Thus, the security is an essential element of any network; therefore, the industrial communications can be possessed through the blockchain networks for enhanced security. The blockchain can revolutionize the working style of the industries. Smart contracts can accelerate the working process and the data management of the industry [55].

The blockchain can be used in the smart industries for tracking the data recorded by the sensor devices and also to prevent the replication of the data in the network. Locating devices and components, authenticating the staff, etc. are not simple in industries due to the mass population. A distributed ledger can provide help in locating the devices, authenticating, and securing the data in transit. The smart sensors used in the industry use the blockchain for exchanging the data between the devices in the network. The blockchain uses the distributed ledger in the smart industries because the data remains protected despite the failure of one node as there is no single point of data storage. With the help of smart contracts in the blockchain, the devices remain anonymous in the smart industry network. The blockchain has simplified business workings and has provided a better customer experience [56]. The various benefits of blockchain in Industrial IoT have been depicted in Figure 13.6.

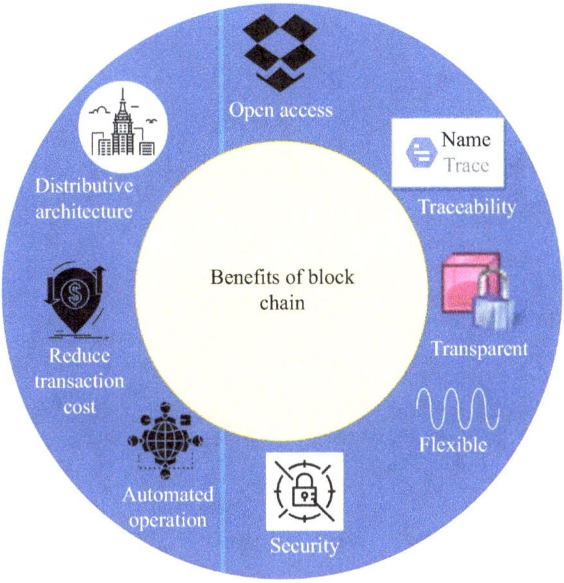

Figure 13.6 Benefits of blockchain in IIoT

13.8.3 Security applications of the blockchain in the IIoT

The blockchain technology finds its applications in diverse areas. The evolution of blockchain started from Bitcoin (Blockchain 1.0). Later, smart contracts were developed (Blockchain 2.0) followed by the efficient and speedier transactions through a Directed Acyclic Graph (Blockchain 3.0). Beyond the transferring of cryptocurrencies and smart contracts, the blockchain technology is finding its use in IoT applications –such as data storage, authentication, time stamping, intelligent transport communication, supply chain management, and security in the mission-critical scenario.

The blockchain technology can be used in IoT agricultural applications for tracking the food supply from worldwide. This tracking system is based on RFID technology and blockchain, which aims to achieve food safety and quality to reduce the loss during logistics. It further helps to grow the chain network and store the large heterogeneous data in an efficient manner [52]. Figure 13.7 represents the various applications of Blockchain in IIoT.

13.8.3.1 Applications in IIoT

IIoT security: The blockchain-based applications increase security and provide a full guarantee that the data published by a user in the IIoT network is neither forged nor altered by third parties.

Data access guarantee: The blockchain distributes the data among peer to peer industrial network in a secure fashion and does not allow any intruder to get access

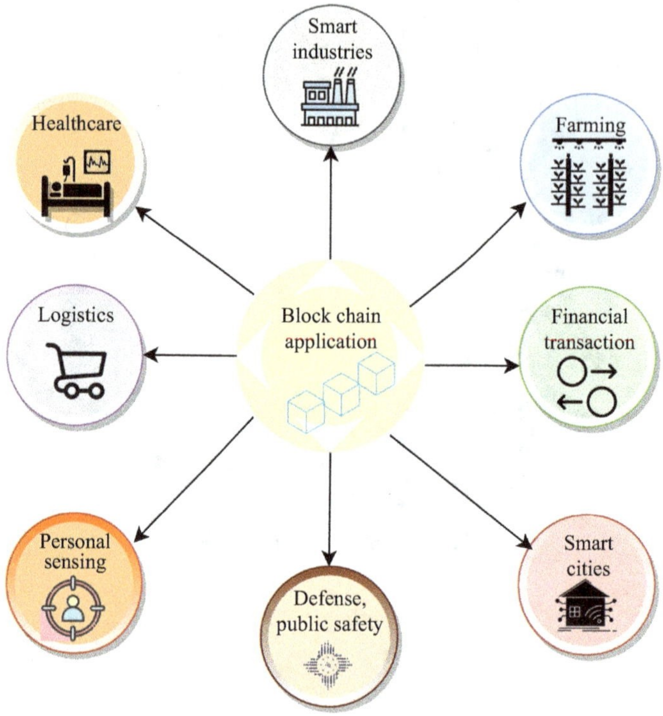

Figure 13.7 Diverse applications of blockchain

to it. The blockchain avoids the denial of service attacks and permits only the legitimate candidates to make or read the transactions.

Integrity of data: It is very important for the IIoT networks to assure that the information collected from the different IIoT devices has not been altered by the intruder, as the modified and malicious data can change the whole meaning of the information that may pose threats to the industry.

13.8.3.2 Applications in big data and data analytics

The ideal smart industry collects a large amount of data from many different devices of the industry. The collected data are big in size and carry the precious information in context to industrial growth. The data analytics is very helpful for industries while predicting future demands. The blockchain can be very helpful in big data and data analytics by eliminating three major problems namely data collection, data trustworthiness, and automatic reliable data circulation. The realization of blockchain networks requires the knowledge of data dissemination sources. The smart contracts can be very helpful for data circulation as they are able to exchange the data securely and transparently. The smart contracts make the content totally transparent to third parties and hence they cannot read the content unless permitted [55].

13.8.3.3 Applications in vertical and horizontal integration systems

The horizontal and vertical integration systems are important to automate data exchanges and to help communicate with suppliers and clients through manufacturing execution system (MES), product lifecycle management (PLM), and IoT platforms. But the smart industry demands higher integration levels. Therefore, the aforesaid platforms may not be shared with the other industrial partners or clients because the additional integration systems are more expensive. The blockchain can serve a good role in the recording, manufacturing, and transactions in the horizontal integration systems. In the supply chain networks, the recording of transactions can help to keep track of goods and also provide transparency to third parties.

13.8.3.4 Applications in autonomous robots and vehicle

Automation is a very special and important key in smart industry applications. The industries are now making ample use of automated products, devices, and machines, for example robots, and driverless autonomous vehicles. Robots and autonomous vehicles are being driven through commands which are exchanged over the unsecured medium. Thus, the blockchain enables robots and other machines to accept legitimate commands through smart contracts [55].

13.8.4 Significant consensus of blockchain

Consensus is an agreement on which nodes of the network agree upon during the processing of a new block. As the new blocks are formed on the basis of consensus, therefore they must be fault-tolerant. Table 13.5 provides an introduction to the two significant consensus algorithms as well as provide a comparison of their working.

13.8.5 Challenges in the realization of blockchain

Scalability: The biggest challenge in the blockchain network is scalability. The architectures supporting the blockchain-based smart industry applications must have to withstand the sufficient amount of traffic that the industrial applications

Table 13.5 Comparison of most commonly used consensus of blockchain

Proof of work (PoW)	**Proof of stake (PoS)**
Miners use their device computing power to solve a puzzle in order to add a block to the chain	Miners are chosen by the algorithm based on the stakes. Miners do not compete with each other to solve the puzzle
The addition of malicious block (attack) is only possible if the attacker device has computational more powerful than 51 percent of the total network	The addition of a malicious block (attack) is only possible if the attacker has more than 51 percent of the total cryptocurrency on the network
A reward is given to the first miner for solving the puzzle	There are no rewards as there was no competition instead the user charges a transaction fee for creating the block

mostly generate. Such amount may create a problem for the centralized cloud-based architectures in the blockchain [57,58].

Privacy and security: Not all devices in the smart industries (Industry 4.0) have the capability to execute the complex security processes due to their restricted computing capabilities and power constraints. Despite being one of the best, RSA-based cryptography may not be the most relevant for certain resource constraint IIoT devices. Since the quantum computers are at the verge of making their existence in large, therefore, the challenge is how the blockchain would still be able to retain the transactions confidential [58].

Energy efficiency, throughput, and latency: Frequent mining, inefficient peer to peer protocols, and computationally complex cryptographic algorithms have a consequence on the energy consumption in every scenario, although such factors are more critical when the battery-operated devices are used. The challenge is regarding blockchain throughput and latency that are based on the consensus algorithm and how blocks are added to the blockchain. In fact, both factors usually increase latency.

Management of multichains: A company may operate its financial transactions using Bitcoin, while the smart contracts are carried out on applications that depend on Ethereum. Therefore, the solutions would have to be formulated and executed to use the different blockchains at the same time.

13.9 Introduction to AI

Artificial intelligence is the technology that deals with the transferring of human behavior and intelligence to the machines. AI provides the framework for solving problems after learning from human intelligence and behavior. The humans must supply all the related information, knowledge, and operations needed for artificial intelligence [59–63].

Artificial intelligence is so flexible that it can be used in any field of learning technology; recently, AI has been attracted to the smart industries as shown in Figure 13.8. The part of AI and deep learning (DL) is widely used for solving the problems of smart industries in a real-world application, such as the smart grid in the smart industries and the processing of data from different devices at different locations. Eventually, the large-scale data generated by the industrial nodes will overburden the traditional industrial mechanisms, resulting in either reduced system efficiency or complete system failure. Therefore, it puts new demands on the large-scale smart industries that how to effectively and collaboratively implement artificial intelligence in industrial applications [64,65].

13.9.1 Types of AI

13.9.1.1 Reactive machine AI

Reactive machine AI is developed for all those machines that run entirely based on present data. The reactive machine can only perform narrow ranged and predefined tasks. For example, the IBM chess program that has beaten the world champion, Gary Kasparov, in chess.

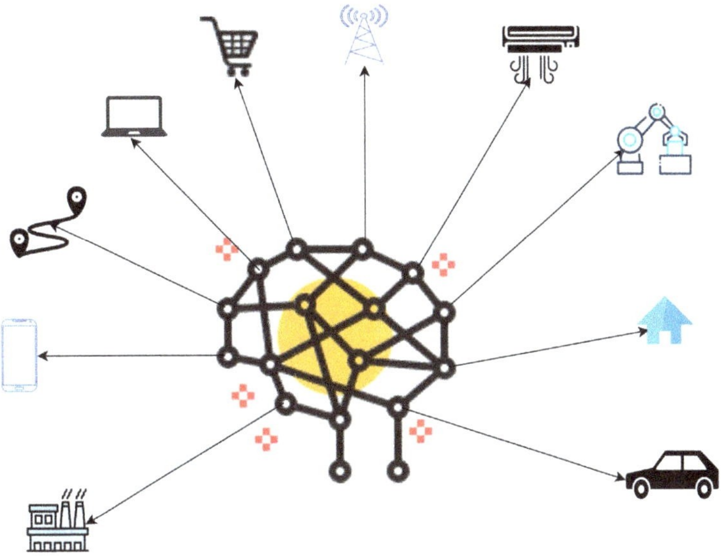

Figure 13.8 Incorporation of artificial intelligence in smart industries and its applications

13.9.1.2 Limited memory AI

Limited memory (LM) AI is developed to improve the performance in machines by incorporating intelligence. Limited memory AI makes the use of previous data that is stored in memory. LM AI has the temporary memory which is used to store the past data that helps the machine to predict future actions. For example, self-driving cars use a number of different sensors to identify different activities on the road and collect the information accordingly and store it in their memory for future actions.

13.9.1.3 Theory of mind AI

This type of AI is a more advanced type of AI. In this type of artificial intelligence, the machines are playing a role in psychology and mainly focus on emotional intelligence so that human thoughts and beliefs can be adequately comprehended. This AI has not fully realized and research is still underway.

13.9.1.4 Self-aware AI

Self-aware AI is the intelligence where the machine has its own intelligence and becomes self-aware of the working and the maintenance of faults in the system. But as of now, self-aware AI has not reached the level of realization [66].

13.9.2 Applications of artificial intelligence

The IIoT applications of AI are Smart Logistics and Predictive Maintenance. The applications are affluent in terms of involved parameters, associated data, and process flow interactions. The user includes suppliers, purchasers, logistics providers, conveyance automobiles, containers, pallets, manufacturing machines, robots, and workforce. In IIoT, the users could be both consumers and producers.

- ***Smart logistics:*** When the stakeholders exchange the goods between each other, then the smart logistic AI is used in the smart industries. Smart logistics also handles the environmental conditions, monitoring, and tracking of goods in industries. IoT brings the smart logistics into a fully automated and adaptive end-to-end process providing optimization across all the areas involved.
- ***Predictive maintenance:*** Predictive maintenance (PM) plays an important role in the smart industries. PM helps to predict the failure and the service requirements of the machine. It also prepares a scheduled maintenance plan to ensure that the production does not get hindered due to the malfunctioning of machines in the industry. The improved operator safety in an industrial environment can be achieved with the pre-emptive detection of unexpected failures in machines [67].

13.10 Introduction to machine learning

Machine learning is latterly the same as artificial intelligence which also refers to the intelligent behavior and actions learned from human beings. Generally, ML includes a number of processes like perception, understanding, learning, judgment, reasoning, planning, designing, and solving different problems. The main motive of ML is to utilize a machine to replicate a human's thinking process and behavior that can be predictable and help to replace the human from the corresponding work in that particular process.

Figure 13.9 shows the numerous new ideas that ML provides for optimized and better performance of IIoT applications. The network interconnection with different devices and data flow analysis in the smart industries using ML methods solidify four features of the smart industry: intelligent production, networked collaboration, personalized customization, and convenient service extension [68].

13.10.1 Types of ML

Machine learning provides different types of support, based on learning models. The three models on which ML works are as follows.

13.10.1.1 Supervised learning

In supervised learning, we make the computer to learn different classifications made by humans. In this type of learning, we give some set of input and output to the computer so that the computer can learn from the input given by the user. Once the input and the output are given to the computer, then the computer starts predicting the results of similar different problems according to the given data.

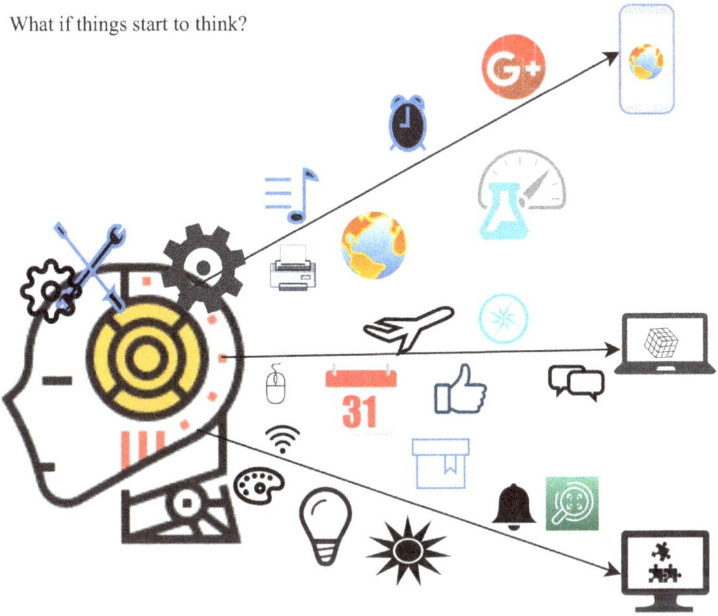

Figure 13.9 Use of machine learning in Industry 4.0

13.10.1.2 Unsupervised learning

Unsupervised learning is the learning in which the system or model learns from the observations and finds the information by themselves. There is no supervised programming before use. The unsupervised learning is based on behavioral learning; it makes inferences from the inputted data and visualizes the output. This type of learning is not effective in the initial stages but becomes more efficient once the computer has gained enough knowledge.

13.10.1.3 Reinforcement learning

In reinforcement learning, the computer tends to work in a dynamic environment in which the computer and the learning model have to complete the task without being told the way to accomplish it. This type of learning goes through the number of failures and learning. If the machine is able to complete the task without any external input, then it can be termed as reinforcement learning [69].

13.10.2 Applications of machine learning

ML is being used in many applications, for example, image processing, fraud detection, pattern reorganization, and many more. The use of decent ML algorithms in IIoT helps many organizations to arrange their big data systematically and securely. The integration of ML in IIoT can help in the improvement of industrial processes, cutting costs, and creating better opportunities for the customer [70]. ML mostly focuses on acquiring knowledge from different parameters in industries, which helps to improve the efficiency of problem-solving.

- ***Intelligent manufacturing:*** Intelligent manufacturing is an important foundation of the industrial IoT. Nowadays in heterogeneous networks, the network data analysis and the network optimization in the smart industries have become more intricate than the social networks. IoT sensors and terminal data in the network can be optimized with the help of graph matching in the smart industries. The allocation of the network resource and the flow size of data can be identified with the help of deep learning reinforcement. Machine learning provides different solutions for the industrial IoT analysis such as heterogeneous equipment for networking, multidata processing, and the discovery of knowledge, decision support, or security of information.
- ***Intelligence traffic guidance (ITG):*** Intelligent traffic guidance helps to control the active traffic with the use of an efficient road network. The road networks collect traffic information and operate the traffic flow accordingly. ML can be used to recover the missing information from the traffic video to improve the quality of service. Based on the digital map, the deep learning methods can be used for predicting the possible congestion time in the network, which will support time customization in real-time for a traffic light.
- ***Smart grid maintenance:*** Smart grid maintenance inspects the machine at regular intervals and evaluates the risk using ML models. The environmental warning information, the operation and maintenance information, and the basic information of equipment can be obtained using a smart grid operated remotely in the smart industries. On the other hand, the smart grid can analyze the relationship between the operation parameters of fault equipment and the network operator. This relationship can build the classification model of the smart grid and later can classify the different fault types of problems in the smart grid while making use of the experiential type classification learning [66].

13.10.3 Challenges in implementing machine learning and artificial intelligence-based security solutions for IIoT

The major challenges faced by the ML technology lies in the integration and the security violations of the data. The analysis of the data will become more cumbersome especially when it is sourced from the different sensors in the different formats. The threat of the security can arrive from any part of the system in IIoT. The major challenges of implementing ML in IIoT are illustrated in Figure 13.10.

ML still requires a different kind of expert-level knowledge in many specific fields. There are still many challenges in the IIoT level of intelligence. The training of a new machine in the new format requires a very large amount of data. Machine learning requires the high quality and quantity of data for better learning which is very time and cost consuming work. The obtained data becomes relatively sparse with the loss of random packet data. Every smart industry must build the ML model to grasp the intelligence of structured and unstructured data which can help the network learning behavior and the network diagnosis behavior to simulate network security [68].

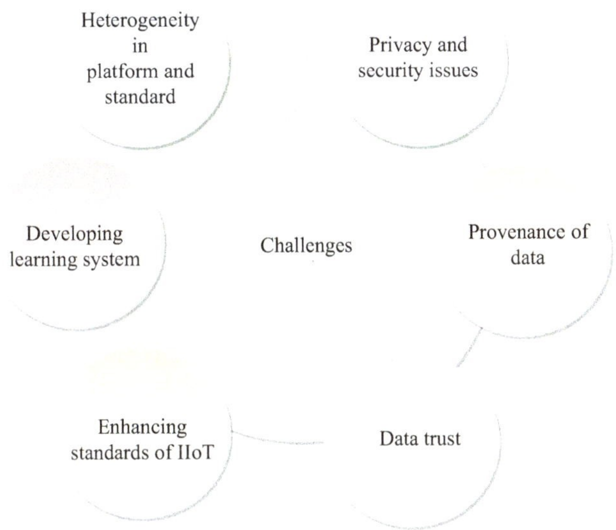

Figure 13.10 Major challenges in implementing ML-based security solutions in IIoT

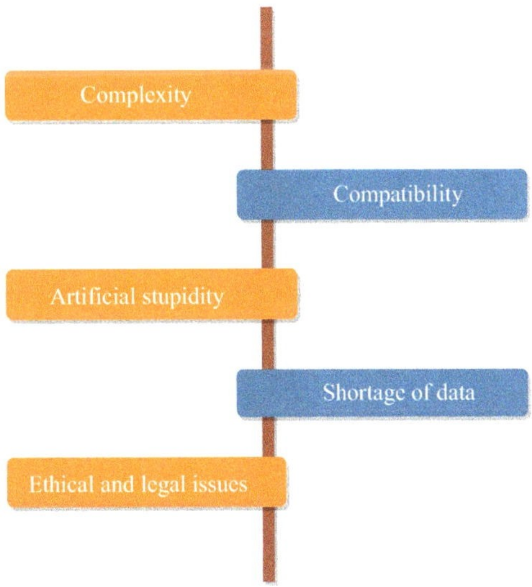

Figure 13.11 Major challenges in implementing AI-based security solutions in IIoT

The main challenges faced by artificial intelligence in IIoT are depicted in Figure 13.11.

Nowadays most of the smart industries have incorporated AI so that they can work more efficiently without the human presence in the industry. Every network system is so smart that they know how to do their work in an optimized manner and more securely in the network. AI also requires a large amount of data for becoming self-efficient in different domains [71].

13.11 Conclusions

This chapter summarizes the advancement of industries over the decades, that is, from Industry 1.0 to Industry 4.0. The chapter also gives insights into how the users are directly connected to their organization without the physical presence in the industry. The industrial IoT equipped with billions of nodes and smart sensors makes the complete industrial operations automatic. These nodes are directly connected to the machines which maintain their operational timings, thereby ensuring efficiency in all aspects. The chapter also highlights the various challenges faced in transforming the present-day industries to smart industries. As the industries are getting connected to the Internet, vulnerabilities, threats, and risks of cyberattacks on the industries are also increasing day by day. For getting protection from these risks, various security considerations have been discussed which may help the industries to secure their personal and official data. Additionally, this chapter discusses various remedies such as blockchain, AI, and ML that can be used to identify vulnerabilities, combat threats, and overcome risks.

References

[1] Waidner M., and Kasper M. 'Security in industrie 4.0: Challenges and solutions for the fourth industrial revolution'. *Design Automation Test in Europe Conference Exhibition*. 2016; 1303–1308.
[2] Esfahani A., Georgios M., Rainer M., *et al.* 'A lightweight authentication mechanism for M2M communications in industrial IoT environment'. *IEEE Internet of Things Journal.* 2019; 6(1): 288–296.
[3] Industrial Internet of things. *En.Wikipedia.Org*.[online]. 2020. Available from https://en.wikipedia.org/wiki/Industrial_internet_of_things [Accessed 23 February 2020].
[4] Jeevitha T., and Ramya. L. N. 2018. [online]. 2020. Available from https://www.researchgate.net/publication/330336790_INDUSTRY_10_TO_40_THE_EVOLUTION_OF_SMART_FACTORIES. INDUSTRY 1.0 TO 4.0: THE EVOLUTION OF SMART FACTORIES [Accessed 23 February 2020].
[5] Industrial Revolution – From Industry 1.0 To Industry 4.0. *Desouttertools. Com.* [online]. 2020. Available from https://www.desouttertools.com/industry-4-0/news/503/industrial-revolution-from-industry-1-0-to-industry-4-0 [Accessed 23 February 2020].
[6] Industry 1.0 to 4.0: The evolution of smart factories. APICS Magazine. *Apics. Org.* [online]. 2020. Available from http://www.apics.org/apics-for-individuals/apics-magazine-home/magazine-detail-page/2017/09/20/industry-1.0-to-4.0-the-evolution-of-smart-factories [Accessed 23 February 2020].

[7] Buntz, B. The Top 20 industrial IoT applications". *IoT World Today*. [online]. 2020. Available from https://www.iotworldtoday.com/2017/09/20/top-20-industrial-iot-applications/ [Accessed 23 February 2020].

[8] Applications of industrial Internet of things (IIoT) - RF page. *RF Page*. [online]. 2020. Available from https://www.rfpage.com/applications-of-industrial-internet-of-things/.[Accessed 23 February 2020].

[9] Team DataFlair. Industrial IoT applications. IoI applications in manufacturing – Dataflair. *Dataflair*. [online]. 2020. Available from https://data-flair.training/blogs/industrial-iot-applications/ [Accessed 23 February 2020].

[10] Autonomous Haulage system optimizes surface mining. Komatsu: Komatsu America Corp. [online]. 2020. Available from *Komatsuamerica.Com*. https://www.komatsuamerica.com/autonomous-haulage-system [Accessed 23 February 2020].

[11] automation.com. 2020. *KUKA introduces KUKA connect robotics software platform at IMTS*. [online] Available at: https://www.automation.com/en-us/products/product04/kuka-introduces-kuka-connect-robotics-software-pla#:~:text=Using%20the%20KUKA%20Connect%20platform,device%2C%20anywhere%20at%20any%20time.&text=Leveraging%20the%20latest%20web%20and,multiple%20layers%20of%20robot%20data [Accessed:23 Feb 2020].

[12] [Online]. 2020. Available from. https://www.maersk.com/?gclid=Cj0KCQiAv8PyBRDMARIsAFo4wK1W2DnI1HjwCbPZdLLrY5IDjWzQDZzaF_ZhfuiLAwEiG1i4qHovPIaAn7WEALw_wcB&gclsrc=aw.ds [Accessed 23 February 2020].

[13] Top 5 challenges for industrial IoT (IIoT) implementation – Finoit. *Finoit Technologies*. [online]. 2020. Available from https://www.finoit.com/blog/5-iiot-industrial-internet-of-things-implementation-challenges/. [Accessed 23 February 2020].

[14] Vavra, C. Six IoI implementation challenges and solutions. *Control Engineering*. [online]. 2020. Available from https://www.controleng.com/articles/six-iot-implementation-challenges-and-solutions/. [Accessed 23 February 2020].

[15] Muscat, I. Cyber threats, vulnerabilities, and risks. Acunetix. *Acunetix*. [online]. 2020. Available from https://www.acunetix.com/blog/articles/cyber-threats-vulnerabilities-risks/ [Accessed 23 February 2020].

[16] Lateef, Z. *What are the types of artificial intelligence?. Branches of AI. Edureka*. 2020. [online] Edureka. Available at: https://www.edureka.co/blog/types-of-artificial-intelligence/ [Accessed 23 February 2020].

[17] Li Q., and Guohong C. 'Multicast authentication in the smart grid with one-time signature'. *IEEE Transactions on Smart Grid*. 2011; 2(4): 686–696.

[18] Chim T. W., Yui S. M., Lucas C. K. H., and Victor O. K. L. 'PASS: Privacy-preserving authentication scheme for smart grid network,' *IEEE 2nd International Conference on Smart Grid Communications, Smart Grid Communication*. 2011; 196–201.

[19] Nicanfar H., Paria J., Konstantin B., and Victor C. M. L. 'Efficient authentication and key management mechanisms for smart grid communications'. *IEEE Systems Journal.* 2014; 8 (2): 629–640.

[20] Kumari S., Xiong L., Fan W., Ashok K. D., Hamed A., and Muhammad. K. K. 'A user friendly mutual authentication and key agreement scheme for wireless sensor networks using chaotic maps'. *Future Generation Computer Systems.* 2016; 63: 56–75.

[21] Chung, Y., Seokjin, C., Youngsook, L., Namje, P., and Dongho, W. 'An enhanced lightweight anonymous authentication scheme for a scalable localization roaming service in wireless sensor networks'. *Sensors.* 2016;16 (10): 1653.

[22] Jan, M. A., Nanda, P., Usman, M., and He, X. 'PAWN: A payload-based mutual authentication scheme for wireless sensor networks'. *Concurrency And Computation: Practice And Experience.* 2016; 29 (17): e3986.

[23] Ruhul, A., and Biswas, G. P. 'A secure light weight scheme for user authentication and key agreement in multi-gateway based wireless sensor networks'. *Ad Hoc Networks.* 2016; 36: 58–80.

[24] Panchal, A. C., Khadse, V. M., and Mahalle, P. N. 'Security issues in IIoT : A comprehensive survey of attacks on IIoT and its countermeasures,' *IEEE Global on Wireless Computing And Networking.* 2018; 124–130.

[25] IoT security considerations (Internet of things security). *Sikich LLP.* [online]. 2020. Available from https://www.sikich.com/insight/iot-security-considerations-internet-of-things-security/ [Accessed 23 February 2020].

[26] How to manage IIoT authentication and protect machine identities - IoT agenda. 2020. *Internetofthingsagenda.Techtarget.Com.* [online]. 2020. Available from https://internetofthingsagenda.techtarget.com/blog/IoT-Agenda/How-to-manage-IIoT-authentication-and-protect-machine-identities [Accessed 23 February 2020].

[27] How encryption is powering the future of IoT. 2020. *IoT for all.* [online]. 2020. Available from https://www.iotforall.com/future-iot-encryption/ [Accessed 23 February 2020].

[28] Lai C., Rongxing L., Dong Z., Hui L., and Xuemin (Sherman) S. 'GLARM: Group-based lightweight authentication scheme for resource-constrained machine to machine Communications'. *Computer Networks.* 2018; 99: 66–81

[29] Cespedes. S., Sanaa T., and Xuemin S. 'A multihop-authenticated proxy mobile IP scheme for asymmetric vanets'. *IEEE Transactions On Vehicular Technology.* 2013; 62 (7): 3271–3286.

[30] Mohammad S. F., Muhamed T., Saru K., and Marko H. 'An efficient user authentication and key agreement scheme for heterogeneous wireless sensor network tailored for the Internet of things environment'. *Ad Hoc Networks.* 2016; 36: 152–176.

[31] Qi J., Neeraj K., Jianfeng M., Jian S., Debiao H., and Naveen C. 'A privacy-aware two-factor authentication protocol based on elliptic curve cryptography for wireless sensor networks'. *International Journal of Network Management.* 2016; 27 (3): e1937.

[32] Digital signature. IoT ONE. *IoT ONE*. [online]. 2020. Available from https://www.iotone.com/term/digital-signature/t784 [Accessed 23 February 2020].

[33] Hess, K. Five security considerations for IoT implementations. ZDNet, *ZDNet*. [online]. 2020. Available from https://www.zdnet.com/article/five-security-considerations-for-iot-implementations/ [Accessed 23 February 2020].

[34] Storozhkova, E., Demush, R., Demush, R., and Demush, R. The future of IoT. Internet of things development. IoT future. Perfectial. *Perfectial*. [online]. 2020. Available from https://perfectial.com/blog/future-of-iot/ [Accessed 23 February 2020].

[35] Future IoT for intelligent society and industry. *Ericsson.Com*. [online]. 2020. Available from https://www.ericsson.com/en/future-technologies/future-iot [Accessed 23 February 2020].

[36] Maria S., Yannis N., Spyridon P., Evangelos P., and Evangelos K. M. "A survey on the Internet of things (IoT) forensics: Challenges, approaches, and open issues". *IEEE Communications Surveys & Tutorials*. 2020; 22 (2), 1191–1221.

[37] Evangelos M., Yannis N., Macro M., and Evangelos p., "Security assessment as a service cross-layered system for the adoption of digital, personalised and trusted healthcare". *2019 IEEE 5th World Forum on Internet of Things (WF-IoT)*. 91–94.

[38] Nikoloudakis Y., Panagiotakis S., Manios T., Markakis E., and Pallis E. 'Composting as a service: A real-world IoT implementation.' *Future Internet*. 2018; 10, 107.

[39] Markakis E., Karras K., Sideris A., Alexiou G., and Pallis E. 'Computing, caching, and communication at the edge: The cornerstone for building a versatile 5G ecosystem'. *IEEE Communications Magazine*. 2017; 55, 152–157.

[40] Sideris A., Markakis E., Zotos N., Pallis E., and Skianis C., 'MPEG-DASH users' QoE: The segment duration effect.' *Seventh International Workshop on Quality of Multimedia Experience (QoMEX), Pylos-Nestoras*. 2015; 1–6.

[41] Georgios G., George X., Evangelos P., and Anastasios K., 'Joint assessment of network- and perceived-QoS in video delivery networks.' *Telecommunication Systems*. 2012; 49, 75–84.

[42] Zotos N., Xilouris G., Pallis E., and Kourtis A., 'An MPLS-DiffServ experimental core network infrastructure for E2E QoS content delivery.' *IEEE/ACS International Conference on Computer Systems and Applications. Doha*. 2008; 947–951.

[43] Evangelos M., Yannis N., Evangelos P., and Barbara M., 'Security assessment as a service cross-layered system for the adoption of digital, personalised and trusted healthcare.' *IEEE 5th World Forum on Internet of Things (WF-IoT)*. 2019; 91–94.

[44] Anapliotis P., Markakis E., Sideris A., Pallis E., and Négru D., 'A novel content-aware multipath routing concept exploiting random utility theory principles.' *International Conference on Telecommunications and Multimedia (TEMU)*. 2012; 49–53.

[45] Evangelos K. M., Yannis N., Evangelos P., *et al.* "The FORTIKA accelerated edge solution for automating SMEs security'. *Challenges in Cybersecurity*

and Privacy: The European Research Landscape, Series in Security and Digital Forensics. 2018; 1–45.

[46] Nagar, T. The future of the Internet of things (IoT) in 2020. *Colocation America*. [online]. 2020. Available from https://www.colocationamerica.com/blog/future-of-iot-in-2020 [Accessed 23 February 2020].

[47] Nikolaos A., Yannis N., Ioannis K., Charalabos S., Evangelos P., and Evangelos K. M., 'Cloud-native application validation & stress testing through a framework for auto-cluster deployment'. *IEEE 24th International Workshop on Computer Aided Modeling and Design of Communication Links and Networks (CAMAD)*. 2019; 1–5.

[48] Rebahi Y., Chiu K.T., Tcholtchev N., Hohberg S., Pallis E., and Markakis E. "Towards a next generation 112 testbed." *The EMYNOS ESInet International Journal of Critical Infrastructure Protection*. 2018; 22: 39–50.

[49] Markakis E. K., Karras K., Zotos N., *et al.* 'EXEGESIS: Extreme edge resource harvesting for a virtualized fog environment.' *IEEE Communications Magazine*. 2017; 55 (7): 173–179.

[50] Yacine R., Majid S. G., Nicolas H., *et al.* 'Virtual network functions deployment between business expectations and technical challenges: The T-NOVA approach.' *Recent Advances in Communications and Networking Technology (Discontinued) Formerly Recent Patents on Telecommunication*. 2016; 5 (1). 49–64.

[51] Gardikis G., Pallis E., and Grafl M. 'Media-aware networks in future Internet media'. In: A Kondoz., and T. Dagiuklas (eds.). *3D future Internet media*. New York, NY: Springer; 2014; 105–112.

[52] Tiago M. F. C., and Paula F. L. "A review on the use of block chain for the Internet of things." *IEEE Access*. 2018; 6: 32979–33001.

[53] Quansi W., Ying G., Zhiling C., and Dapeng. W. (eds.). "A block chain-based data sharing scheme in the supply chain by IIoT." *IEEE International Conference on Industrial Cyber Physical Systems (ICPS)*. 2019.

[54] Li B., Mi H., and Min L., and Wang, J. 'BPIIoT: A light-weighted block-chain-based platform for industrial IoT." *IEEE Access*. 2019; 7: 58381–58393.

[55] Tiago M. F. C., and Paula F. L. "A review on the application of block chain to the next generation of cyber secure Industry 4.0 smart factories." *IEEE Access*. 2019; 7: 45201–45218.

[56] Kim K. R. C., Zheng Y., and Weizhi. M. "Block chain in Industrial IoT applications: Security and privacy advances, challenges, and opportunities." *IEEE Transactions on Industrial Informatics*. 16(6): 4119–4121.

[57] Ahmed A. M., Olov S., and Karl A. "A survey of block chain from the perspectives of applications, challenges, and opportunities." *IEEE Access*. 2019; 7: 117134–117151.

[58] Tanweer A. "Blockchain and its role in the Internet of things (IoT)." *International Journal of Scientific Research in Computer Science, Engineering and Information Technology*. 2019; 50 (1), 151–157.

[59] Nikolaos A., Yannis N., Ioannis K., Charalabos S., Evangelos P., and Evangelos K. M., 'Cloud-native application validation & stress testing through a framework for auto-cluster deployment.' *IEEE 24th International Workshop on Computer Aided Modeling and Design of Communication Links and Networks (CAMAD)*. 2019; 1–5.

[60] Rebahi Y., Chiu KT., Tcholtchev N., Hohberg S., Pallis E., and Markakis E. "Towards a next generation 112 testbed." *The EMYNOS ESInet International Journal of Critical Infrastructure Protection*. 2018; 22: 39–50.

[61] Markakis E. K., Karras K., Zotos N., *et al.* 'EXEGESIS: Extreme edge resource harvesting for a virtualized fog environment.' *IEEE Communications Magazine*. 2017; 55 (7): 173–179.

[62] Yacine R., Majid S. G., Nicolas H., *et al.* 'Virtual network functions deployment between business expectations and technical challenges: The T-NOVA approach.' *Recent Advances in Communications and Networking Technology (Discontinued) Formerly Recent Patents on Telecommunication*. 2016; 5 (1). 49–64.

[63] Gardikis G., Pallis E., and Grafl M. 'Media-aware networks in future Internet media'. In A. Kondoz, and T. Dagiuklas (eds.). *3D future Internet media*. New York, NY: Springer; 2014, 105–112.

[64] Simmons, A.B., and Chappell, S.G. "Artificial intelligence: Definition and practice." *IEEE Journal of Oceanic Engineering.* 1988; 13(2): 14–42.

[65] Meng H., Hongwei L., Xizhao L., Guowen Xu, Haomiao Y., and Sen L. "Efficient and privacy-enhanced federated learning for industrial artificial Intelligence." IEEE Transactions on Industrial Informatics. 2019; 1–11.

[66] Lateef, Z. What are the types of artificial intelligence? Branches of AI. Edureka. *Edureka*. [online]. 2020. Available from https://www.edureka.co/blog/types-of-artificial-intelligence/ [Accessed 23 February 2020].

[67] Senthamiz S. A., Ramamurthy B., Aitor H. H., *et al.* "Accelerating industrial IoT application deployment through reusable AI components." *2019 Global IoTSummit (GIoTS)*. Aarhus, Denmark. 2019; pp. 1–4.

[68] Baotong C., and Jiafu W. "Emerging trends of ML-based intelligent services for industrial Internet of things (IIoT)." *Computing, Communications and IoT Applications (ComComAp)*. Shenzhen, China. 2019; pp. 135–139.

[69] Das, S., and Nene, M. J. "A survey on types of machine learning techniques in intrusion prevention systems." *2017 International Conference on Wireless Communications, Signal Processing and Networking (WiSPNET)*. Chennai. 2017; pp. 2296–2299.

[70] Reddy, R. R., Mamatha C., and Reddy. R. G. "A review on machine learning trends, application and challenges in Internet of things." *International Conference on Advances in Computing, Communications and Informatics (ICACCI)*. Bangalore. 2018; pp. 2389–2397.

[71] Rana,. A. K., Krishna, R., Dhwan, S., Sharma, S., and Gupta, R. "Review on artificial intelligence with Internet of things: Problems, challenges and opportunities." *2nd International Conference on Power Energy, Environment and Intelligent Control (PEEIC)*. Greater Noida, India. 2019; pp. 383–387.

Chapter 14

Improved gain vector-based recursive least squares for smart antenna applications

Peter N. Chuku[1], *Thomas O. Olwal*[1] *and Karim Djouani*[1]

In future, the improvement of the utilization of the frequency spectrum for a more desirable quality of service (QoS) hinges upon the use of smart antenna (SA) systems in wireless and cellular networks. A signal's output power can be enhanced by using SA techniques. These techniques enhance the signal power toward expected directions within the network. However, long-range communication applications still face some unsolved problems such as signal fading and cochannel interference. This chapter provides solutions to the problems of signal fading and cochannel interference by enhancing the already conventional recursive least squares (RLS) algorithms in SA arrays over long-range communication channels. This chapter further provides performance-based comparisons between the enhanced RLS algorithm and other beamforming algorithms such as the conventional RLS algorithm and the least mean squares (LMS) algorithm. In the conventional RLS algorithm, matrix inversion computations are not required. This is because the conventional RLS algorithm already determines the inverse correlation matrix directly. For this reason, the RLS algorithm can save computational power. The RLS algorithm is enhanced through the introduction of a constant denoted as $\boldsymbol{m}$ to the gain factor of the algorithm and the reason for this introduction is to yield an improved gain vector. Results from our simulation results indicate that the enhanced RLS decreases mean square error (MSE) of the signals, which makes the output of the filter smoother. These results also show that our enhanced RLS improves SNR when compared to the conventional methods. The results of this study show the benefits of improving the gain of the SA, which yields an increase in the sharpness and range of the SA over a long-range communication channel.

14.1 Introduction

Recently, there is an increased desire for the proliferation of mobile services that will cater to the needs of wireless and mobile communications. Some of these

[1]Department of Electrical Engineering, FSATI, Tshwane University of Technology, Pretoria, South Africa

multimedia services include multimedia data transfer, Internet access, and audio-visual communications [1]. The ever increasing demand for the utilization of multimedia services has preempted the reliance of millimeter wave communication. This communication is highly required in fifth-generation (5G) and Internet of things (IoT) [2]. IoT connects multiple electronic devices such as fridges, televisions, cars, and cell phones to each other through the use of the Internet. Furthermore, there is a recent use of cloud-based technology for communications and storage purposes. This technology utilizes servers that store high volumes of data files. These files can be transferred from a user/location to another user/location [3]. For the above-mentioned demands, the utilization of multiple element antennas (MEAs) will serve as a formidable solution [4]. MEAs have the ability to prevail over the limited capacity challenges that the single input single output (SISO) systems encounter. MEAs are termed as adaptive or directional antennas because of their ability to adapt their directionality to vary interfering signals. Directional antennas have many advantages over omnidirectional antennas. Some of these advantages are that they have a wider transmission range, have less interference, an enhanced network capacity, and they can reuse spatial frequencies [5]. Hence, the utilization of directional antennas in wireless networks cannot be overemphasized. We will classify directional or adaptive antennas as smart antennas for the purpose of this study. For definitive purposes, a smart or adaptive antenna is a combined system of antenna arrays, which utilizes digital signal processing techniques for the purpose of transmitting and receiving signals in an adaptive, sensitive, and spatially approach [6]. Some properties are associated with SAs of which one of them is spatial filtering. Spatial filtering enables the SA to collect transmit-energy coming from a desired direction and simultaneously hinders the transmit energy from getting to another direction.

SAs are adaptive array antennas, which imply that they have multiple input and multiple output (MIMO) antenna arrays with adaptive signal processing algorithms. These adaptive algorithms are utilized to recognize the direction of arrival (DOA) of the signal [7]. SAs are used to determine beamforming vectors that continuously locate and track the beams of the antenna arrays located in a communication network. SAs primarily have two functions. These are to perform direction of arrival (DOA) of a signal and beamforming, respectively [7]. It is important to note that SA designs involve signal processing of antenna arrays. SAs utilize beamforming to enhance the capacity and range of a signal that is transmitted or received within a communication network.

The vector weights of an adaptive antenna require a specific technique in which they have to be updated and that technique is known as adaptive filtering [8]. However, in recent years, studies [9] have shown that adaptive algorithms have been utilized as a technique for adaptive filtering in SAs. The manner in which most beamforming algorithms (least men squares, normalized least mean squares, etc.) function is highly dependent on the state of their signal and filter order. However, the functionality of the recursive least squares (RLS) algorithm (which is also an adaptive algorithm) relies on the forgetting factor of the algorithm. The RLS algorithm also consists of a complex weight that is comprised of the relative amplitude and phase shift for each antenna [9]. The RLS algorithm is more superior to other beamforming algorithms such as the least mean squares (LMS), constant modulus algorithm (CMA), etc., because it possesses a quicker convergence rate and has a lesser mean square error (MSE) [8]. Recent studies [8,9] and [10] have proposed several solutions that suggest significant enhancements in the performance of the conventional RLS algorithm. However, literature review showed that these improvements required a high computational cost and complexity [11].

This study is aimed at presenting an enhanced RLS-based SA design. This design will improve the gain and directivity of SA in a long-range communication network. Our approach is derived from the knowledge that the gain vector $k(n)$ plays an important role in the functionality of the RLS algorithm. In theory, output computations of the RLS filter involve a combination of the input sample, tap coefficients $u(n)$, and the weight vector $w(n)$. On the other hand, the weight vector is continuously updated by taking the product of the gain vector $k(n)$ and the error signal $e(n)$ [9]. The value of the weights vector $w(n)$ is dependent on the gain vector $k(n)$ and the error signal $e(n)$. Our work is based upon the fact that the value of the gain vector $k(n)$ will affect the outcome of the weights vector $w(n)$. This enhancement of the gain vector $k(n)$ will lead to a significant improvement in the rate of convergence of the RLS algorithm.

The rest of the chapter is organized as follows: a literature review on related work is presented in Section 14.2, system modeling is discussed in Section 14.3, the performance evaluation and results of the enhanced RLS algorithm are discussed in Section 14.4. Section 14.5 concludes the chapter.

14.2 Related work

Studies in [12] show that smart antenna arrays imbibe approaches of intelligent signaling into their functionality when compared to the switched-beam array approach. Smart antenna array model makes use of digital algorithms that form beams in a communication link that comprises of interfering signals. Figure 14.1 illustrates the beamforming signal of a SA [13]. This signal also consists of interfering signals that operate on the same frequency as the beamforming signal. The adaptability of SAs enables them to null interfering signals in order to enable an enhanced signal. The interfering signals mentioned above are nulled when the weights of each antenna elements in the array are varied.

Literature studies in [10] focused on the benefits of SA transceiver architecture. Authors in [10] also discussed the usage progressions of SAs in future technologies. These progressions include cross-layer optimization, reconfigurability to changing channel propagation and network conditions, and multiuser diversity. Several challenges however accompanied this presentation such as the design of a suitable method of simulation and the accurate modeling of interference, channel characteristics, and implementation losses which were presented in [10]. Studies in [14] proposed a delay-aware scheme which has a high energy efficiency. This scheme is utilized in resource management and energy

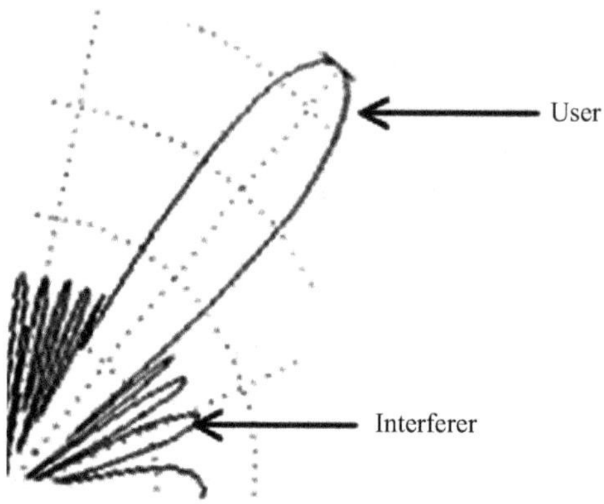

Figure 14.1 Adaptive array of an antenna. Adapted from [13]

conservation in the 5G radio network architecture. Authors in [14] based their scheme on the outcome from comparing the delays of a radio's access points (AP) when a delay-sensitive transmission is required. This scheme results in the provision of an optimal TV-white space through the use of a spectrum broker. The use of the spectrum broker is to manage the consumption of energy in a 5G communication network which takes into account the delay-aware scheme and the evaluation of the queuing delays that are being compared.

Another characteristic of SAs is that they use beamforming techniques to improve the range of the transmitted signal and they also enhance the signal capacity in a communication channel. Beamforming, as defined in [15], is the process of differentiating the spatial properties between noise and signal. Chuku *et al.* in [16] further defined beamforming as a SA technique used to enhance the range and directivity of a transmitted signal and this technique provides a better signal-to-noise ratio (SNR). The beamformer is the system that is usually used to derive beamforming in an antenna array. The beamformer receives a desired signal that is beamed from a specific direction and then attenuates or nulls signals originating from other directions [15]. The block diagram in Figure 14.2 illustrates adaptive beamforming with different antenna elements (x) and their respective weights (w) [9].

Beamforming in SAs can be achieved by some adaptive algorithms such as the least mean squares (LMS), recursive least squares (RLS), constant modulus algorithm (CMA), and other hybrid versions consisting of all or some of these algorithms. These adaptive algorithms are known to adjust the weights of incoming signals, sample these signals, and then keep updating the signals so these signals eventually converge to an optimal solution. Adaptive algorithms are very vital in the functionality of SAs. The SA confines the broadcast energy of the transmitted and received signals by narrowing the resultant beam. After the beam has been narrowed, the SA then optimizes the frequency in which the signals are distributed through space in a multivector network environment on a real-time basis. The SA

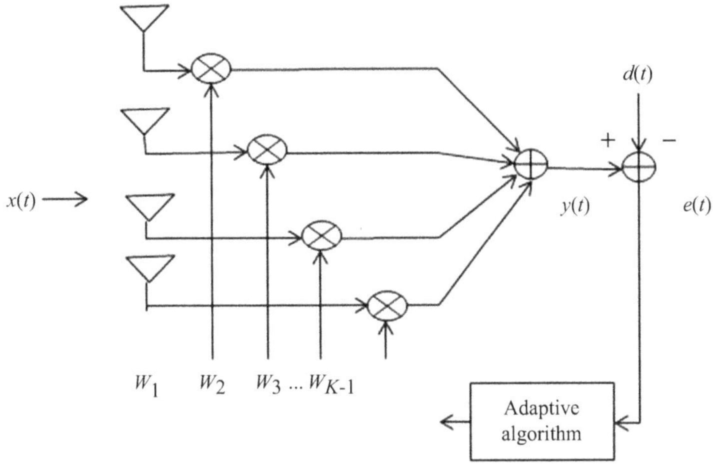

Figure 14.2 Adaptive beamforming. Adapted from [9]

distributes a signal by directing the signal to an intended user while redirecting the signal from unintended users within channels that share the same communication cells with the intended user [17].

Authors in [18] presented an analytical performance comparison between some digital signal processing techniques that are used for beamforming. These digital signal processing algorithms are used in a SA system. The study in [18] took into consideration, the advantages and disadvantages of the three adaptive algorithms that were discussed. The algorithms that were taken into consideration were the RLS, LMS, and the CMA. The authors in [18] also proposed that high error rates would be experienced when the smart antenna engages in the coverage of signals when the number of subscribers is increased. In most cases, this signal coverage is more than the antenna's capacity, thereby increasing the rate of errors in the antenna element. Results of the study in [18] showed that the CMA algorithm rejects interference more than other algorithms because the CMA algorithm has the ability to block unwanted signals which emanate from neighboring networks. Furthermore, results from [18] showed that the LMS algorithm makes use of techniques that iteratively updates the weights vector in an opposing direction to that of the gradient vector. The benefit of this technique is that it causes the LMS to reduce the mean square error (MSE) of a signal in comparison to other digital signal processing techniques. Studies conducted by authors in [8] show that the RLS algorithm has a faster convergence rate when compared to the LMS algorithm because matrix inversion computation is not a requirement of the RLS algorithm. In the RLS algorithm, the inverse correlation matrix is computed directly into the filter. Authors in [19] further carried out some studies about the RLS algorithm. These studies proposed a methodology that depends on adding an RLS-based adaptive linear equalizer to an antenna on its receiver end. The studies in [19] further investigated the performance of the RLS-based adaptive linear equalizer

(ALE) in a ZigBee system. These tests were carried out in varying multipath fading systems. The authors then compared the performance of the RLS algorithm with that of the LMS algorithm. Performance results received from their simulations showed that the RLS-based ALE is very efficient in a ZigBee system having a minimum bit error rate (BER). These results further showed that the RLS-based ALE has a high adaptation response. Results gotten from the studies in [2, 11, and 12] demonstrated the unique advantages of using the RLS algorithm for beamforming purposes. These results also suggest that the RLS algorithm is the best algorithm that can be used at the base station (BS) of a SA system. However, the study in [19] came with a limitation. This limitation results from the proximity of the angles between the interfering and desired signals. This proximity of angles is because of the reduction of the SNR from 10 dB to 2 dB.

Concluding remarks from studies carried out in [18], [19], and [9] showed that the RLS algorithm is the preferred adaptive technique that can produce the best beamforming signals in the direction of desired users. It was noted clearly that RLS algorithms designed in those studies demonstrated critical limitation concerning failure to completely nullify cochannel interference (CCI) in a signal [20]. As a result of this shortcoming, we seek to improve the gain vector of the RLS algorithm for beamforming over long-range communications by further introducing a constant demoted as m to the gain factor of the conventional RLS algorithm.

14.3 System model

An enhanced RLS-based model algorithm is modeled in this section. We achieve this by adapting the weights $w(n)$ of the RLS filter recursively. For the weights of the RLS algorithm to be adapted, values of the error signal $e(n)$, and the gain factor $k(n)$ first have to be determined. Error signal value can be determined by calculating the variance of the desired signal $d(n)$ and the output signal $y(n)$ of the RLS filter. We have dedicated this section to an enhanced gain factor $k(n)$ value that is aimed at yielding improved weights $w(n)$ for the RLS algorithm. A diagrammatic illustration for the computation of the weights $w(n)$ of the RLS filter is illustrated in Figure 14.3.

Due to the inefficiencies of the correlation matrix $\Phi^{-1}(n)$ which lacks some distinct variables that constitute the RLS algorithm, it is computationally still used to derive the RLS algorithm. For the computational inefficient calculations of $\Phi^{-1}(n)$ to be avoided, authors in [21] suggested that it is advisable to deploy a matrix inversion technique. The matrix inversion technique is constantly implemented to avoid the computationally inefficient calculations of the correlation matrix $\Phi^{-1}(n)$.

Due to the matrix inversion techniques, the autocorrelation of the tap input vector $u(n)$ of order M-by-M is identified by

$$\Phi^{-1}(n) = \sum_{i=1}^{n} \lambda^{n-1} u(i) u^{H}(i) + \delta\lambda^{n} I \tag{14.1}$$

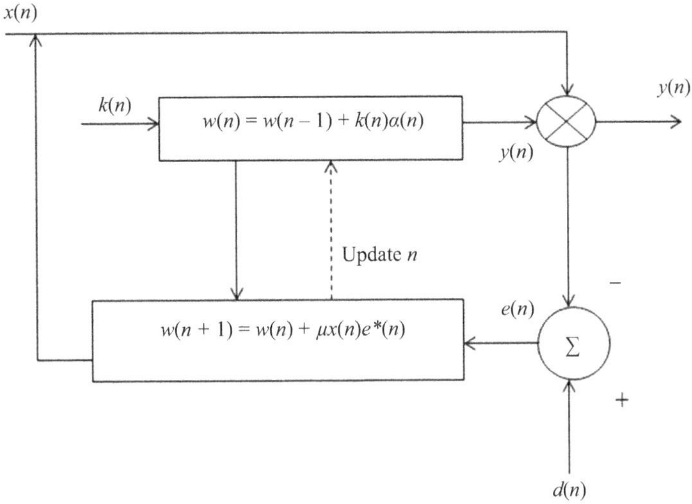

Figure 14.3 Computation of adaptive weight

Here, δ denotes the regularizing parameter of the RLS algorithm and λ (lambda) is known as the forgetting factor. The M-by-1 cross-correlation matrix of the tap input vector $u(n)$ and $d(n)$ which is the expected or desired response, computed as

$$\mathrm{z(n)} = \sum_{i=1}^{n} \lambda^{-1} u(i) d(i) \tag{14.2}$$

The tap weight vector $w(n)$ of the RLS algorithm is calculated as

$$\Phi(n) w(n) = z(n) \text{ OR}$$

$$w(n) = \Phi^{-1}(n) z(n) \tag{14.3}$$

After applying the matrix inversion method cited by [6], we arrive at the equation below:

$$\Phi^{-1}(n) = \lambda^{-1}\Phi^{-1}(n-1) - \frac{\lambda^{-2}\Phi^{-1}(n-1)u(n)u^{H}(n)\Phi^{-1}(n-1)}{1+\lambda^{-1}u^{H}(n)\Phi^{-1}(n-1)u(n)} \tag{14.4}$$

where $k(n)$ utilizes the M-by-1 vector and is identified as the gain vector of the RLS algorithm. The gain factor $k(n)$ is defined by the tap input vector $u(n)$, which is then converted by the inverse of the correlation matrix $\Phi^{-1}(n)$. The gain factor $k(n)$ is given as

$$k(n) = \Phi^{-1}(n) u(n) \tag{14.5}$$

To generate a weights vector for the RLS from (14.5) we compute $w(n)$ as

$$w(n) = \Phi^{-1}(n)z(n-1) + \Phi^{-1}(n)d(n) - k(\text{n})u^H(n)\Phi^{-1}(n-1)z(n-1) \quad (14.6)$$

The simplified equation for the weight vector after simplifying the weight vector $w(n)$ in (14.6) is given as

$$w(n) = w(n-1) + k(\text{n})e(n) \quad (14.7)$$

where the error signal $e(n)$ is calculated as

$$e(n) = d(n) - y(n) \quad (14.8)$$

$P(n)$ which is the inverse correlation matrix at step n is denoted as

$$P(n) = \Phi - 1(n) \quad (14.9)$$

We have finally arrived at our enhanced gain factor $k(n)$ which is given as

$$k(n) = \frac{\pi(n)}{m \times \lambda + u^H(n)\pi(n)} \quad (14.10)$$

where $\pi(n)$ is denoted as

$$\pi(n) = P(n-1)u(n) \quad (14.11)$$

Our enhanced RLS algorithm is initialized by observing the following steps:

1. The weights vector initial value, $\text{w}(0) = 0$

   $$\text{P}(0) = \delta^{-1}\mathbf{I}$$

2. And the value for δ is dependent on the value of the signal-to-noise ratio (SNR), that is

   $$\delta = \begin{cases} \text{small positive constant for high SNR} \\ \text{large positive constant for low SNR} \end{cases}$$

3. Verification of δ is based on regularization grounds
4. And instantaneously, $n = 1, 2, 3, 4, \ldots$
5. Final computations:

$$\pi(n) = P(n-1)u(n),$$

$$k(n) = \frac{\pi(n)}{m \times \lambda + u^H(n)\pi(n)} \text{ and}$$

$$P(n) = \frac{P(n-1) - k(n) * \pi(n)}{\lambda}$$

The peculiarity of our enhanced RLS algorithm is based on the introduction of the enhancing constant m to the gain factor $k(n)$ of the RLS algorithm. We have

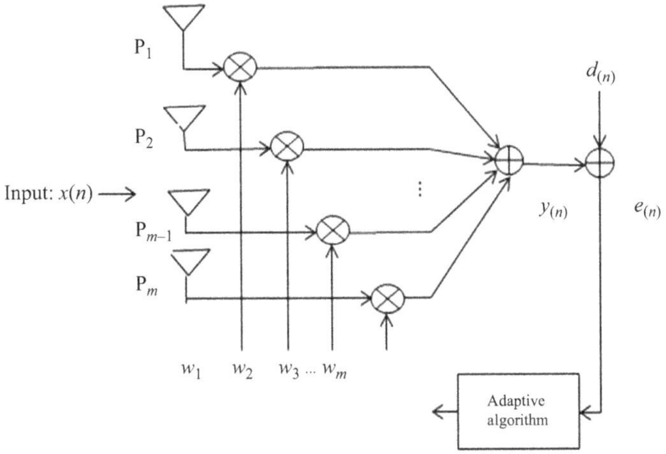

Figure 14.4 Weights summation of adaptive filtering

introduced the enhancing constant m to the RLS algorithm in order to improve the directivity and gain of the smart antenna required for long-range communication networks. The enhancement of the gain factor $k(n)$ influences the entire performance of the weights vector $w(n)$ at the output of the adaptive filter. This enhancement is illustrated in Figure 14.4. Figure 14.4 illustrates the weight summation of adaptive filtering.

It is important to know that the RLS algorithm is a deterministic algorithm. This means that the performance index of the RLS algorithm is a combination of the weighted error squares for known data that are presented. The weights vector is iteratively updated after every sample.

The constant m that has been introduced to the gain factor $k(n)$ in this study will be assigned a value. In order to determine the best possible value for m that can be introduced to $k(n)$, we conducted some mathematical computations and several simulation tests. These computations were conducted to determine the best value of the weights output. They also showed that the weights output of our enhanced RLS filter outperforms the weights of other adaptive algorithms that were previously used for smart antenna design. Studies conducted by authors in [22] stated that the convergence rate of beamforming algorithms is dependent on the value of the step-size parameter μ. The value of the step-size parameter μ lies between 0 and 1. An optimized value for the constant m was initially obtained by mathematical computations. However, the simulation results further confirmed that the value for m obtained through our mathematical computation was an optimized value. The constant m was assigned numerical values that ranged between 0.1 and 1.9. Arithmetic computations were applied to the equation of our gain constant $k(n)$ in (14.10). The arithmetic computations resulted to the optimization of the gain factor $k(n)$ when the constant m was given a value of 0.25. It is for this reason that the

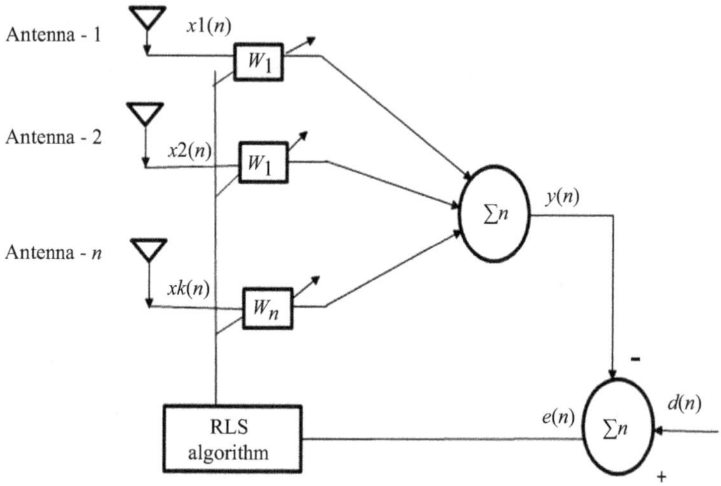

Figure 14.5 Computation of the enhanced RLS

constant m has been given the value of 0.25. Figure 14.5 below illustrates how the enhanced RLS algorithm is computed.

14.4 Performance evaluation and results

The performance of RLS-based model for smart antenna design was conducted with MATLAB® software. Three cases were considered for simulations. The following are the cases we chose to consider for simulations: the error curve of the RLS algorithm and the system output of the RLS algorithm. Comparisons between the filter weights and the estimated weights for three adaptive algorithms were conducted. These algorithms are the LMS algorithm, the conventional RLS algorithm, and our enhanced RLS algorithm. For simulation purposes, the following parameters are going to be considered (Table 14.1).

Simulation outcomes of the six experiments conducted in this work are represented in Figures 14.6–14.11. The purpose for these tests is to confirm that the enhancement we have made on the gain factor $k(n)$ positively affects the performance of the RLS algorithms. These effects will further signify that the enhanced RLS algorithm outperforms other adaptive algorithms because of its improved gain factor $k(n)$.

In our first simulations, we compare the system output results of the conventional RLS and our enhanced RLS algorithms; these results are displayed in the graphs below (Figure 14.6). The results in Figure 14.6 both display the output curve of the RLS system by matching the true and estimated output against the number of samples that are being tested. The result in Figure 14.6(a) shows that the density of the noisy output gradually begins to fade after the 600th sample. Remember that one of the justifications we made for our choice of the RLS algorithm is that it has

Table 14.1 Simulation parameters

S. No.	Parameters	Specifications	Reasons
1	Number of antenna elements	5	Five elements are used so that our simulations will be based on an antenna with few elements
2	Number of system points	2000	It is advisable to base your tests using a wide range of variables
3	Forgetting factor	0.99	A forgetting factor of 0.99 is a perfect value for the RLS algorithm
4	Number of training points	50	Gives a large number of iterative tests
5	Spacing between elements	1	This indicates the distance between each antenna element and it is measured in meters

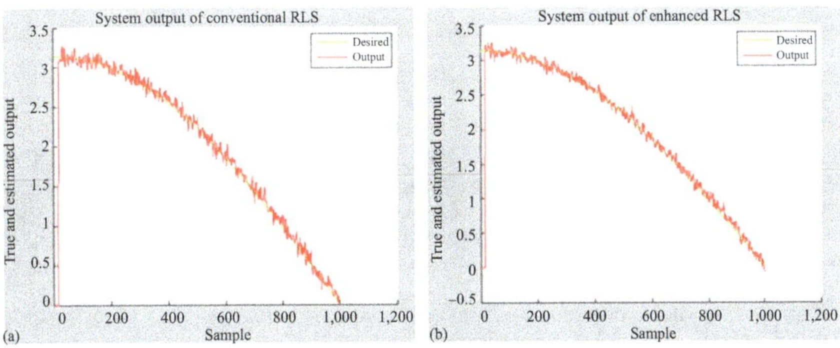

Figure 14.6 System output comparison between Conventional RLS and Enhanced RLS algorithms

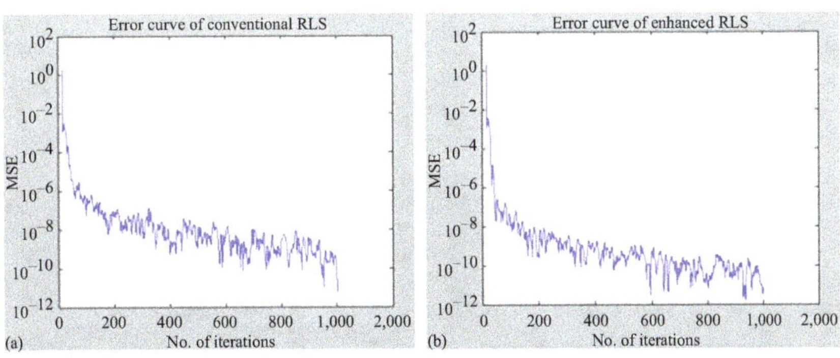

Figure 14.7 MSE comparison between conventional RLS and Enhanced RLS algorithms

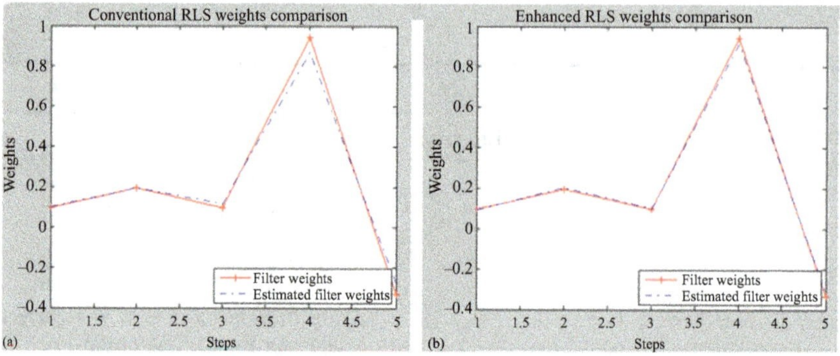

Figure 14.8 Filter weights comparison between conventional RLS and Enhanced RLS algorithms

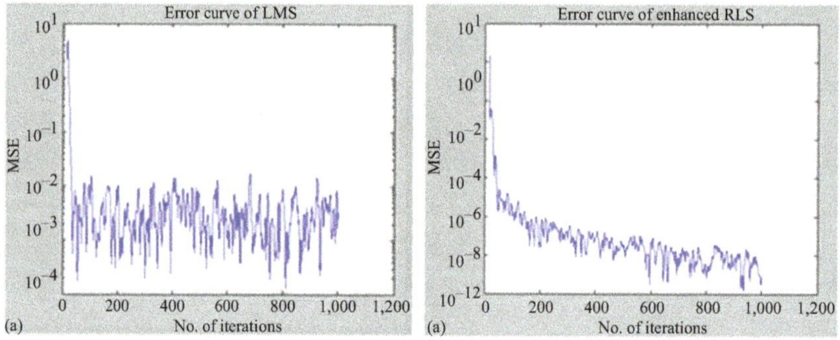

Figure 14.9 MSE comparison between LMS and enhanced RLS algorithms

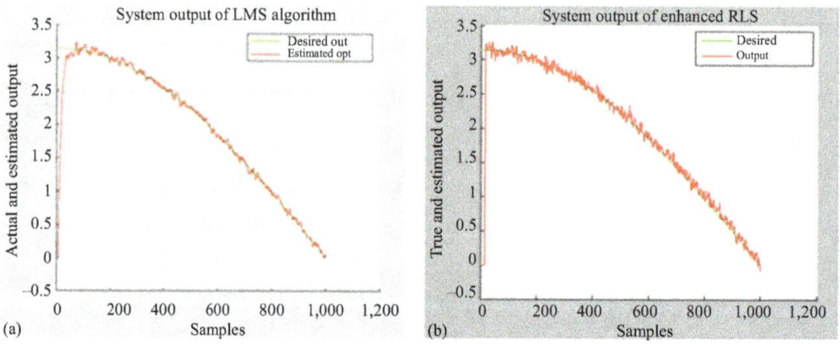

Figure 14.10. System output comparison between LMS and enhanced RLS algorithms

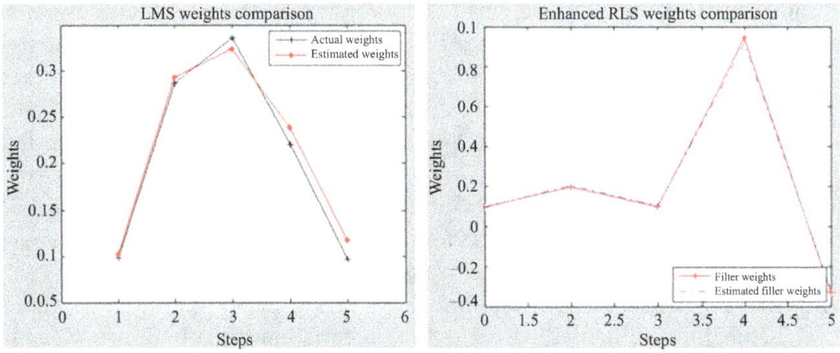

Figure 14.11 Filter weights comparison between LMS and enhanced RLS algorithms

the ability to cancel out the effects of noise at a faster rate when compared to other adaptive algorithms. Therefore, in Figure 14.6(a), the output of the conventional RLS algorithm started to incrementally become less noisy at an exponential rate. Whereas in the system output of our enhanced RLS algorithm (Figure 14.6(b)), it can be noticed that the density of the noise signal begins to fade away when it reaches the 400th sample. This result further indicates that the enhancement we made to the conventional RLS algorithm enables it to cancel out the effect of noise on a communication channel at a quicker rate; thus, signifying that the enhanced RLS algorithm is more efficient in dealing with a noisy communication channel. Therefore, simulation results in Figure 14.6 clearly show that our enhanced RLS algorithm has a system output with lesser noise density when compared to the conventional RLS algorithm.

A minimum square error (MSE) comparison is conducted in Figure 14.7. The aim of this simulation is to graphically show the error tolerance the RLS algorithm has. This directly has an effect on the overall performance of the SA because its signal output will have less errors. Simulation results in Figure 14.7(a) and (b), respectively, display the error curves of both the conventional RLS and enhanced RLS algorithms. We see from the error curve results of the error curve of our enhanced RLS algorithm that the MSE stays below 10^{-9} as it approaches the 1000th iteration. The initial results were high but began to decrease with the incremental number of iterations. However, the error curve of the conventional RLS algorithm slightly falls below the 10^{-8} point even at the 1000th iteration. Thus, enhancing the gain factor $k(n)$ of the RLS algorithm yields as enhanced RLS algorithm that has lesser square errors as illustrated in the results below (Figure 14.7(a) and (b)) which confirm that our enhanced RLS algorithm has fewer errors than the conventional RLS algorithm.

Our third simulation in Figure 14.8 takes into account the filter weights comparison between conventional RLS and enhanced RLS algorithms. Adaptive algorithms act as filters because one of their functions is to filter the output of their corresponding signals. Therefore, we want to show through simulation results, what effect an improved gain factor $k(n)$ has on the filter weights of the RLS algorithm.

The comparison of the filter weights is between the conventional RLS algorithm and our enhanced RLS algorithm. In Figure 14.8, when the number of steps is increased (on the vertical axis), the estimated weights are supposed to track or keep up with the actual filter weights of the adaptive algorithm. Hence, we observe that the estimated weights of our enhanced RLS algorithm consistently track the filter weights. This is shown in Figure 14.8(b). The significance of this result is to indicate that by improving the gain factor $k(n)$ of the RLS algorithm, the efficiency of the RLS algorithm is increased. On the other hand, simulation results of Figure 14.8(a) show that the estimated weights can be seen to lose track of the filter weights in some instances. This variation, however, is not so significant because of the unique qualities of the RLS algorithm. The variation actually occurs when the number of samples is increased. This result signifies the superiority of the enhanced RLS algorithm over the conventional RLS algorithm.

In the simulations shown in Figures 14.9–14.11, a comparison is made between our enhanced RLS algorithm and the least mean square (LMS) algorithm. This comparison is done in order to confirm which of the two algorithms performs better in the presence of interfering signals. Figure 14.9 compares the MSE results between the LMS algorithm and our enhanced RLS algorithm. We can see in Figure 14.9(a) that the MSE curve of the LMS algorithm does not exponentially decrease but linearly stays constant at 10^{-3} root-mean-square deviation. Also observed in Figure 14.9(a) is how the MSE dropped from 10^{1} to 10^{-2} within the 50th iteration. Afterward, the error curve remains constant. Thus, the LMS algorithm displayed a 30% decrease in error (as shown in Figure 14.9) on average. On the other hand, our enhanced RLS algorithm showed an exponential decrease in the MSE from the first iteration to the 1000th iteration. Also, observed in Figure 14.9(b) is a 40% error correction between the 200th and 1000th iteration. Therefore, our enhanced RLS algorithm is more superior to the LMS algorithm in canceling out errors. This quality makes our enhanced RLS algorithm to be preferred over other beamforming algorithms in a long-range communication channel as it cancels out the interfering or unwanted signals.

From the error curves shown in Figure 14.9, we observe that the results from the error curve of our enhanced RLS algorithm has a reduced value when compared to that of the LMS and the conventional RLS algorithms. We arrived at this conclusion due to the respective error levels of the LMS and conventional RLS algorithms. From Figure 14.9, we can also see that the error curve of our enhanced RLS algorithm reduces when there is an increment in the number of iterations. The mean square error is fundamental in the optimization of the weights vectors, which in turn, increases the performance of a communication channel as it improves the directivity of an antenna array.

Figure 14.10 represents the system outputs for the LMS algorithm and enhanced RLS algorithm, respectively. We observe in Figure 14.10(a) that the LMS algorithm has quite a noisy system output (shows greater variability on the output). Furthermore, it can be seen that its actual output does not exactly track its desired output. The output of the enhanced RLS algorithm (Figure 14.10(b)) is seen to possess a smooth system output with less noise. The exact output is very similar to the desired output. Secondly, as the number of samples increases in Figure 14.10 (b), the system output becomes less noisy.

The zigzag representation of the Red curve indicates the presence of noise from the system's output. As the number of samples increase, the noise on the output reduces. Hence, the noise ratio decreases with time. Based on the results of the LMS algorithm, we see in Figure 14.10(a) that the density of the actual system output is initially saturated which indicates a very noisy signal. However, the noise in the system output gradually begins to fade only after the 600th sample. Between the 200th sample and the 800th sample, the noise level decreased by 3 dB and between the 800th sample and the 1000th sample, the noise level decreased by another 1 dB. Whereas, in the enhanced RLS algorithm (Figure 14.10(b)), the density of the noise signal begins to fade away almost immediately. On the graph, between the 200th sample and the 800th sample, we notice that the noise level dropped by 3 dB. However, after the 800th sample of our enhanced RLS algorithm, there was a negligible (close to 1 dB) drop in the noise level. We also see in Figure 14.10 below that our enhanced RLS algorithm has a better tolerance to noise than the LMS algorithm of more than 6 dB. These results show that our enhanced RLS algorithm has a less noisy output than the LMS and the conventional RLS algorithm. This attribute makes the enhanced RLS algorithm very suitable for long-range wireless networks because it will be able to filter out a high level of noise from the signal. As we had earlier mentioned, this design will benefit wireless and cellular operators immensely. Also because of this quality, the enhanced RLS will be a resourceful tool in the design of 5G antennas because of its high resistance to noise.

From Figure 14.11, the filter weights and the estimated weights are being compared using the LMS and enhance RLS algorithms, respectively. We can see in Figure 14.11(a) that as the number of samples increases, the distance between the filter weights and estimated weights broadens. However, in Figure 14.11(b), which illustrates our enhanced RLS algorithm, we see that the estimated weights still maintain pace with the filter weights even when there is an increase in the number of samples.

In Figure 14.11(a) (the LMS algorithm), at steps 2, 3, and 4, the convergence rate is at an average of 65%. Also on steps 2.5, 3.5, and 4.5, the convergence rate is at an average of 80%. As can be observed in the weights comparison, the convergence rate at step 4 of the enhanced RLS algorithm is faster by 25% when compared to that of the conventional RLS algorithm. The result in Figure 14.11 thus shows the superiority of the enhanced RLS algorithm over the LMS algorithm in terms of its efficiency.

Results from Figures 14.6–14.11 lead us to arrive at the conclusion that our enhanced RLS algorithm displays a more superior outcome when it is compared to the other beamforming algorithms that were tested in this work. These results further show the following:

- The filter weights of the enhanced RLS algorithm have an efficiency of 90%.
- The convergence rate of the enhanced RLS filter as shown by the filter weights is fast.
- The enhanced RLS algorithm has less errors.

Therefore, the enhanced RLS algorithm is a very efficient algorithm for beamforming over a long-range wireless communication.

14.5 Conclusions

A performance-based evaluation of our enhanced RLS algorithm and other beamforming algorithms are discussed in this chapter. The work done in this chapter was to enhance the gain factor $k(n)$ of the RLS algorithm. The aim of this enhancement was to improve the performance of the RLS algorithm resulting in a considerable improvement in the value of the gain factor. We introduced a constant m to the gain factor $k(n)$ of the RLS algorithm. The introduction of the constant m is aimed at improving the output performance of the adaptive antenna. The benefit of this enhancement is that it does not increase the output power of the RLS filter, rather it increases the gain and directivity of the RLS filter. The aim of the simulation tests conducted in Figures 14.6–14.11 was to verify the following outcomes: the comparison of the RLS system output, the comparison of the MSE between the conventional RLS algorithm, the LMS algorithm, and the enhanced RLS algorithm. Comparison of the filter weights between the three algorithms (conventional RLS, LMS, and our enhanced RLS) was carried out. The results of these simulations show that our enhanced RLS algorithm has an improved performance in the level of MSE, which leads to a quicker rate of convergence and directivity of the adaptive algorithm. The output of our enhanced RLS algorithm displayed a smoother waveform, which indicates that it is the best adaptive filter when compared to other adaptive algorithms. This property makes our enhanced RLS algorithm to eliminate unwanted noise from surrounding communication networks. When we compared our enhanced RLS algorithm to the conventional RLS algorithm, its filter weights output is almost accurate to the estimated weights. Therefore, the problems of weight complexity that are experienced with adaptive filters have been significantly reduced by the emergence of our enhanced adaptive algorithm. As seen from simulation results, the design proposed in this chapter reduces SNR, updates weights speedily, and increases the convergence rate of the RLS algorithm for SAs over long-range communication networks. Emerging applications such as the 4IR, 5G, and IoT require antennas with the ability to cater for a wider transmission range, have less interference, enhanced network capacity, and spatial reuse. Simulation results show that our enhanced RLS algorithm is the preferred adaptive algorithm that can be used in these emerging applications alike.

Acknowledgments

This work was supported by Tshwane University of Technology under FSATI. The authors seize this opportunity to appreciate Tshwane University of Technology and the National Research Foundation (NRF) for granting us the funding for research, study equipment, and necessary facilities for the purpose of this study.

References

[1] Jain, R. K., Katiyar, S., and Agrawal, N. K. "Smart antenna for cellular mobile communications." *Networking and Internet Architecture*, arXiv:1204.1790 (cs). VSRD-IJEECE. 2012; 1(9): 530–541.

[2] Mavromoustakis, C. X., Mastorakis, G., and Batalla, J. M. "Internet of things (IoT) in 5G mobile technologies." *Technological Educational Institute of Crete*. Greece, National Institute of Telecommunications. 2016.

[3] Papanikalaou, K., Mavromoustakis, C. X., Mastorakis, G. and Bourdena, A. "Energy consumption optimization using interaction in the mobile cloud." *6th International Conference on Mobile Networks and Management MONAMI*. 2014.

[4] Thompson, J., Xiaohu, G., Hsiao-Chun, W., *et al.* "5G Wireless communication systems: prospects and challenges." *IEEE Communications Magazine*. 2014; 52(2): 62–64.

[5] Hong-Ning, D., Kam-Wing, N., Minglu, L., and Min-You, W. "An overview of using directional antennas in wireless networks." *International Journal of Communication Systems*. 2013; 26: 413–448.

[6] Surenda, L., Shameem, S., and Khan, H. "Performance comparison of LMS, SMI and RLS adaptive beamforming algorithms for smart antennas." *IJCST*. 2012; 3(2).

[7] Mavromoustakis, C. X., Pallis, E., and Mastorakis, G. *Resource management in mobile computing environments*. Springer International Publishing AG; 2014.

[8] Oluwole, A. S. and Srivastava, V. M. "Smart antenna for wireless communication systems using spatial signal processing." *Journal of Communications*. 2017; 12(6).

[9] Zaharis, Z. D., Lazaridis, P. I., and Mastorakis, G. "Implementation of antenna beamforming by using a novel neural network structure." *Conference: 2016 International Conference on Telecommunications and Multimedia (TEMU)*. July 2016.

[10] Anzhong, H. "DOA-based beamforming for multi-cell massive MIMO systems." *Journal of Communications and Networks*. 2016; 18(5).

[11] Ramineni, B., Sagar, G. C., Jain, K. A., Ganga Prasad, M. S., Ramakrishna, T. V., and Sarat Kumar, K. "Comparison and performance evaluation of different adaptive beamforming algorithms in wireless communications with smart antenna." *International Journal of Engineering Research and Applications (IJERA)*. May–June 2012: 630–633.

[12] Jyothi Lakshimi, T. S. and Sivvam, S. "Smart antennas for wireless communication." *International Journal of Applied Engineering Research*. 2017; 12(1); ISSN0973-4562.

[13] Joshi, R. and Dhande, A. "Adaptive beamforming using LMS algorithm." *International Journal of Research in Engineering and Technology*. 2014; 3(5): 589–593.

[14] Mavromoustakis, C. X., Bourdena, A., Mastorakis, G., Pallis, E., and Kormentzas, G. "An energy-aware scheme for efficient spectrum utilization in a 5G mobile cognitive network architecture." *Telecommunication Systems*. 2015. doi: 10.1007/s11235-014-9885-4.
[15] Divya, R. and Jayasimha, T. "Adaptive beamforming algorithm for smart Wimax antenna." *International Journal of Advance Research in Computer Science and Management Studies*. 2014; 2(5).
[16] Chuku, N. P., Olwal, T. O. and Djouani, K. "Enhanced RLS in smart antennas for long range communication networks." *9th International Conference on Ambient Systems, Networks and Technologies (ANT 2018)*.
[17] Renukadas, B. B. "Smart antennas for mobile communications." *IOSR Journal of Electronics and Communication Engineering (IOSR-JECE)*. Ver. 1 2016; 11(4): 69–72.
[18] Basha, T. S. G., Aloysius, G., Rajakumar, B. R., Prasad, M. N. G., and Sridevi, P. V. "A constructive smart antenna beamforming technique with spatial diversity." *IET Microwave, Antennas Propagation*. 2012; 6(7): 773–780.
[19] Romia, A. M., Ali, H. S., and Abdalla, M. I. "Optimization of recursive least square-based adaptive linear equalizer for ZigBee transceiver." *Wireless Personal Communications*. 2018; 103: 2758–2772.
[20] Mallaparapu, U., Nalini, K., Ganesh, P., *et al.* "Non-blind adaptive beamforming algorithms for smart antennas." *International Journal of Recent Research and Applied Studies (IJRRAS)*. 2011; 495–495.
[21] Martinek, R., Vanus, J., Kelnar, M., Bilik, P., and Zidek, J. "Application of recursive least square algorithm to adaptive channel equalization." *World Congress Measurement in Research and Industry*. September 2015.
[22] Khan, A. "Beamforming using adaptive algorithms." Telecom Department Hazara University Mansehra. Session 2010–2014.

Chapter 15

Forecast of electricity consumption: a comparison of ARIMA and neural networks

Theodoros Pseftelis[1], Constandinos Mavromoustakis[1], George Mastorakis[2], Periklis Chatzimisios[3], Evangelos K. Markakis[4], Evangelos Pallis[4] and Jordi Mongay Batalla[5]

Electricity consumption is a critical factor in the climate change problem. The in time and reliable prediction of future consumption can help experts take the appropriate measures to eliminate electricity production side effects on the planet. Experts also can use forecasts to design suitable renewable energy systems. In this chapter, we analyze two well-known forecasting models. The first is the autoregressive integrated moving average (ARIMA), which has been used in many real-life cases in the previous years, and the second one is the neural network forecasting method which, is based on human's brain function. Each method is analyzed with its implementation and steps. The last section is a head-to-head comparison of the two methods.

15.1 Introduction

Climate change is considered one of the most severe problems. Scientists around the globe guess that human activity will have hazardous consequences in the next generations. One of the main culprits is electricity production plants [1,2]. Therefore, the need to control energy resources and the advent of alternative energy compels the use of high-level tools to forecast electricity consumption [3–15]. It is crucial to find ways to eliminate their usage or at least to be used in more productive ways.

[1]Department of Computer Science, University of Nicosia, Makedonitissas Avenue Nicosia, Nicosia, Cyprus
[2]Department of Management Science and Technology, Hellenic Mediterranean University, Agios Nikolaos, Crete, Greece
[3]Department of Information and Electronic Engineering, International Hellenic University, Thessaloniki, Greece
[4]Department of Electrical and Computer Engineering, Hellenic Mediterranean University, Heraklion, Greece
[5]Department of Computer Network and Services, National Institute of Telecommunications and Warsaw University of Technology, Warsaw, Poland

In recent years with the blooming of the Internet [16], Internet of things (IoT) technology [17,18] has become a game-changer ally to this purpose. Correctly, using sensors, monitors, and other IoT devices [19–21], we can retrieve useful data about house electricity consumption. In most cases, we can process them in real-time. So the question is, "How can we exploit data collected by IoT devices to produce handy solutions for electricity consumption optimization?" An efficient way to use these data is to predict future electricity consumption. Decision-makers having knowledge with relevant precision of the future electricity consumption can utilize better the electricity power plants. From a data science point of view, there are different forecasting methods [30]. Different forecasting techniques are applied to various real-life problems. So it is essential to clarify which methods produce better results for the inevitable problem of electricity consumption. As a result, scientists will not be spending time to select the appropriate method [31–41] but will use these valuable time for the forecasting procedure itself.

The remainder of this chapter is organized as follows: Section 15.2 contains details about the used dataset. In Section 15.3, we analyze the autoregressive integrated moving average (ARIMA) forecasting method and its procedure. In Section 15.4, we discuss the neural network (NN) forecasting method and its step by step implementation. In Section 15.5, we compare the ARIMA against the NN method citing the pros and cons of each method for electricity consumption prediction.

15.2 Data overview

The used dataset [19] contains 11 different minutely measurements for the electricity, water, and natural gas consumption of a residential house in Canada. Captured period starts from 2012 to 2014. Data were collected with industry-standard equipment and stored with a high degree of precision and integrity. For the aforementioned reasons, this dataset constitutes the ideal choice for this work.

15.2.1 Import data

```
# Import data from csv file
raw_data <- read_csv("data/Electricity_P.csv")

# Display sample
sample_data <- sample_n(raw_data, size = 10)
kable(sample_data[,1:10], "latex",
      caption = "Raw Data",
      booktabs = T,
      linesep = "") %>%
kable_styling(latex_options = c("striped", "hold_position"))
```

In this chapter, we use only the whole house electricity (WHE) consumption measurement for modeling and forecasting (Tables 15.1 and 15.2).

Table 15.1 Raw data

UNIX_TS	WHE	RSE	GRE	MHE	B1E	BME	CWE	DWE	EQE
1375020840	667	180	0	487	0	1	0	0	40
1384971120	516	43	0	473	0	1	0	0	41
1387460820	539	173	0	366	0	0	0	0	38
1372626780	3162	1855	2	1305	0	0	0	769	40
1360085100	2563	41	2	2520	0	0	0	0	45
1340447520	594	171	2	421	0	5	0	0	40
1384790760	1389	470	0	919	22	1	0	0	46
1370683740	434	43	0	391	0	3	0	0	41
1382546580	1142	43	2	1097	0	334	0	0	41
1367630820	1334	447	2	885	0	2	0	7	41

Table 15.2 Electricity consumption (W)

Min	Max	Mean	SD
0	12260	1112.327	946.2197

15.2.2 Data information

From the daily–monthly electricity consumption as shown in Figures 15.1 and 15.2, it seems to be an increasing electricity waste. Precisely, the high consumption period starts in October and ends in April. On the other hand, there is a lower consumption rate period in the summer months. We can assume that a significant part of electricity consumption is being spent on heating.

15.3 ARIMA

ARIMA represents the autoregressive integrated moving average, which is a generalization of an ARMA model. Arima models are applied to data mainly to analyze their behavior or to predict their future points. ARIMA achieves these goals by trying to describe the autocorrelations between time series data. The AR part implies that the time series data are correlated with their prior values. The "I" shows that data have been replaced with their computed differences. Last, the MA parameter suggests that the regression error is a linear combination of error terms whose values coincided and at several moments in the past. The full model can be written as:

$$y'_t = c + \phi_1 y'_{t-1} + \cdots + \phi_p y'_{t-p} + \theta_1 \varepsilon_{t-1} + \cdots + \theta_q \varepsilon_{t-q} + \varepsilon_t \tag{15.1}$$

where y'_t is the differenced series (it may have been differenced more than once). The "predictors" on the right-hand side include both lagged values of y_t and lagged errors. It is worth mentioning that there are two kinds of ARIMA models: seasonal and nonseasonal.

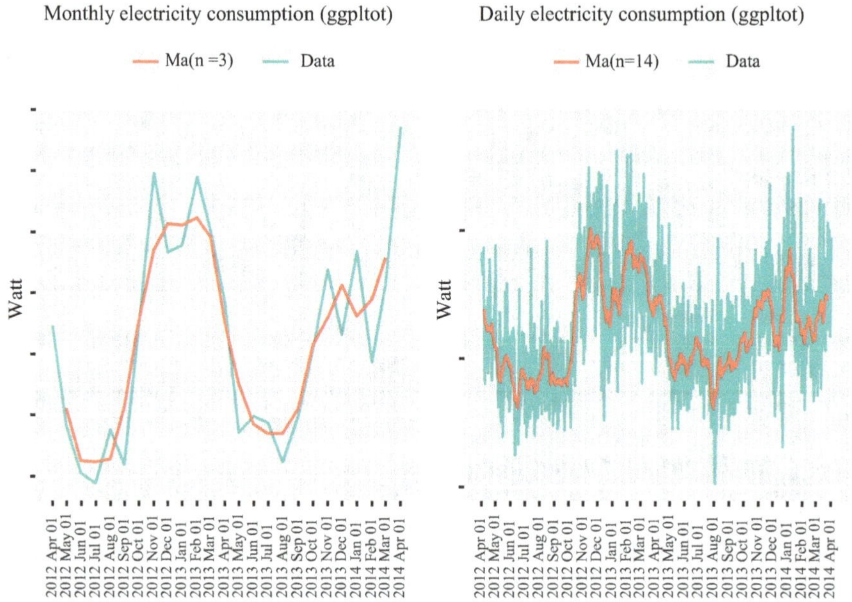

Figure 15.1 Monthly–daily consumption (2012–2014)

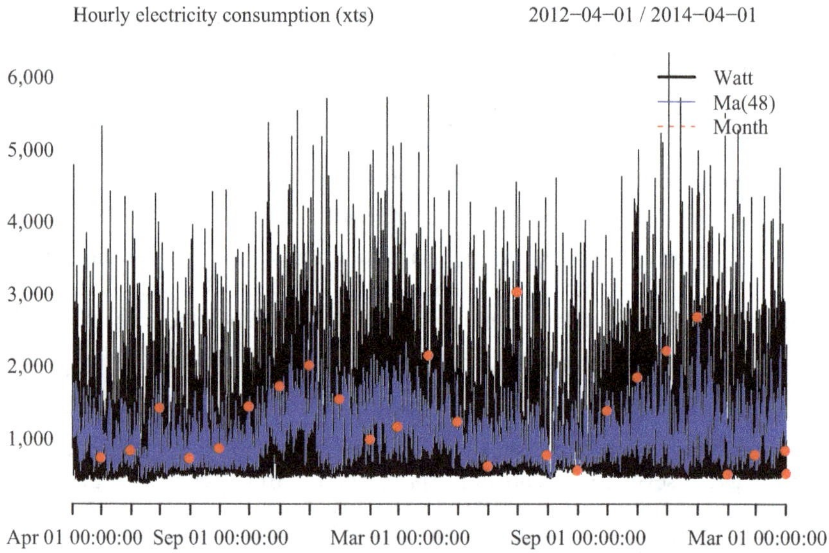

Figure 15.2 Hourly consumption (2012–2014)

15.3.1 Data preparation

It is common practice to separate data into two parts: a training set and a test set. The training set is used to fit the model and test for evaluating the forecast accuracy, precisely, because the test dataset is unknown to the fitting model, it is a reliable indication of how well the model can forecast new data. In the below *R* code, the 90% of the data becomes the training set, and the remaining 10% becomes the test set.

In Figure 15.3, the black and the red line represent the training and the test dataset, respectively.

15.3.2 Modeling

The very first step is to visualize the data to check for outliers or other unexpected behavior. The next step is to identify if the time series is stationary. More accurately, if $\{y_t\}$ is a stationary time series, then for all ss, the distribution of $(y_t, \ldots, y_{t+s})$ does not depend on t [30]. Stationarity in data is determined as it produces forecasts accurately. One way to look for stationarity is to examine autocorrelation function plot (ACF) and partial autocorrelation plot (PACF).

```
# R code for Autocorrelation Function Plot
  # library(forecast)
  Acf(diff(ts_m))
```

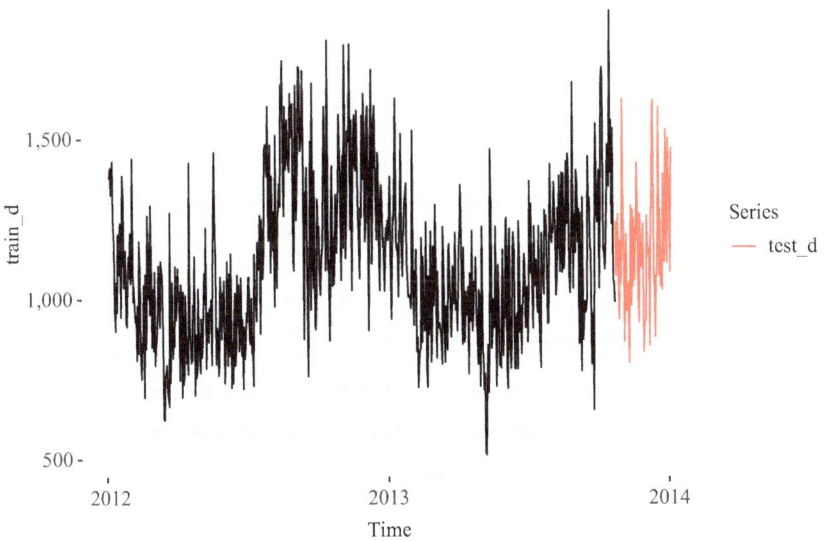

Figure 15.3 Training and test daily data

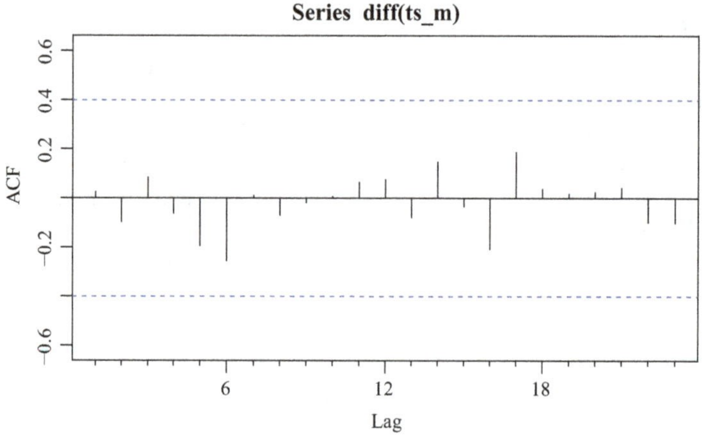

An alternative solution is unit root tests. Two widely used unit root tests are Kwiatkowski–Phillips–Schmidt–Shin (KPSS) and augmented Dickey–Fuller (ADF) [42].

```
# R Code for Dickey-Fuller test
library(tseries)
adf.test(ts_m)
```

```
##
## 	Augmented Dickey-Fuller Test
##
## data:  ts_m
## Dickey-Fuller = -1.7785, Lag order = 2, p-value = 0.6568
## alternative hypothesis: stationary
```

p-value < 0.05 means that we reject the null hypothesis and keep the alternative hypothesis, so the time series is stationary.

One way to transform a nonstationary time series to stationary is differencing. Differencing is the calculation of the difference between two sequential chronological data. One of the most crucial modeling parts is to define p, d, and q parameters:

- p is the lag order
- d is the differencing degree
- q is the order of the moving average

So the question is how to know what is the appropriate values for the parameters. Autocorrelation function and partial autocorrelation plots are used to define AR(p) and MA(q) values. On the other hand, $I(d)$ stands for the number of differences. The right approach is to try different model combinations, for example, Arima (1,1,2) and Arima (1,0,1):

```
# R Code for Fitting the model
# Set up Hourly Time Series Object
data <- ts(whe_h$watt, frequency = 24 )
# Difference data for stationarity
data <- diff(data)
# Fit AR(1)I(0)MA(0) to data
fit <- Arima(data, order = c(1,0,0))
```

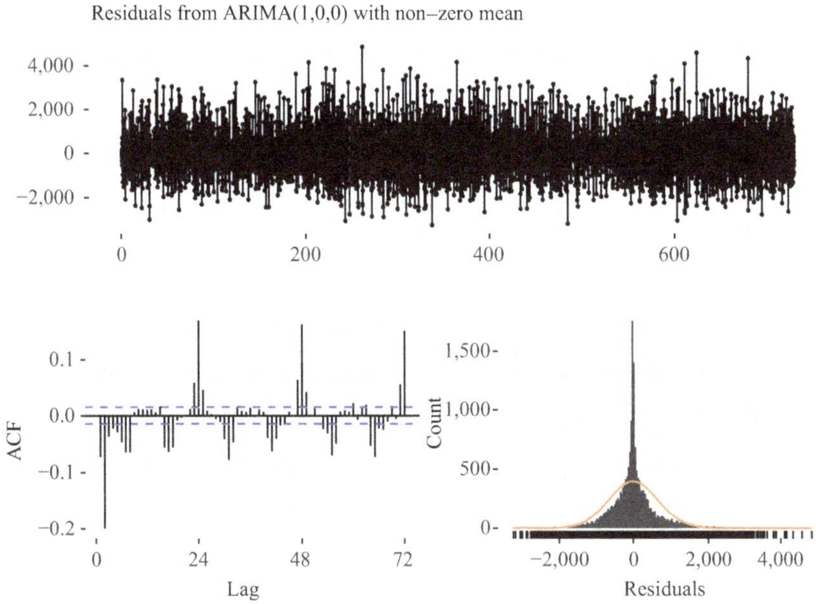

Figure 15.4 Residuals examination plots

Then an information criterion is used to keep the model that minimizes the criterion's value. Information criterion measures the quality of the model, taking into account the complexity of the model and how well it fits in data. Standard criteria are Akaike's information (AIC) criterion [43] and Bayesian information criterion (BIC) [44].

Furthermore, an efficient way to see how well the model fits the data is the residuals examination. As clearly defined by Hyndman, "The residuals in a time series model are what is left over after fitting a model. For many (but not all) time series models, the residuals are equal to the difference between the observations and the corresponding fitted values":

$$e_t = y_t - \widehat{y}_t \tag{15.2}$$

Residuals help to check whether a model has sufficiently caught the information in the time-series. Residuals that come from adequate forecasting methods are uncorrelated between them and also have zero mean. One way to check for correlations is by plotting residuals ACF. For example, in Figure 15.4, specifically, in the bottom left, the ACF plot shows that there is a correlation between lags. Consequently, the testing model might not be appropriate for forecasting. Residuals must look like white noise.

```
# R Code for Residuals Examination
# library(forecast)
checkresiduals(fit)
```

```
##
## Ljung-Box test
##
## data:  Residuals from ARIMA(1,0,0) with non-zero mean

## Q* = 2669, df = 46, p-value < 2.2e-16
##
## Model df: 2.   Total lags used: 48
```

Another solution to identify correlations is by portmanteau test. A test for a group of autocorrelations is named a portmanteau test. A well-known such test is the Ljung-Box test [20] based on the following:

$$Q^* = T(T+2)\sum_{k=1}^{h}(T-k)^{-1}r_k^2 \tag{15.3}$$

15.3.3 Forecast

After fitting the models, we can make forecasts to predict future values. This procedure in R language is quite straightforward using **forecast** package.

```
# Forecast using auto.arima function
fit_h <- auto.arima(train_h, stepwise = FALSE)
fcast <- forecast(fit_h, h = 876)
autoplot(fcast)
```

In Figure 15.5, we use the training set to fit the model. From the graph, it is visible that this model does not produce reliable predictions. In **forecast()** function with h parameter, the number of future periods to forecast is defined.

15.3.4 Evaluate results

To be sure that the model produces solid forecasts, forecasts must be validated against the test dataset. *R* **accuracy()** method from **forecast** package evaluates forecasts using a large number of different error measurements:

```
# Fit Model
fit_h <- auto.arima(train_h,stepwise = FALSE)
# Forecast next 876 periods
fcast <- forecast(fit_h,h = 876)
# Evaluate the accuracy against test data set
accuracy(fcast, test_h)
```

```
##                       ME     RMSE      MAE       MPE     MAPE      MASE
## Training set    4.157792 591.7108 400.7914 -19.68889 38.95056 0.5767752
## Test set     -115.363350 717.4950 562.0815 -47.17203 66.17354 0.8088862
##                     ACF1 Theil's U
## Training set -0.02577784        NA
## Test set      0.47809353  1.349497
```

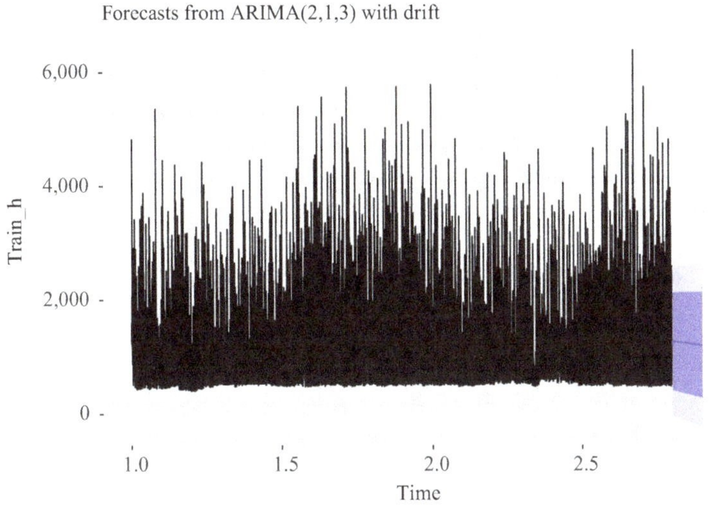

Figure 15.5 Forecast plot

There are a variety of measurements for forecasting data accuracy [30]. Root-mean-square-error (RMSE):

$$\text{Root mean squared error: RMSE} = \sqrt{\text{mean}(e_t^2)}$$

and mean percentage error (MAPE):

$$\text{Mean absolute percentage error: MAPE} = \text{mean}(|p_t|)$$

are two of the most applied measures of prediction accuracy in forecasting methods.

15.3.4.1 Cross-validation

Cross-validation, also known as out-of-sample testing, is a model validation technique for assessing how the results of a statistical analysis will generalize to an independent dataset. In other words, one can determine how accurately a forecasting model will perform in practice. The central concept of this technique is to split data into two subsets: The training set and the test set. The training set is used to train the model, and the test set is used to evaluate it. Taking into account that the test set is unseen to the trained model, it constitutes a functional counterpart of real-life problems. The main drawback of this approach is that usually there not enough data for the training and the test phase. Thus, the need for a method that utilizes the whole dataset arises.

K-fold cross-validation solves this issue. This method splits dataset into K folds. Then it holds out one dataset and uses the remaining datasets to train the model. In the next step, the "hold out" sample is used to evaluate the trained model. These steps are repeated for every one of the K-fold sets. In the end, the results from every model's evaluation are summarised by cross-validation.

15.4 Neural networks

Neural networks are mathematical models that simulate the function of biological neurons, so they are capable of automatically discovering and extracting features from unprocessed data. They may be consist of multiple layers and can produce forecasts even for models with a nonlinear relationship. Furthermore, NNs are robust to noise and missing values.

15.4.1 Different types of neural networks

There is a variety of different NN models [32]. The most well known among these algorithms are as follows:

- Perceptrons were first-generation NNs, first introduced by F. Rosenblat [45]. There are single-layer and multilayer perceptron algorithms. They are used for supervised learning problems and they are exceptionally efficient for linearly separated problems. However, they have severe limitations to confront real-life challenges.
- Convolutional neural network(CNN) is consists of one or more convoluted layers, as well as by one or more connected layers. They use convolution algorithms that applied like filters to extract specific predefined characteristics. For example, by applying a vertical Sobel filter [46] to a particular image, you can obtain the image's vertical lines. Combining different extracted features, CNN can recognize patterns. CNN is widely applied in various fields from health [21] diabetics to natural language processing [47] and more [48].
- Vanilla neural networks have limited capabilities; they cannot handle vectors with no predefined size. This issue is alleviated by recurrent neural network (RRN) that can process sequences of vectors efficiently as inputs and produce outputs on nonfixed-sized vectors [49]. Besides that, RNN remembers "things" learned from previous inputs, allowing information to persist.

15.4.2 Time series forecasting with neural network

In time-series forecasting with NN model, lagged values are used as inputs. So far different kinds of NNs such as recurrent [50] or convolutional have been used for time-series forecasting. However, it should be mentioned that these networks sometimes add complexity and suffer from a lack of interpretability. To this work, we use a feed-forward autoregressive NN in favor of its intuitive nature. The feed-forward network consists of one input layer, one or more hidden layers, and the output layer. When a NN contains no hidden layer of neurons, it is equal to a linear regression model. With the addition of one or more hidden layers, the model can recognize and nonlinear relationships between data. It is called feed-forward because one layer forwards its output as input to the next layer and so on: "A NNAR$(p, P, k)_m$ model has inputs $(y_{t-1}, y_{t-2}, \ldots, y_{t-p}, y_{t-m}, y_{t-2m}, y_{t-Pm})$ and k neurons in the hidden layer" [30].

The main parameters of an autoregressive neural network (NNAR) are:

- The number of nonseasonal lags used as input
- The number of seasonal lags used as input(if there is a seasonal effect in time series)
- The number of neuron (nodes) in the hidden layer

15.4.2.1 Data preparation

A typical time series characteristics are seasonality, trend, and stationarity. While in classic autoregression models, time-series must be deseasonalized and stationary, in NN forecasting, this is not a necessity. On the other hand, it is recommended to separate data into a training set and a test like ARIMA models.

15.4.2.2 Modeling

In R fit model implementation with **R**, the modeling is achieved with $nnetar(p, d, size)$ function:

- *p* represents the *non-seasonal* lags number.
- *P* represents the *seasonal* lags number.
- *size* represents the *nodes* of hidden layer.

  If parameters are not defined they selected automatically!

```
# Neural Network hourly forecasting
nfit <- nnetar(ts_d)
```

15.4.2.3 Forecast

Following the fitting of the model, we move to predict future values. Unlike classical statistical forecasting methods, NNs do not provide a direct way to define prediction intervals. Nevertheless, there are indirect techniques [51,52] for deriving the prediction intervals.

In Figure 15.6, we can see that model has recognized the yearly seasonal effect 365. Also it used 14 lagged values as input, *1* seasonal lagged value, and 8 neurons

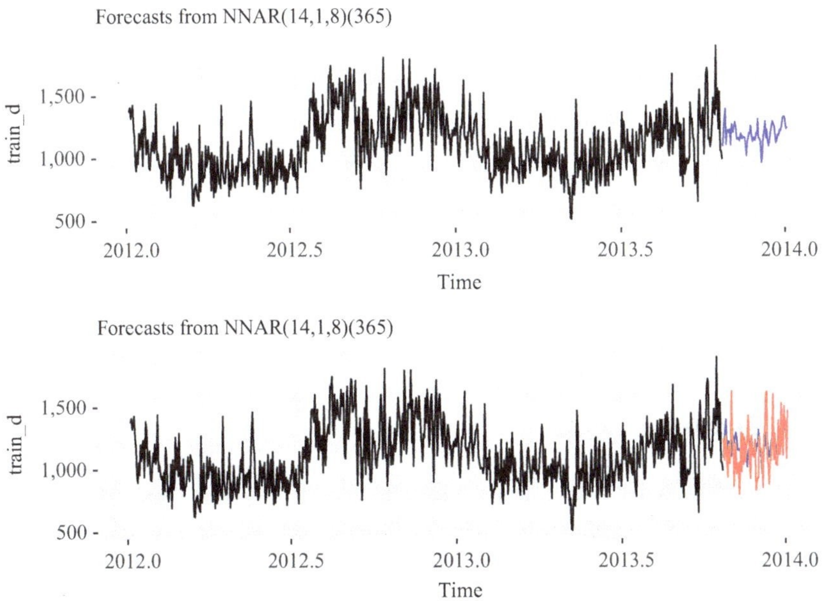

Figure 15.6 Neural network daily forecast against test data

in the hidden layer (the number of nodes in the hidden network is usually half of the lagged inputs $k = (p + P + 1)/2$ where k is the number of hidden neurons.

15.4.2.4 Evaluate results

To evaluate the results, we train the network with the training dataset. The model evaluation procedure usually follows the below steps:

1. Separate data to training and test set

```
# Training set 90%
train_d <- head(ts_d, n = round(length(ts_d)*0.9) )
h <- length(ts_d) - length(train_d)
# Test set the remaining 10%
test_d <- tail(ts_d, h)
```

2. Training the NN with the training dataset

```
# Fit model with training set
nfit <- nnetar(train_d)
```

3. Forecast the same period as the test set

```
# Forecast is the same period with test set
fcast <- forecast(nfit, h = length(test_d))
```

4. Compare Forecasted values against actual values(test dataset).
5. Statistical Performance Evaluation of the Model. In R the statistical evaluation is accomplished with **accuracy()** function from *forecast* package. With this method we RMSE, MAPE and other model performance measures for the model. Specifically these measures indicates how well the model performs.

```
##                       ME      RMSE       MAE        MPE      MAPE
## Training set   0.09830957  48.43538  35.76721 -0.6796142  3.513744
## Test set     -33.51829125 188.33577 151.62149 -5.2934000 13.586374
##                   MASE       ACF1 Theil's U
## Training set 0.1670059 0.09838935        NA
## Test set     0.7079581 0.13976786 0.8304667
```

In Figure 15.6, in the upper plot, the blue line depicts the forecasted data; in the lower plot, the red line represents the actual data(test data) and the blue line represents the forecasted data. We notice that the forecasted line not entirely reflects the real data. This is maybe a not captured seasonal effect.

15.5 Compare methods

NNs and ARIMA have their own advantages and disadvantages. NNs is data-driven, and it can be applied to a variety of different datasets. ARIMA demands data to meet specific criteria. Furthermore, NNs can solve nonlinear problems, something which is unfeasible for ARIMA. From this point of view, NNs can efficiently face real-life challenges. On the other hand, NNs demand big datasets for their efficient training. ARIMA can produce reliable results with a relatively small number of historical data. Another advantage of ARIMA compared to NNs is the fact that its implementation consists of robust and transparent steps. In contrast, NNs sometimes acts as a "black box" with many hidden layers of neurons with complicated interaction between them. It is worth mentioning that no method constitute a panacea for all kinds of forecasting challenges. Thus, it is advisable to use a combination of different forecasting techniques, and after evaluating and compare their results, then choose the right one.

References

[1] Mahlia, T. M. I. "Emissions from electricity generation in Malaysia." *Renewable Energy*. 2002; 27(2): 293–300.

[2] Pacca, S. and Horvath, A. "Greenhouse gas emissions from building and operating electric power plants in the Upper Colorado River Basin." *Environmental Science & Technology*. 2002; 36(14): 3194–3200.

[3] Poznyak, T. I., Oria, I. C., and Poznyak, A. S. (eds.). *Ozonation and biodegradation in environmental engineering*. Elsevier, 2019. Available from http://www.sciencedirect.com/science/article/pii/B9780128128473000081.

[4] Triebe, O., Laptev, N., and Rajagopal, R. *AR-Net: A simple auto-regressive neural network for time-series*. 2019. arXiv: 1911.12436 [cs.LG].

[5] Palomares-Salas, J. C., De La Rosa, J. J. G., Ramiro, J. G., Melgar, J., Aguera, A., and Moreno, A. "ARIMA vs. Neural networks for wind speed forecasting." *IEEE International Conference on Computational Intelligence for Measurement Systems and Applications*. 2009; 129–133.

[6] Mousicou, P., Mavromoustakis, C. X., Bourdena, A., Mastorakis, G., and Pallis, E. "Performance evaluation of dynamic cloud resource migration based on temporal and capacity-aware policy for efficient resource sharing." *Proceedings of the 2nd ACM workshop on High performance mobile opportunistic systems*. 2013; 59–66.

[7] Mavromoustakis, C. X., Mastorakis, G., Bourdena, A., and Pallis, E. "Energy efficient resource sharing using a traffic-oriented routing scheme for cognitive radio networks." *IET networks*. 2014; 3(1): 54–63.

[8] Mavromoustakis, C. X., Dimitriou, C. D., Mastorakis, G., and Pallis, E. "Real-time performance evaluation of F-BTD scheme for optimized QoS energy conservation in wireless devices." *IEEE Globecom Workshops (GC Wkshps)*. 2013; 1151–1156.

[9] Papanikolaou, K., Mavromoustakis, C. X., Mastorakis, G., Bourdena, A., and Dobre, C. "Energy consumption optimization using social interaction in the mobile cloud." *International Conference on Mobile Networks and Management*. 2014; 431–445.

[10] Papadopoulos, M., Mavromoustakis, C. X., Skourletopoulos, G., Mastorakis, G., and Pallis, E. "Performance analysis of reactive routing protocols in mobile ad hoc networks." *International Conference on Telecommunications and Multimedia (TEMU)*. 2014; 104–110.

[11] Mavromoustakis, C. X., Bourdena, A., Mastorakis, G., Pallis, E., and Kormentzas, G. "An energy-aware scheme for efficient spectrum utilization in a 5G mobile cognitive radio network architecture." *Telecommunication Systems*. 2015; 59(1): 63–75.

[12] Mavromoustakis, C. X., Kormentzas, G., Mastorakis, G., Bourdena, A., Pallis E., and Dimitriou, C. "Joint energy and delay-aware scheme for 5G mobile cognitive radio networks." *IEEE Global Communications Conference*. 2014; 2624–2630.

[13] Mavromoustakis, C. X., Kormentzas, G., Mastorakis, G., Bourdena, A., Pallis, E., and Rodrigues, J. "Context-oriented opportunistic cloud offload processing for energy conservation in wireless devices." *IEEE Globecom Workshops (GC Wkshps)*. 2014; 24–30.

[14] Skourletopoulos, G., Mavromoustakis, C. X., Mastorakis, G., Batalla, J. M., and Sahalos, J. N. "An evaluation of cloud-based mobile services with limited capacity: A linear approach." *Soft Computing*. 2017; 21(16): 4523–4530.

[15] Skourletopoulos, G., Mavromoustakis, C. X., Mastorakis, G., *et al*. "Big data and cloud computing: A survey of the state-of-the art and research challenges." *Advances in mobile cloud computing and big data in the 5G Era*. 2017; 23–41.

[16] Batalla, J. M., Mastorakis, G., Mavromoustakis, C. X., and Zurek, J. "On cohabitating networking technologies with common wireless access for home automation system purposes." *IEEE Wireless Communications*. 2016; 23(5): 76–83.

[17] Batalla, J. M., Gajewski, M., Latoszek, W., Krawiec, P., Mavromoustakis, C. X., and Mastorakis, G. "ID-based service-oriented communications for unified access to IoT." *Computers & Electrical Engineering*. 2016; 52: 98–113.

[18] Mavromoustakis, C. X., Mastorakis, G., and Batalla, J. M. *Internet of Things (IoT) in 5G mobile technologies*. 2016; 8.

[19] Makonin, S., Ellert, B., Bajic, I. V., and Popowich, F. "Electricity, water, and natural gas consumption of a residential house in Canada from 2012 to 2014." *Scientific Data.* 2016; 3.160037: 1–12.

[20] Ljung, G. M. and BOX, G. E. P. "On a measure of lack of fit in time series models." *Biometrika*. 1978; 65(2): 297–303. doi: 10.1093/biomet/65.2.297.

[21] Lian, C., Liang, Y., Kang, R., and Xiang, Y. "Deep convolutional neural networks for diabetic retinopathy classification." *Proceedings of the 2nd International Conference on Advances in Image Processing (ICAIP'18)*; Chengdu, China. Association for Computing Machinery, 2018, pp. 68–72. doi: 10.1145/3239576.3239589.

[22] Mastorakis, G., Kormentzas, G., and Pallis, E. "A fusion IP/DVB networking environment for providing always-on connectivity and triple-play services to urban and rural areas." *IEEE network*. 2007; 21(2): 21–27.

[23] Mavromoustakis, C. X., Dimitriou, C. D., and Mastorakis, G. "Using real-time backward traffic difference estimation for energy conservation in wireless devices." *Proceedings of the Fourth International Conference on Advances in P2P Systems (AP2PS 2012)*. 2012; 18–23.

[24] Mastorakis, G., Mavromoustakis, C. X., Bourdena, A., and Pallis, E. "An energy-efficient routing scheme using backward traffic difference estimation in cognitive radio networks." *2013 IEEE 14th International Symposium on "A World of Wireless, Mobile and Multimedia Networks" (WoWMoM)*. 2013; 1–6.

[25] Mastorakis, G., Bourdena, A., Mavromoustakis, C. X., Pallis, E., and Kormentzas, G. "An energy-efficient routing protocol for ad-hoc cognitive radio networks." *Future Network & Mobile Summit*. IEEE. 2013; 1–10.

[26] Mastorakis, G., Mavromoustakis, C. X., Bourdena, A., Kormentzas, G., and Pallis, E. "Maximizing energy conservation in a centralized cognitive radio network architecture." *IEEE 18th International Workshop on Computer Aided Modeling and Design of Communication Links and Networks (CAMAD)*. 2013; 175–179.

[27] Mavromoustakis, C. X., Pallis, E., Mastorakis, G., *et al. Resource management in mobile computing environments*. 2014.

[28] Kryftis, Y., Mavromoustakis, C. X., Mastorakis, G., Pallis, E., Batalla, J. M., and Skourletopoulos, G. "Resource usage prediction for optimal and balanced provision of multimedia services." *IEEE 19th international workshop on computer aided modeling and design of communication links and networks (CAMAD)*. 2014; 255–259.

[29] Markakis, E. K., Karras, K., Zotos, N., *et al.* "EXEGESIS: Extreme edge resource harvesting for a virtualized fog environment." *IEEE Communications Magazine*. 2017; 55(7): 173–179.

[30] Hyndman, R. J., Athanasopoulos, G., and OTexts.com. *Forecasting: principles and practice*. English. Print edition. OTexts.com [Heathmont?, Victoria], 2014; 291 pages.

[31] As'ad, M. "Finding the best ARIMA model to forecast daily peak electricity demand." *Applied Statistics Education and Research*. 2012.

[32] Burke, L. and Ignizio, J. P. "A practical overview of neural networks." *Journal of Intelligent Manufacturing*; 1997; 8: 157–165.

[33] Graves, A., Mohamed, A., and Hinton, G. "Speech recognition with deep recurrent neural networks." *IEEE International Conference on Acoustics, Speech and Signal Processing*. 2013; 6645–6649. doi: 10.1109/ICASSP.2013.6638947.

[34] Dimitriou, C. D., Mavromoustakis, C. X., Mastorakis, G., and Pallis, E. "On the performance response of delay-bounded energy-aware bandwidth allocation scheme in wireless networks." *IEEE International Conference on Communications Workshops (ICC)*. 2013; 631–636.

[35] Bourdena, A., Mavromoustakis, C. X., Kormentzas, G., Pallis, E., Mastorakis, G., and Yassein, M. B. "A resource intensive traffic-aware scheme using energy-aware routing in cognitive radio networks." *Future Generation Computer Systems*. 2014; 39: 16–28.

[36] Kryftis, Y., Mavromoustakis, C. X., Mastorakis, G., *et al.* "Resource usage prediction algorithms for optimal selection of multimedia content delivery methods." *IEEE international conference on communications (ICC)*. 2015; 5903–5909.

[37] Batalla, J. M., Kantor, M., Mavromoustakis, C. X., Skourletopoulos, G., and Mastorakis, G. "A novel methodology for efficient throughput evaluation in virtualized routers." *IEEE International Conference on Communications (ICC)*. 2015; 6899–6905.

[38] Kryftis, Y., Mastorakis, G., Mavromoustakis, C. X., Batalla, J. M., Pallis, E., and Kormentzas, G. "Efficient entertainment services provision over a novel network architecture." *IEEE Wireless Communications*. 2016; 23(1): 14–21.

[39] Batalla, J. M., Mastorakis, G., Mavromoustakis, C. X., and Pallis, E. *Beyond the Internet of Things*. 2017.

[40] Batalla, J. M., Krawiec, P., Mavromoustakis, C. X., *et al.* "Efficient media streaming with collaborative terminals for the smart city environment." *IEEE Communications Magazine*. 2017; 55(1): 98–104.

[41] Ciobanu, R.-I., Negru, C., Pop, F., Dobre, C., Mavromoustakis, C. X., and Mastorakis, G. "Drop computing: Ad-hoc dynamic collaborative computing." *Future Generation Computer Systems*. 2019; 92: 889–899.

[42] Palachy, S. *Detecting stationarity in time series data*. 2019. Available from https://towardsdatascience.com/detecting-stationarity-in-time-series-data-d29e0a21e638.

[43] Akaike, H. "A new look at the statistical model identification." *IEEE Transactions on Automatic Control*. 1974; 19(6): 716–723. doi: 10.1109/TAC.1974.1100705.

[44] Schwarz, G. "Estimating the dimension of a model." *The Annals of Statistics*. 1978; 6(2): 461–464. Available from http://www.jstor.org/stable/2958889.

[45] Rosenblatt, F. "The perceptron: A probabilistic model for information storage and organization in the brain." *Psychological Review*. 1958; 65–386.

[46] Sobel, I. "An isotropic 3x3 image gradient operator." *Presentation at Stanford A.I. Project 1968*. 2014.

[47] Yu, D. and Deng, L. "Deep learning and its applications to signal and information processing [Exploratory DSP]." *IEEE Signal Processing Magazine*. 2011; 28(1): 145–154. doi: 10.1109/MSP.2010.939038.

[48] Indolia, S., Goswami, A. K., Mishra, S. P., and Asopa, P. "Conceptual understanding of convolutional neural network – a deep learning approach." *Procedia Computer Science*. 2018; 132. International Conference on Computational Intelligence and Data Science, pp. 679–688. doi: https://doi.org/10.1016/j.procs.2018.05.069.

[49] Gregor, G., Danihelka, I., Graves, A., Rezende, D. J., and Wierstra, D. *DRAW: A recurrent neural network for image generation*. 2015; arXiv: 1502.04623 [cs.CV].

[50] Cai, X., Zhang, N., Venayagamoorthy, G. K., and Wunsch, D. C. "Time series prediction with recurrent neural networks trained by a hybrid PSO–EA algorithm." *Neurocomputing*. 2007; 70(13). Selected papers from the 3rd International Conference on Development and Learning (ICDL 2004) Time series prediction competition: the CATS benchmark, pp. 2342–2353. Available from http://www.sciencedirect.com/science/article/pii/S0925231207000380.

[51] Yadavalli, V. S. S. and Boraine, H. "Electricity load forecasting with artificial neural networks." *South African Journal of Industrial Engineering.* 2003; 14(2): 23–35.

[52] Ungar, L. H., De Veaux, R. D., and Rosengarten, E. "Estimating prediction intervals for artificial neural networks." *Proceedings of the 9th Yale Workshop on Adaptive and Learning Systems.* Citeseer. 1996.

Chapter 16
Smart interoperability public safety wireless network

Adil Akasha[1], *Rashid A. Saeed*[1,2] *and Elmustafa Sayed Ali*[1,3]

Smart public safety network is one of the hot topics in wireless communication. This chapter presents the design and development of first responders network (First Net) which provides broadband spectrum to the emergency service personnel at the time of disaster in a national level. In the same direction, the proposed design solves the problems of radio networks' interoperability. The smart public safety network uses incompatible radio systems and technologies in different frequencies to coordinate and collaborate works without any interferences. Smart radio network technologies aim to optimize the connections between different public safety organizations. An adaptive controlled system for different two-way radio networks interoperability is designed to increase the trust and reliability at the disaster and crisis times. The design usually depends upon the infrastructure and environmental condition of radio networks that already exist in the country. For simplicity, the presented design is focused on voice communications by accessing audio through commonly featured radio sockets. On the other hand, the existence of the third parties in the system such as the Internet, cellular, or even a console in large two-way radio systems reduces the reliability of the whole design and increases the cost and lack of time.

16.1 Introduction

The services of public safety organizations are valuable to any society to create a stable and secure environment, which include the protection of people and assets addressing a large number of disasters both natural and man-made, such as acts of terrorism, technological, radiological, or environmental accidents [1,2]. The ability of exchanging information such as voice or data is very important for the

[1]Department of Electronics Engineering, Sudan University of Science and Technology (SUST), Khartoum, Sudan
[2]Department of Computer Engineering, Taif University (TU), Al-Taif, Saudi Arabia
[3]Department of Electrical and Electronics Engineering, Red Sea University (RSU), Portsudan, Sudan

coordination of public safety organizations during emergency events. Wireless communication such as two-way radio is important in many fields in order to serve the mobility of first responders [3]. Public safety works in a tough environment, but it is an important task needed for human society to be sustained. The working conditions are risky and subject to time-pressure, and workers must collaborate to achieve shared goals [4].

The ability of emergency first responders, their equipment, and response system to respond to emergency and continue working throughout regional, local, and/or nationwide for huge natural or artificial tragedy is called public safe grade (PSG). First responder network (First Net) should be envisioned by PSG to afford supervision to control the nationwide public safety broadband network (NPSBN) along with the national public safety telecommunications council (NPSTC) [5].

Public safe grade is also one of the key recommendations and requirements for land mobile radio (LMR) network to support the response for emergencies and huge natural disasters [6]. A PSG network and emergency system are constructed to cope with the failures due to artificial or natural disaster, which is globally considered as the first defense and cornerstone for safe, secure, and reliable environment that leads to validated and verified performance [7,8]. PSG is used by public safety responders and assessed by public safety bureaucrats to provide dependable and safe processes in the event of natural or artificial disasters or events.

Resilience network is one of the main pillars for public and wealth safety in local incidents and disasters. The personnel commands, routines, and protocols during that time are vital for the situation to be under control and all safety teams to be connected and communicate covering a larger area [9]. Land mobile radio (LMR) public safety (PS) voice networks should be designed to provide the most resilient, reliable, and trusted communication channels for the team. Emergency responders and commanders respond that the LMR PS networks to work at any time and under all circumstances to be approved by the PS regularity committee. The Nationwide Public Safety Broadband Network (NPSBN) cannot be compromised by any reason, that is, cost, regulation, policies, or anything else. Sometimes, the PS regulations go beyond more than the conventional public safe grade (PSG) procedure which needs to distinguish, accommodate, and join many wireless communication sites without permission from the operators of these sites. These include many other functions and adding values that are not usually available for the conventional customer-provided equipment (CPE).

A conventional base station of the wireless network might not be equipped to meet the demand of these high technologies and functions. However, the central hub of the PSG can activate these critical features with the blueprint provided by the manufacturers and agreed by the regulatory and standardization bodies for functioning based on all these recommendations [10]. First Net, having a bylaw which is in agreement with all local authorities, should be able to measure the status or criticality of each base station (BS) and recommend how to balance between cost and risk [11].

16.2 Public safety and emergency networks

Studies on the same topic such as the coalition peering domain (CPD) is relatively old but it is still used as a tactical solution. In this system, first, the interoperation is not always direct, and a network in-between is used. These terminals (responders) have no choice to select a connection because they are already built as fixed connections [7]. Second, recent technology widely used in the United States is Project 25 (P25) which is a digital trunked radio system that provides both voice and data [12]. Although the system is reliable but still the existence of a third party as a console increases the massage delivery time.

The third recent technology is software-defined radio (SDR) which contains efficient techniques for radio design with many advantages in terms of adaptability and reconfigurability to support new waveforms and multifunctionality, such as bands, modes of operation, and air interfaces [13]. Although this technology makes a benefit for interoperability issues, it joins different bands of radio in one radio, rather than joining networks. The fourth is the wireless gateways and directional antenna (WGDA). Although the gateway design can be simple, the directional antenna may include a very complicated control system, and the system is limited by the number of directions the antenna can be positioned [14]. The fifth method is cellular and Internet (CI). Cellular and Internet services offer data transfer in a higher speed than other radio networks.

However, the existence of the third party limits the continuity and flexibility of the network. Radio networks prefer as private networks that are not affected by cellular or Internet. Private networks choose the area of interest which may not be covered by cellular and Internet [15–17]. The last type discussed is the advanced control unit (ACU), which is a large switching system, has an expanding capability to include any number of networks, the connection between networks controlled by internal microprocessor which in turn controlled by user interface software, and the manual operation of the ACU form a third party between first responder and the concerned agency that makes the system less benefits for time loss considerations [18].

16.3 The need for public safety and emergency collaboration

First responders and other emergency organizations arrived on the scene without the ability to interconnect between them while relevant and timely information is very important for making tactical, strategic, and planning decisions when responding to emergencies. Many of the large radio equipment suppliers manufacture their own interoperability solutions which may not work with other manufacturer's radio equipment and can more expensive. Other advanced systems try to control the interconnections manually (console), but they delay the response to an emergency by a considerable amount of time. Hence time is a considerable factor in public safety and the research focuses on the delay caused by the connections control rooms or consoles by designing an automatic switching system that is reliable with less time loss and cost.

16.4 Cooperative wireless communication

The point-to-point channel is the simplest communication topology. Its model consists of only one transmitter and one receiver existing in an otherwise empty space. Even with such a simple topology, communications depend on the properties of the environment and the problem is not always trivial. The capacity of a communications system composed of more than those two essential devices is even harder to determine, due to their interactions, and it is impossible to calculate the capacity region as the mere union of independent point-to-point scalar problems.

The broad term “cooperative wireless communications” does not refer to a specific, isolated field of knowledge. It is more like a transversal concept that covers any type of communication problem in a wireless network where one transmitter desires to deliver information to a receiver, and where one or several cooperative nodes nearby help in the communication so that the transmitter can achieve a different (possibly higher or more stable) rate than with a point-to-point transmission.

16.4.1 Spectrum scarcity

Since the public safety wireless network depends on radio networks’ interoperability, it founds that spectrum plays an important role especially in such cooperative communications. Moreover, to facilitate the smart public safety services, the network must sustain high-speed, ultra-reliable wireless data transmissions. The increase in data traffic exchange due to the growth of demands in mobile broadband networks significantly increased the spectrum shortfall for commercial wireless radio services, which will turn to reduce the performance of public safety wireless networks. Spectrum scarcity complicates challenges at several levels in public safety communications (PSC), for example, in capacity, interoperability, and functionality [16].

For emergency services, the scarcity in capacity will high congestion in resource allocation especially for urban areas. The lack of spectrum will face difficulties of interoperability for PSC systems because of multiple frequency bands and standardization, in addition to difficulties of functionality to support efficient QoS. Because of these challenges, there is an urgent need for models that motivate users to share their spectrum at multiple stages to smoothly exploit it for public safety applications. The software defines radio (SDR) and cognitive radio (CR) technologies provide a smart solution for the mentioned spectrum scarcity levels. Communication systems based on SDR and CR technologies have to validate the technical requirements of the technologies used in the public safety networks [17].

16.4.2 Software defined radio

Recently, software-defined radio (SDR) technology was developed rapidly, which significantly changed the abilities of wireless communication capabilities and used in a variety of application areas, such as military and commercial services. For PSC application, SDR provides a solution to address communications and interoperability requirements for public safety. It can facilitate interoperability between public safety radio systems and other systems, as well as multimedia devices that provide the user with access to land mobile radio networks, the commercial cellular system, relay systems, and wireless data system. Many research groups studied the use of SDR in PSC.

The public safety special interest group (SIG) is one of these groups was put rules to bring together developers, users, regulators, and educators to address issues specific to the application of SDR technology for public safety communication services.

16.4.3 Cognitive radio

Cognitive radio (CR) is an exciting technology for solving the artificial spectrum scarcity. It will improve the spectrum efficiency and spectrum sharing methods by using interference cancellation (IC) technique. In cooperation spectrum sharing, the devices under different administrative control must communicate and cooperate with each other to avoid mutual interference, which requires a means of common protocol for sharing. CR technology provide a solution for this challenge in order to reduce spectrum underutilization phenomenon resulting from the legacy spectrum licensing system based on command and control. For PSC, CR enables the systems intelligently access underutilized spectrum bands. Moreover, CR can offer network resilience, flexible network topology, and security.

Cognitive radio in PSC will provide access to the commercial spectrum in times of emergency, and the commercial licensee will have preemptible, secondary access to the public safety broadband spectrum. CR will also provide shared infrastructure to achieve significant cost efficiencies while maximizing public safety access to interoperable broadband spectrum.

16.5 Radio systems in public safety and emergency

Radio systems are complex terrestrially based wireless communications systems. Commonly, the most used radio system type is two-way radio systems to transmit and receive simultaneously. In general, radio systems are not the same as all applications since they are governed by many elements involved. The primary elements that should be considered are capacity, frequency, and coverage. The two-way radio systems can be operated by many frequencies depending on the licenses (see Figure 16.1) [4,6]. Practically, operators need frequency bands for the radio system to enable channelization with a particular frequency with an ability to switch between more than one frequency [13,14].

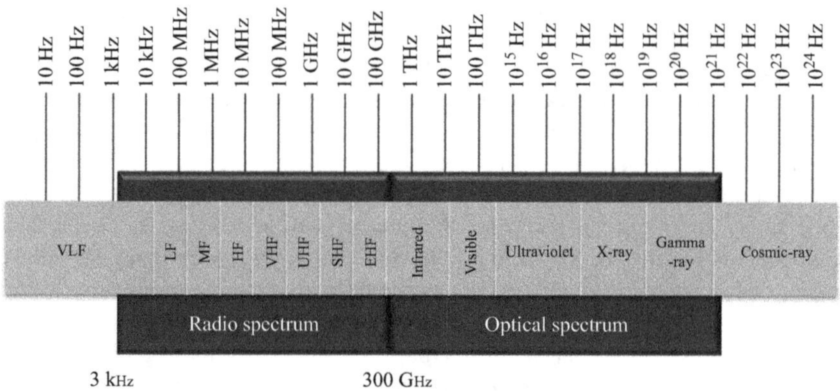

Figure 16.1 Radio frequency bands

For public safety applications, two-way radio systems require to be operated in high performance and ensure that the characteristics of the selected frequency band must fulfill the requirements of efficiency [18]. These requirements are related to many aspects, that is, license and regulations, local congestion, propagation area, and ability to reduce the presence of noise, interference. In public safety communications, the mentioned aspects have more impact according to the application. Public safety organizations must identify the radio services and frequency channels according to the frequency bands used. Public service agencies have an interest in maintaining some common frequencies for interarea or interservice coordination in emergencies [19,20].

Generally, public safety radio communications systems are limited range, enabled communication services within a particular group and are incompatible. Due to the independence of public safety systems agencies, the design of such systems presents challenges because of the need to incorporate adaptiveness and appropriate representation. This what is known as the need for interoperability among public safety agencies. Cognitive radio has been mentioned as a key technology in solving interoperability issues [21,22]. The use of full cognitive radio provides a smart solution that takes into consideration every possible observable parameter of a node or wireless network in order to address interoperability problems.

In addition, cognitive radio focuses on the importance of exploiting the capabilities of technical communication between devices and determines the limitations of interoperability of these applications. From these mentioned considerations, it is necessary to understand the characteristics associated with how to operate dynamic communication groups devoted to public safety smartly in the light of recent developments in communications technology and redefine the concept of interoperability with consideration to organizational, operational, and contextual aspects in order to enhance awareness of radio capabilities in the field of public safety [23,24].

16.6 Characteristics of smart radios

Two-way radios have the ability to provide many communication services in important and vital areas. These devices are relied on by many government agencies, customers, and public safety agencies according to the needs of each sector, as this depends on reliability and efficiency to support communication needs. Therefore, two-way radio communication systems must be characterized by common smart features such as the followings:

1. **Digitization:** All system information such as date, network, battery status, and menu options are displayed in a digital manner. Digital microphones and amplifiers provide volume and enhanced noise cancellation. Automatic voice control allows the adjuster to adjust the volume to the preferred level, which will apply the radio to all incoming and outgoing transmissions.
2. **Intelligent screen control:** Must support a powerful touch screen designed to recognize and respond to deliberate touch using the smart touch interface. This technology reduces the time it takes to scroll through a range of options using analog radio. It also provides great touch operation and a high-speed user interface design to help the officer's ability to quickly find what he needs.
3. **Smart programming:** Communication technology is rapidly developing, often through software updates. With the built-in broadband coverage provided by LTE, radio management teams can send comprehensive updates to identify radios or all radios via the smart programming feature.
4. **Intelligent positioning and communication:** The built-in safety system must support the smart positioning application, which transmits the position of the emergency agent with a faster relay to enable transmission and tracking of the site for more voice traffic. In addition, the system should introduce a new feature for embedded systems, which is smart communication that allows automatic switching between network coverage based on signal strength and availability. This automatic switching from one network to another ensures that communication continues without interruption.

16.6.1 Radio interoperability

Interoperability of public safety communications is the ability to share information via voice and data signals on demand. This level of interoperability is available using equipment from many manufacturers of public safety communications systems. Radio interoperability provides solutions to enable users to participate in the same group of conversations so that they can collaborate and better coordinate their response, providing a high level of response to disaster prevention. [25]. In general, radio interoperability is a wide and complex issue. It depends on the convergence of issues such as governance through the cooperation of all public safety agencies, standard operating procedures for different types of use cases, in addition to training employees to ensure full employment.

The ability of public safety agencies to communicate on radio depends on demand in real time. This is according to the specific needs of the situation and any

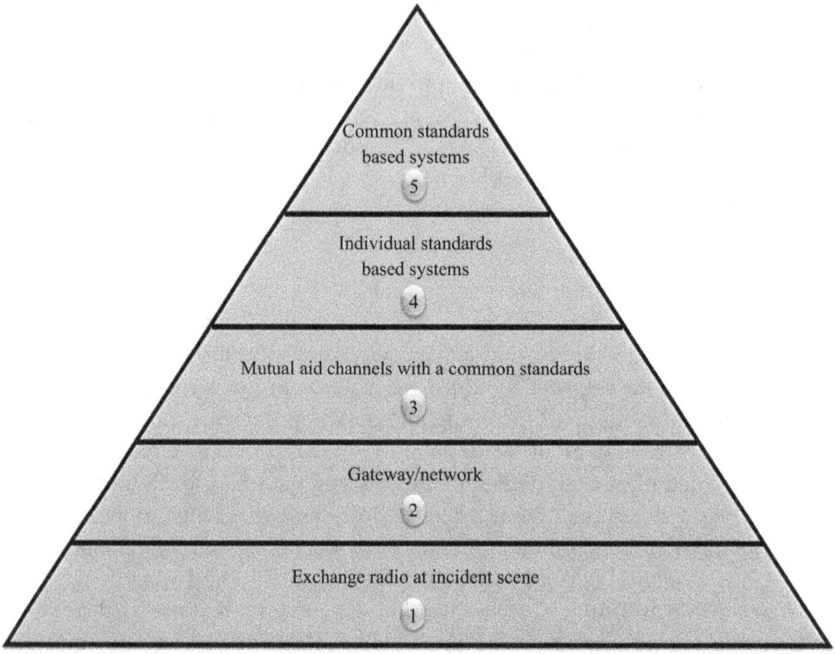

Figure 16.2 Radio interoperability levels

operational procedures and policies that may exist between the agencies involved. In public safety, the communications can be structured in dependent or independent infrastructure. In independent infrastructure, communications link occurs between mobile radio terminals over a direct radio frequency path. In dependent infrastructure, the communications link requires the use of equipment, other than mobile radio terminals, for the establishment of the link and complete operation.

Since there is a specific technology to facilitate radio interoperability according to the used range, the methods for providing this capability are universal, and therefore a general hierarchy of levels of radio interoperability can be established based on operational and technical means (see Figure 16.2). Through this framework, comprehensive testing networks that benefit from existing independent infrastructures can be built through a collaborative process between public safety agencies

There are some challenges in formulating the framework for radio interoperability in general. This challenge is related to access, video surveillance cameras, mobile data terminals in law enforcement vehicles, which operate over push-to-talk (PTT) networks, instant messages of computers and handheld personal digital assistants (PDAs) carried by field personnel, geographic information systems, and hazardous materials databases [24,26]. Despite these challenges, the use of this collaborative approach enables them to be understood in addition to the complex challenges of interoperability that each agency faces [27].

16.6.2 Interoperable communications for first responders

In 1970, the first interoperable communication challenge for respondents was addressed by developing the FIRESCOPE system of accident events in Southern California as a result of a catastrophic wildfire. It is difficult to plan for future disasters and develop action plans to coordinate with neighboring communities in the face of the responsibility to meet safety demands today. This will be one of the most important challenges the public safety agencies are facing. For this purpose, interoperability became a more urgent issue. Recently, the issue of interagency interoperable communications was highlighted as an operational weakness for the first responder community and defined according to different kinds of interoperability as shown in Figure 16.3 [28].

16.6.2.1 Syntactic interoperability

If two or more systems are able of connecting and interchanging data, then they are unveiling syntactic interoperability. Certain data presentations and formats are needed and communication protocols and the like are essential. HTTP or Python languages are amongst the protocols of syntactic interoperability. This is also correct for the lower layer of data presentations and formats, such as guaranteeing numerical characters are kept in the same disparity of ASCII or a Unicode format in all the transmission systems. Syntactical interoperability is an essential state for other interoperability, that is, semantic interoperability [23]. Syntactic interoperability is a requirement for semantic interoperability. Syntactic interoperability refers to the packet and mechanisms for communicating data.

16.6.2.2 Semantic interoperability

Semantic interoperability is not about packaging data exchange (syntax) only but the exchange with understood, unambiguity, and meaningful with common code of protocol between the two or more parties and systems. Semantic interoperability is usually adding some extra data as a vocabulary and interpret the main data. The extra data is sent along with the basic data. A data vocabulary interprets the contents of the packets, the data ruins worthless.

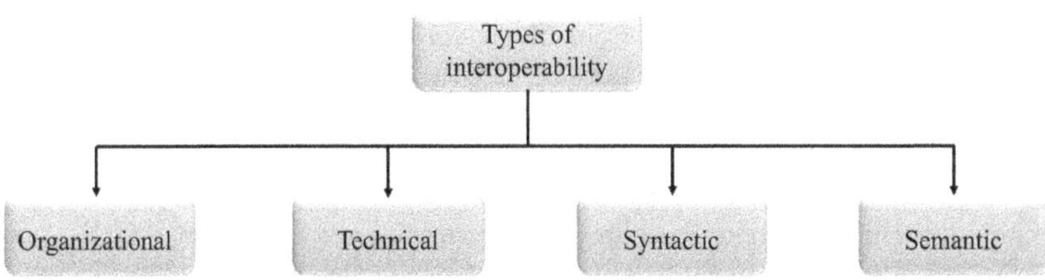

Figure 16.3 Types of interoperability

Syntactic interoperability includes a mutual data format and mutual protocol to build any information so that the method of handling the data to be understandable from the format. It also permits the discovery of syntactic mistakes, thus letting receipt schemes to request resending of any packets that look to be distorted or imperfect. However, data formatted in one composition may sometimes be exactly interpreted into various arrangements. Where precise interpretation of syntaxes is probable, systems using various syntaxes may also collaborate precisely. In sometimes, the capability to precisely interpreted data format amongst formats using various syntaxes might be inadequate in one way, when the data utilized have various layers of fluency (capability to express data).

16.6.2.3 Interoperability organizations

Internationally many bodies have been dedicated to guarantee compatibility between various products. All have in the mutual goal that they need to thrust toward world-wide understanding and devices cooperate for the sake of integrated network which usually reduces the monopoly, cost reduction, and resource sharing. Some focus on the government's interest, or businesses/companies' interest for voice, video, and or data exchange with barriers [29]. This is distributed at three layers: the core network layer, the distributed network layer, and the customer-provided equipment (CPE) layer.

Internationally, network-centric actions industry group simplifies global network compatibility across boundaries, linguistic, and practical barriers. In Europe, for instance, the Interoperable Delivery of European Public Services Administrations, Businesses and Citizens (IDABC) worked on issues for the European collaborated Framework. IDABC is succeeded by the Interoperability Solutions for Public Administrations (ISA) force task. IDABC created the semantic interoperability center Europe (SEMIC). International Telecommunication Union (ITU) formerly the International Telegraph Union is also similarly working for common understanding for the regulations and policies for the international wide and bilateral and regional agreements among member state [11]. ITU contains ITU-Radio Committee (ITU-R), ITU-Standardization Committee (ITU-T), and ITU-Development Committee (ITU-D).

ITU-R was formerly called International Radio Advisory Committee (CCIR). In the United States, the Institute of Electrical and Electronics Engineers (IEEE) founded in 1884 and it has a wide band of information technology and communication technology standards to allow quick response for new innovations, that is, IEEE 255 up to IEEE 12207. One of the widely spread technology all over the world, IEEE technology is the IEEE 802.11 (wireless local area network) or commercially called Wireless Fidelity (Wi-Fi) [30].

16.6.2.4 Cross-domain interoperability

Cross-domain interoperability enables different institutions, departments, sectors, and other technical people working together for a common standard, rules, or policies, in addition to bush their ability of sharing the resource and information exchange [5]. The ability of cross-domain interoperability (CDI) is pretty progressively becoming vital for the private and public sector, where connection and linkage becoming more universal and codependent. CDI enables interaction, spreads product value, and empowers customers to be more innovative and effective within their own spheres and the mutual effort.

Cross-domain interoperability is categorized by mutual empathy and contracts on both sides of a domain border that allow separate institutions to modify or change their connections, devices, or services collaboratively within the bigger community. Each user admits and applies the use of common, boarded-wide, or national-wide policies and regulation protocols. Therefore, cross-domain communication and interoperability may not be under the administration of one operator or service provider.

16.6.2.5 Interjurisdictional interoperability

Recent trends have been envisioned the insistence of communications cooperation and compatibility in public safety. For instance, lately, the standard and international regulatory bodies concern much about that enforced legal decisions and judgment committee to standardize multiple devices to interoperate in incident response. The reason behind that is to produce a smart and quick response to prevent or thwart terrorism. In addition to that, there is much concern about securing critical substructure in the public and private sector, the firms and buildings and the access to it, and emergency response systems. One of the most concern lately was the susceptibility of society to unembellished storms and other nature-related tragedies [31].

Interjurisdictional cooperation also spread over day-to-day public safety procedures as well as emergency response. For instance, the FIFA World Cup Games and other major crowded events, that is, religious and music events, and conventions. Likewise, agencies gradually expect to cooperate during crimes that span agency authorities, such as car accidents, terrorism, and serial crimes [32–34].

16.7 Vision of new communications generation for emergency

Lately, some communications technologies are presented to address and intensify interoperability and compatibility issues. When alliances of many organizations produce joint communications devices that might include legacy ultra-high frequency (UHF), very high frequency (VHF) radios, and push-to-talk (PTT) radios with outdoor wireless equipment using new community radio frequency allocations at 2.4 and 5 GHz, new trunking systems preserve spectrum and create a queue to handle the demand for voice or data channels.

Another example is personal digital assistants (PDAs) with instant messaging (IM) and capability of PTT-over-cellular phones, especially valuable for conveying arithmetical information such as location coordinates. If the interoperability and compatibility solutions are not provided for such devices, these diversities cause incompatible communications technologies to efficiently detach first responders even if they operate adjacently [35–37].

16.7.1 First responders communications systems (FRCS)

First responder public safety system is an emergency radio communication system that provides safety services to many cities. This network sends emergency wireless frequencies from outside the building, where the signal is supposed to be stronger, to the building, where it is usually weaker or absent. The system shown in Figure 16.4, proposed by Akasha, A. *et al.* [38], interconnects first responders (FRs), that is, (FR1), (FR2), and (FR3), directly using the same frequency of the other agency (AG) in case of (FR3), or indirectly for (FR1) and (FR2) to call other agencies through their own agencies as they have shared frequencies.

Such FRCS provides an automated incident control system to support inter/intra agency communications among first responders. Agencies may include fire, police, border patrol, emergency service, and safety. FRCS integrated any components of different communications channels, radio systems, microwave, cellular equipment, satellite, and even public telephony networks. In addition, positioning systems and sensors enhance the C3 capabilities [39]. The improvement of remote locations by using the 4G LTE cellular data networks, particularly provides a great development in FRCS [40]. The development in 5G networks promises new opportunities to make significant solutions to first responder communications.

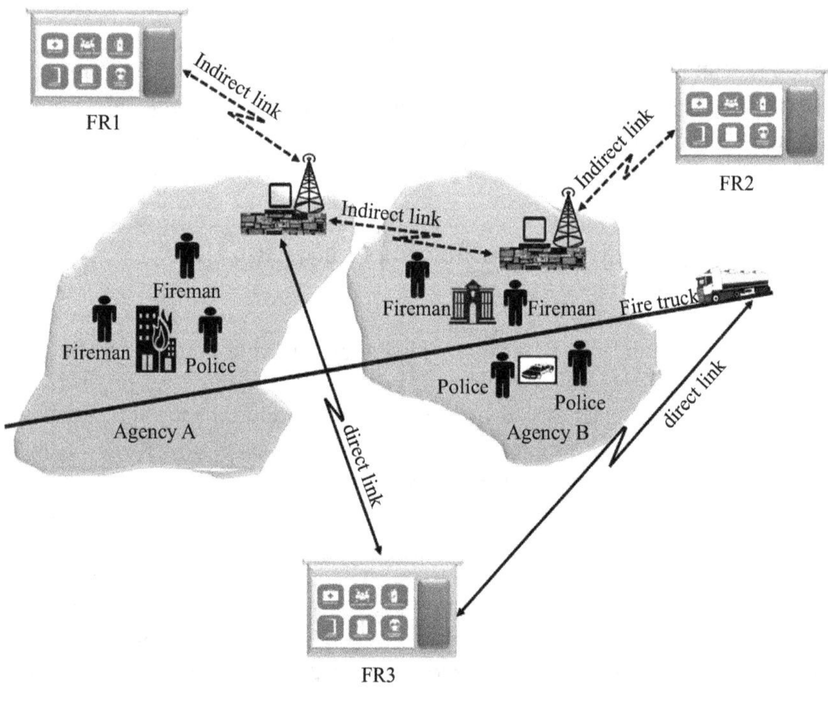

Figure 16.4 First responder system with direct and indirect links

The 5G networks can provide utilizing more public safety broadband spectrum, which is better able to penetrate physical objects.

16.7.2 Decentralized trunking system

Legacy emergency networks are usually separated by frequency bands intend to serve various group of users, that is, policy, military, special weapons and tactics (SWAT) team, etc. Each group of users are able to be served by half-duplex, full-duplex channels, or both. In this way, alliances can guarantee both resource sharing and security. In this system, each channel needs its own A base transceiver station (BTS) where a wireless network can be facilitated between the central station and end-users [41].

BTS is coordinating the share of multichannel radios switches between end-users by scanning and switching among all authorized channels. Legal and registered users can be able to connect and access other BTS and groups on the same or neighbor networks dynamically by their RF equipment to connect to other channels/frequencies. One of the vital challenges to this method and scenario is the interoperability of the radio components that are manufactured by different vendors or if networks operate in mismatched standards, that is, 25 kHz versus narrowband (NB), frequency-division multiple access (FDMA) versus time-division multiple access (TDMA), or digital circuitry versus analog circuitry [42].

Telecommunication Union (TU) is doing a great job on compatibility and interoperability interfaces for radio standards to make sure that analogous features can be given to all groups of users, that is, TIA/EIA. For such cases, if there is no air compatibility, audio cross-patching between dispatch panels can be simply modeled and used. For the same level of technology, trunked communication networks and protocols are one of the best scenarios to be adopted for interoperable over conventional networks. Devices with different trunking networks cannot cowork or communicate with each other as in the case of legacy networks. Users from different trunked protocols cannot communicate or connect if they use different radio frequencies or adopt different RF technologies [43].

Nevertheless, trunked protocols are constructed to maintain interoperability control and management features. One of the great features for the trunked networks is that one can initiate in real time a new talk user groups or adding new members to an existing talk user group; and also it permits the creation of talk user groups across the same institutional lines on-demand basis.

16.7.3 Trust and risk in collaborative ventures

Trust and risk in collaborative ventures are sophisticated and complicated function for the different parties that has strategic alliances. Hence, to figure out how different parties can properly and effectively eliminate and deal with this risk, it requires to verify the relationships and interfaces between different parties in terms of the trust, risk, and interoperability. In [44], the authors proposed a complete and unified solution for the three pillars (trust, risk, and control) in the background of strategic alliances of departments challenging that trust and control are the two fundamental predecessors of risk. First, the authors proposed three structures and each of them is included in certain key layers.

Risk is proposed to be separated into two parameters as relational risk and performance risk. Trust has been separated into two parameters and they are considered as goodwill trust and competence trust. Finally, control is distributed into three different dimensions: behavior control, output control, and social control. Then, the authors negotiated the different interfaces between the various types of trust, risk, and control in alliances. Accordingly, many approaches to risk vulnerability improvement and elimination are discussed. A couple of combinations were introduced, that is, through goodwill trust, behavior control, and social control, where the relationships are minimized and reduced; while competence trust, output control, and social control are related with minimized performance risk [45].

A number of trust-structure methods and control techniques have been proposed to eliminate risk in various sorts of strategic alliances. Finally, this chapter conducted plans for an experimental test of the proposed integrated above-mentioned model and gave brief recommendations for the research vision and directions and management requirements.

16.8 Recent public safety interpretability systems and future directions

The most recent works are based on consoles or control room and this may be helpful in storing and retrieving of data for a given period of time for sake of further studies or analysis. But at the same time, these systems increase hardware crowding hence increases cost; on the other hand, the console is manually operated and this increases the time spend for a massage to reach its target. Other systems use public networks such as cellular networks as a third party in the whole system, and this affects system dependency as well as reduces system reliability. A comparison between selected types of first responder systems is stated in Table 16.1.

The new directions of public safety interpretability system depend on developing ways to use existing infrastructure such as antennas, radios networks, communication cables, power supplies, transceiver devices to interact with public safety systems, in addition to intelligent interaction with humans, such as the ability to integrate voice interaction to ensure rapid understanding and transmission of information. Due to the significant expansion in the use of smart systems and AI in communication networks, especially in smart cities, machine learning in public safety networks will provide a noticeable development in the interaction between public safety networks and other communication networks [46]. The large amount of heterogeneous data that can be exchanged in public safety communication systems can be served by machine learning (ML) systems [47]. ML systems have a great performance in dealing with the most complex communications and offer unique solutions in interacting with humans and cognitive radio interoperability, in addition to radio spectrum sharing, especially in emergency situations.

Table 16.1 Public safety interoperability systems comparison

	Coverage area	Relative no. of users	Reliability	Complexity	Main disadvantage
CI	Cellular and Internet (CI) according to the public network	Limited to the no. of subscribers	Reliable per single and group connection	Simple terminals	Limited coverage area
CPD	Limited by the network fixed terminals	Equal to no. of terminals	Less reliability per single connection	Simple connections	Limited connection paths
P25	Can be expanded relative to radio type	Up to the no. of the radios	Reliable per single and group connection	Complicated console	Manual console (lack of time) and needs infrastructure
SDR	Can be expanded relative to radio type	Up to the no. of the radios	Reliable per single and group connection	Simple (single) radio	Type of frequency sharing
WGDA	Limited by the no. of directions the antenna can take	Limited to antenna positions	Not reliable either per single or group connection	Complicated mechatronics	Mechanical parts sensitivity
ACU	Can be expanded relative to radio type	Up to the no. of the radios	Reliable per single and group connection	Complicated console	Manual console (lack of time)

16.9 Conclusion

The exchange of voice messages between different radio receivers and transmitters in networks is a model of interaction between radio networks regardless of frequency bands, which provide a greater possibility to manage communications for public safety systems. This method enables the implementation of automatic procedures instead of manual procedures to reduce time loss between different network connections. The system can also operate an interactive, updateable program to control the public safety system communication process. In general, the manual procedures for voice messaging take more time, but in case of relying on automatic procedures, they can be given a gain of time, which may provide efficient or direct value. Interactive systems can provide a number of advantages over other two-way radio systems, such as reducing cost and time and increasing control efficiency through compatibility and interoperability of devices in simple and less crowded infrastructure.

References

[1] Bourdena, A., Mavromoustakis, C. X., Kormentzas, G., Pallis, E., Mastorakis, G., and Yassein, M. B. "A resource intensive traffic-aware scheme using energy-aware routing in cognitive radio networks." *Future Generation Computer System*. 2014; 39: 16–28.

[2] Mavromoustakis, C. X., Mastorakis, G., Bourdena, A., and Pallis, E. "Energy efficient resource sharing using a traffic-oriented routing scheme for cognitive radio networks." *IET Networks*. 2014; 3: 54–63. Available from https://doi.org/10.1049/iet-net.2013.0132.

[3] Yuksel, M., Guvenc, I., Saad, W., and Kapucu, N. "Pervasive spectrum sharing for public safety communications." *IEEE Communications Magazine*. 2016; 54(3): 22–29. doi: 10.1109/MCOM.2016.7432143.

[4] Kader, M. F., Irfan, M., Shin, S. Y., and Chae, S. "Cooperative spectrum sharing in cognitive radio networks: An interference free approach." *Physical Communication*. 2017; 25(1): 66–74. Available from https://doi.org/10.1016/j.phycom.2017.09.001.

[5] Bourdena, A., Pallis, E., Kormentzas, G., and Mastorakis, G. "Efficient radio resource management algorithms in opportunistic cognitive radio networks." *Transactions on Emerging Telecommunications Technologies*. 2014; 25: 785–797. Available from https://doi.org/10.1002/ett.2687.

[6] Bourdena, A., Pallis, E., Kormentzas, G., and Mastorakis, G. "A prototype cognitive radio architecture for TVWS exploitation under the real time secondary spectrum market policy." *Physical Communication*. 2014; 10: 159–168. Available from https://doi.org/10.1016/j.phycom.2013.11.003.

[7] Gorcin, A. and Arslan, H. "Public safety and Emergency case communications: Opportunities from the aspect of cognitive radio." *3rd IEEE Symposium on New Frontiers in Dynamic Spectrum Access Networks*; Chicago, IL, 2008, pp. 1–10. doi: 10.1109/DYSPAN.2008.68.

[8] Bourdena, A., Pallis, B., Arvanitis, A., Mastorakis, G., and Kormentzas, G. "A spectrum aware routing protocol for public safety applications over cognitive radio networks." *International Conference on Telecommunications and Multimedia (TEMU)*; Chania, 2012, pp. 7–12. doi: 10.1109/TEMU.2012.6294737.

[9] Mokhtar, R., Saeed, R., and Abdallah, Y. "Study on energy detection-based cooperative sensing in cognitive radio networks." *Journal of Networks*. 2013; 8(6): 1255–1261. doi:10.4304/jnw.8.6.1255-1261

[10] Saeed, R. A., Ismail, A. F., Kamrul Hasan, M., Mokhtar, R., Salih, S. K. A., and Wahida, H. "Throughput enhancement for WLAN TV white space in coexistence of IEEE 802.22." *Indian Journal of Science and Technology*. 2015; 8(11): 1–7. doi: 10.17485/ijst/2015/v8i11/71783.

[11] Mastorakis, G., Bourdena, A., Mavromoustakis, C. X., Pallis, E., and Kormentzas, G. "An energy-efficient routing protocol for ad-hoc cognitive radio networks." *Future Network and Mobile Summit*. 2013; 1–10.

[12] Ann, S. and Kim, H. S. "Relay association method for optimal path in IEEE 802.16j mobile multi-hop relay networks." *European Transaction on Telecommunications*. 2010; 21: 624–631.

[13] Ermolov, V., Heino, M., Karkkainen, A., *et al.* "Significance of nanotechnology for future wireless devices and communications." *18th Annual IEEE International Symposium on PIMRC'07*. Athens, 2007, pp. 1–5. doi: 10.1109/PIMRC.2007.4394126.

[14] Xie, R., Yu, F. R., and Ji, H. "Joint power allocation and beamforming with user's selection for CR networks via discrete stochastic optimization." *ACM Wireless Networks*. 2011; 18: 481–493. doi:10.1007/s11276-012-0413-8.

[15] Bian, K. and Park, J. M. "MAC-layer misbehaviours in multi-hop cognitive radio networks." *International Conference on Science, Technology, and Entrepreneurship (UKC)*. Korea, August 2006; 1–8.

[16] Xu, X., Liang, W., Wark, T., and Jeong, J. "Maximizing network lifetime via 3G gateway assignment in dual-radio sensor networks." *IEEE 37th Conference on Local Computer Networks (LCN)*; Clearwater, FL, 2012, pp. 479–486. doi: 10.1109/LCN.2012.6423664.

[17] Benmohamed, L., Contestabile, J. M., Desimone, A., Doshi, B., and Forte, J. "Communications architecture for national security and public safety: Multijurisdictional, multitechnology network of interoperable networks." *IEEE Military Communications Conference (MILCOM)*; Baltimore, MD, 2014, pp. 1201–1206. doi: 10.1109/MILCOM.2014.201.

[18] Saeed, R. A., Mokhtar, R. A., Chebil, J., and Abdallah, A. H. "TVBDs coexistence by leverage sensing and geo-location database." *IEEE International Conference on Computer & Communication Engineering (ICCCE)*; Kuala Lumpur, 2012, pp. 33–39. doi: 10.1109/ICCCE.2012.6271147.

[19] Hamadna Allah, O. A., Mokhtar, R. A., Alsaqour, R. A., Salih, S. H. O., and Saeed, R. A. "Spectrum requirements for IMT-advanced: Sudan case study." *International Journal of Applied Engineering Research*. 2014; 9(24): 24425–24435.

[20] Mastorakis, G., Mavromoustakis, C. X., Bourdena, A., and Pallis, E. "An energy-efficient routing scheme using backward traffic difference estimation in cognitive radio networks." *2013 IEEE 14th International Symposium on "A World of Wireless, Mobile and Multimedia Networks" (WoWMoM)*; Madrid, 2013, pp. 1–6. doi: 10.1109/WoWMoM.2013.6583446.

[21] Gupta, P. and Patil, P. "4G-a new era in wireless telecommunication." *Magister Program in S/W Engineering*; Malardalen University, 2009.

[22] Ouyang, Y. and Hosein Fallah, M. "A performance analysis for UMTS packet switched network based on multivariate KPIs." *IEEE Wireless Telecommunications Symposium (WTS)*; Tampa, FL, 2010, pp. 1–10, doi: 10.1109/WTS.2010.5479629.

[23] Huang, J. H., Wang, L. C., Chang, C. J., and Su, W. S. "Design of optimum relay location in two-hop cellular systems." *ACM/Springer Wireless Networks*. 2010; 16(8): 2179–2189.

[24] Raluca, M-E., Razvan, M-E., and Terzis, A. "Gateway design for data gathering sensor networks." *5th Annual IEEE Communications Society Conference SECON'08 on Sensor*, Mesh and Ad Hoc Communications and Networks. 2008.

[25] Saeed, R. A., Mokhtar, R., and Khatun, S. "Spectrum sensing and sharing for cognitive radio and advanced spectrum management." *ICGST International Journal on Computer Networks and Internet Research (CNIR)*. 2009; 9(2): 87–97.

[26] Sun, Z. and Wang, W. "Investigation of cooperation technologies in heterogeneous wireless networks." *Journal of Computer Systems, Networks, and Communications*. 2010. Article ID 413987.

[27] Zhang, G., Li, A., Zhou, Y., and Du, Y. "Towards underlying radio technologies for LTE based public safety networks." *JCP*. 2017; 12(2): 97–104.

[28] Shin, J., Lee, D. K., and Cho, H. S. "Is a multi-hop relay scheme gainful in an IEEE 802.22-based cognitive radio system." *IEICE Transactions on Communications*. 2009; 92: 3528–3532.

[29] Dorfinger, P., Panholzer, G., von Tüllenburg, F., Christaldi, M., Tusa, G., and Böhm, F. "Self-aligning wireless communication for first responder organizations in interoperable emergency scenarios." *Proceedings of the 2014 International Conference on Wireless Networks (ICWN 2014)*. Las Vegas Nevada, USA, 2014.

[30] Alsaqour, O., Alsaqour, R., Alahdal, T., Saeed, R. A., and Al-Hubaishi, M. "A comparative study of simulation based performance evaluation of routing protocol for ad-hoc networks." *Innovations and Advances in Computing, Informatics, Systems Sciences, Networking and Engineering*. Springer International Publishing. 2015.

[31] Mastorakis, G, Mavromoustakis, C. X., Bourdena, A., Pallis, E., and Sismanidis, G. "Energy-efficient routing in cognitive radio networks." *Resource Management in Mobile Computing Environments*. 3: 323–340.

[32] Net-Hobby, P. L. and David E. P. "Emergency radio communications system incorporating integral public safety radio bridging capability." U.S. Patent No. 8,934,934, 13 January 2015.

[33] Mastorakis, G., Bourdena, A., Kormentzas, G., Pallis, E., Zacharopoulos, V., and White, T. V. "Spaces exploitation utilizing a cognitive radio system based on DVB-H." *2011 International Symposium on Computer Networks and Distributed Systems (CNDS)*. 2011.

[34] Min, B. C., Parasuraman, R., Lee, S., Jung, J. W., and Matson, E. T. "A directional antenna-based leader–follower relay system for end-to-end robot communications." *Robotics and Autonomous Systems*. 2018; 101: 57–73.

[35] Hobby, P. L. and Petty, D. E. SAFECOM 911 Inc. "Emergency radio communications system incorporating integral public safety radio bridging capability." U.S. Patent 9,860,923, 2018.

[36] Bohlander, M. J., Fortna, R. T., Huang, A. G., Jolma, J. P., Weiss, J. C., and Deluca, A. K. Axon Enterprise Inc. "Event-based responder dispatch." U.S. Patent 9,642,131, 2017.

[37] Mavromoustakis, C. X., Mastorakis, G., Bourdena, A., and Pallis, E. "Efficient multimedia services provision over cognitive radio networks using a traffic oriented routing scheme." In F. Hu, and S. Kumar (eds.), *Multimedia over Cognitive Radio Networks*. 2015: 119–147.

[38] Akasha, A., Saeed, R. A., and Odeh, N. "Public safety telecommunication interoperability two-way radio systems." *Joint International Conference on Information and Communication Technologies for Education and Training and International Conference on Computing in Arabic (ICCATICET)*. Khartoum, Sudan, August 2017.

[39] Bourdena, A., Makris, P., Skoutas, D. N., Skianis, C., Kormentzas, G., and Pallis, E. "Joint radio resource management in cognitive networks: TV white spaces exploitation paradigm." In *Evolution of cognitive networks and self-adaptive communication systems*. IGI-Global: 2013.

[40] Mukherjee, A., Saeed, R. A., Dutta, S., and Naskar, M. K. "Fault tracking framework for software-defined networking (SDN)." *Resource allocation in next-generation broadband wireless access networks*. IGI Global: 2017; 247–272.

[41] Mastorakis, G., Bourdena, A., Kormentzas, G., and Pallis, E. "Spectrum aware routing in ad-hoc cognitive radio networks." *21st EEE Future Network & Mobile Summit 2012 (FUNEMS)*. Berlin, Germany, 2012.

[42] Alsamhi, S. H., Ma, O., Ansari, M. S., and Gupta, S. K. "Collaboration of drone and Internet of public safety things in smart cities: An overview of QoS and network performance optimization." *Drones*. 2019; 3(1): 13.

[43] Kapucu, N., Haupt, B., and Yuksel, M. "Spectrum sharing policy: Interoperable communication and information sharing for public safety." *Risk, Hazards & Crisis in Public Policy*. 2018; 9: 39–59. Available from https://doi.org/10.1002/rhc3.12129.

[44] Mavromoustakis, C. X., Kormentzas, G., Mastorakis, G., Bourdena, A., and Pallis, E. "Joint energy and delay-aware scheme for 5G mobile cognitive radio networks." *IEEE Global Communications Conference*. 2014; 2624–2630.

[45] Bourdena, A., Kormentzas, G., Pallis, E., and Mastorakis, G. "Radio resource management algorithms for efficient QoS provisioning over cognitive radio networks." *IEEE International Conference on Communications (ICC)*; Budapest, 2013, pp. 2415–2420. doi: 10.1109/ICC.2013.6654893.

[46] Carvalho, D. V., Pereira, E. M., and Cardoso, J. S. "Machine learning interpretability: A survey on methods and metrics." *Electronics*. 2019; 8: 832.

[47] Gilpin, L. H., Bau, D., Yuan, B. Z., Bajwa, A., Specter, M., and Kagal, L. "Explaining explanations: An overview of interpretability of machine learning." *IEEE 5th International Conference on Data Science and Advanced Analytics (DSAA)*; Turin, Italy, 2018, pp. 80–89. doi: 10.1109/DSAA.2018.00018.

Index

www.ingramcontent.com/pod-product-compliance
Lightning Source LLC
LaVergne TN
LVHW011558200926
848484LV00001B/328
9781839530951